Lecture Notes in Computer Science 10079

Commenced Publication in 1973
Founding and Former Series Editors:
Gerhard Goos, Juris Hartmanis, and Jan van Leeuwen

Editorial Board

David Hutchison
  Lancaster University, Lancaster, UK
Takeo Kanade
  Carnegie Mellon University, Pittsburgh, PA, USA
Josef Kittler
  University of Surrey, Guildford, UK
Jon M. Kleinberg
  Cornell University, Ithaca, NY, USA
Friedemann Mattern
  ETH Zurich, Zurich, Switzerland
John C. Mitchell
  Stanford University, Stanford, CA, USA
Moni Naor
  Weizmann Institute of Science, Rehovot, Israel
C. Pandu Rangan
  Indian Institute of Technology, Madras, India
Bernhard Steffen
  TU Dortmund University, Dortmund, Germany
Demetri Terzopoulos
  University of California, Los Angeles, CA, USA
Doug Tygar
  University of California, Berkeley, CA, USA
Gerhard Weikum
  Max Planck Institute for Informatics, Saarbrücken, Germany

More information about this series at http://www.springer.com/series/7407

Paola Festa · Meinolf Sellmann
Joaquin Vanschoren (Eds.)

Learning and Intelligent Optimization

10th International Conference, LION 10
Ischia, Italy, May 29 – June 1, 2016
Revised Selected Papers

 Springer

Editors
Paola Festa
University of Naples Federico II
Naples
Italy

Joaquin Vanschoren
Eindhoven University of Technology
Eindhoven
The Netherlands

Meinolf Sellmann
Thomas J. Wastson Research Center
Yorktown Heights, NY
USA

ISSN 0302-9743   ISSN 1611-3349   (electronic)
Lecture Notes in Computer Science
ISBN 978-3-319-50348-6   ISBN 978-3-319-50349-3   (eBook)
DOI 10.1007/978-3-319-50349-3

Library of Congress Control Number: 2016958517

LNCS Sublibrary: SL1 – Theoretical Computer Science and General Issues

Printed on acid-free paper

This Springer imprint is published by Springer Nature
The registered company is Springer International Publishing AG
The registered company address is: Gewerbestrasse 11, 6330 Cham, Switzerland

Preface

The large variety of heuristic algorithms for hard optimization problems raises numerous interesting and challenging issues. Practitioners are confronted with the burden of selecting the most appropriate method, in many cases through an expensive algorithm configuration and parameter tuning process, and subject to a steep learning curve. Scientists seek theoretical insights and demand a sound experimental methodology for evaluating algorithms and assessing strengths and weaknesses. A necessary prerequisite for this effort is a clear separation between the algorithm and the experimenter, who, in too many cases, is "in the loop" as a crucial intelligent learning component. Both issues are related to designing and engineering ways of "learning" about the performance of different techniques, and ways of using past experience about the algorithm behavior to improve performance in the future. This is the scope of the Learning and Intelligent Optimization (LION) conference series.

This volume contains the papers presented at LION 10: Learning and Intelligent Optimization, held during May 29 – June 1, 2016 in Ischia, Italy. This meeting, which continues the successful series of LION events (see LION 5 in Rome, Italy; LION 6 in Paris, France; LION 7 in Catania, Italy; LION 8 in Gainesville, USA; and LION 9 in Lille, France), explores the intersections and uncharted territories between machine learning, artificial intelligence, mathematical programming, and algorithms for hard optimization problems. The main purpose of the event is to bring together experts from these areas to discuss new ideas and methods, challenges, and opportunities in various application areas, general trends, and specific developments. The International Conference on Learning and Optimization is proud to be the premier conference in the area.

A total of 47 papers were submitted to LION 10: 28 submissions of long papers, 13 submissions of short papers, and six abstracts for oral presentation only. Each manuscript was independently reviewed by at least three (usually four) members of the Program Committee. The final decision was made based on a meta-reviewing phase where each manuscript's reviews were shown to five other members of the Program Committee, who then voted to either accept or reject the manuscript. Only long papers that received four or five votes in favor were accepted as long papers. Papers that received at least three votes in favor were accepted as short papers (papers submitted as long papers had to be shortened).

In total, 14 long papers and nine short papers were accepted. The selection rate for long papers was 34%.

These proceedings also contain two papers submitted to the generalization-based contest in global optimization (GENOPT). These were reviewed independently by the GENOPT Organizing Committee.

During the conference, the keynote talk was delivered by Bistra Dilkina, on "Learning to Branch in Mixed Integer Programming."

In addition, we were pleased to have two tutorial talks:

- "Learning Algorithms and Bioinformatics," by Giovanni Felici and Emanuel Weitschek
- "Automatic Algorithm Configuration," by Meinolf Sellmann

Finally, we gratefully acknowledge the support of our sponsors and partners in organizing this conference:

- IBM Research
- University of Naples Federico II, Italy
- Department of Mathematics and Applications "R. Caccioppoli", University of Naples Federico II, Italy

We hope that these proceedings may serve you well in your endeavors in learning and intelligent optimization.

May 2016   Paola Festa
                                                        Meinolf Sellmann
                                                     Joaquin Vanschoren

Organization

Program Committee

Carlos Ansótegui	Universitat de Lleida, Spain
Bernd Bischl	LMU Munich, Germany
Christian Blum	IKERBASQUE, Basque Foundation for Science, Spain
Mauro Brunato	University of Trento, Italy
André Carvalho	University of São Paulo, Brazil
John Chinneck	Carleton University, Ottawa, Canada
Andre Cire	University of Toronto Scarborough, Canada
Luca Di Gaspero	University of Udine, Italy
Bistra Dilkina	Georgia Institute of Technology, USA
Paola Festa (Co-chair)	University of Naples Federico II, Italy
Tias Guns	K.U. Leuven, Belgium
Frank Hutter	University of Freiburg, Germany
Eyke Hüllermeier	Philipps-Universität Marburg, Germany
George Katsirelos	INRA, Toulouse, France
Lars Kotthoff	University of British Columbia, Canada
Dario Landa-Silva	The University of Nottingham, UK
Hoong Chuin Lau	Singapore Management University, Singapore
Jimmy Lee	The Chinese University of Hong Kong, SAR China
Marie-Eléonore Marmion	Université Lille 1, France
George Nemhauser	Georgia Tech/School of ISyE, USA
Barry O'Sullivan	4C, University College Cork, Ireland
Claude-Guy Quimper	Université Laval, Canada
Helena Ramalhinho Lourenço	UPF, Spain
Francesca Rossi	University of Padova, Italy and Harvard University, USA
Ashish Sabharwal	AI2, USA
Horst Samulowitz	IBM Research, USA
Marc Schoenauer	Projet TAO, Inria Saclay Île-de-France, France
Meinolf Sellmann (Co-chair)	IBM Research, USA
Bart Selman	Cornell University, USA
Yaroslav Sergeyev	University of Calabria, Italy
Peter J. Stuckey	University of Melbourne, Australia
Thomas Stützle	Université Libre de Bruxelles (ULB), Belgium
Eric D. Taillard	University of Applied Science Western Switzerland, HEIG-Vd, Switzerland

Michael Trick   Carnegie Mellon, USA
Joaquin Vanschoren   Eindhoven University of Technology, The Netherlands
  (Co-chair)
Petr Vilím   IBM Czech, Czech Republic

Additional Reviewers

Akgun, Ozgur
Cardonha, Carlos
Gunawan, Aldy
Khalil, Elias
Klein, Aaron

Lindauer, Marius
Lindawati, Lindawati
Mayer-Eichberger,
  Valentin
Meseguer, Pedro

Misir, Mustafa
Teng, Teck-Hou
Yap, Roland

Contents

Short Papers

GENOPT Papers

Full Papers

Learning a Stopping Criterion for Local Search

Alejandro Arbelaez$^{(\boxtimes)}$ and Barry O'Sullivan

Insight Centre for Data Analytics, Department of Computer Science,
University College Cork, Cork, Ireland
{alejandro.arbelaez,barry.osullivan}@insight-centre.org

Abstract. Local search is a very effective technique to tackle combinatorial problems in multiple areas ranging from telecommunications to transportations, and VLSI circuit design. A local search algorithm typically explores the space of solutions until a given stopping criterion is met. Ideally, the algorithm is executed until a target solution is reached (e.g., optimal or near-optimal). However, in many real-world problems such a target is unknown. In this work, our objective is to study the application of machine learning techniques to carefully craft a stopping criterion. More precisely, we exploit instance features to predict the expected quality of the solution for a given algorithm to solve a given problem instance, we then run the local search algorithm until the expected quality is reached. Our experiments indicate that the suggested method is able to reduce the average runtime up to 80% for real-world instances and up to 97% for randomly generated instances with a minor impact in the quality of the solutions.

1 Introduction

Local search is a popular technique to solve many combinatorial problems. These problems can be classified as either decision or optimisation problems. A combinatorial decision problem involves finding a solution that satisfies the constraints of the problem. The satisfiability (SAT) problem is perhaps one of the most important decision problems and involves determining whether a given Boolean formula in conjunctive normal form (i.e., conjunction of clauses) is satisfiable or not. In combinatorial optimisation the goal is to find a solution that satisfies the constraints and proving that the solution is optimal. For example, the Travelling Salesman Problem (TSP) is a well-known combinatorial optimisation problem that involves finding the shortest Hamiltonian circle in a complete graph.

A local search algorithm starts with a given initial solution and iteratively improves it, little by little, by performing small changes while a given stopping criterion is not met. In the context of SAT the algorithm typically starts with a random assignment for the variables and the algorithm flips one variable at a time until either a solution is obtained or a time limit is reached. Alternatively, to solve a TSP the algorithm heuristically builds a starting solution and then applies the k-opt move, i.e., k-links of the current solution are replaced with k new links by improving the current solution until the optimal solution is reached,

© Springer International Publishing AG 2016
P. Festa et al. (Eds.): LION 2016, LNCS 10079, pp. 3–16, 2016.
DOI: 10.1007/978-3-319-50349-3_1

no improvement is observed after applying the k-opt move, or a given time limit is reached.

In this paper, we focus our attention in local search to tackle combinatorial optimisation problems. Ideally the algorithm is executed until a given target solution is reached, (i.e., optimal or near-optimal). In this paper we exploit the use of instance features to predict the expected quality of the solution for a given problem instance. We then execute the local search algorithm until the desired quality is reached. A proper stopping criterion might considerably reduce the runtime of the algorithm. Therefore accurate predictions are very important, in particular with the increasing availability of computational power in cloud systems, e.g., Amazon Cloud EC2, Google Cloud, and Microsoft Azure. Typically, these systems charge for usage by processor-hour. Therefore reducing the computational time and still providing high quality solutions is expected to be very valuable in the near future.

This paper is structured as follows. Sections 2 and 3 provide background material of the two target problems, i.e., CRP and TSP. Section 4 details the machine learning methodology to design the stopping criterion. Section 5 reports on the experimental validation. Related work is discussed in Sect. 6 before final conclusions are presented in Sect. 7.

2   The Cable Routing Problem

The *Cable Routing Problem* (CRP) that we are tackling in this paper relates to the bounded spanning tree problem with side constraints. In particular we focus on a network design problem arising in optical networks where a bound is given on the number of carriers. Each selected carrier is connected to a set of clients using a network that follows a tree topology that respects distance, degree, and capacity constraints.

A long-reach passive optical network (LR-PON) architecture consists of three subnetworks: (1) an Optical Distribution Network (ODN) for connecting customers to facilities; (2) a backhaul network for connecting facilities to metro-nodes; and (3) a core network connecting pairs of metro-nodes. We focus our attention on the ODN part, where the fibre cable is routed from a facility to a set of customer forming a tree distribution network. A PON is a tree network associated with each cable and the signal attenuation in a PON is due the number of customers in the PON and the maximum length of the fibre between the facility and the customer. Additionally, non-root nodes are limited to maximum branching of p nodes. In [1] the authors show the relationship between the size and the maximum length of the PON.

Informally speaking, in the CRP we want to determine a set of spanning trees with side constraints, i.e., distance, degree, and capacity, such that each tree is rooted at the facility, each customer is present in exactly one tree and the total cost of the solution is minimised. The number of trees denotes the number of optical fibres that run from the exchange-site to the customers. Typically, the distance is limited to up 10 km, 20 km, and 30 Km for respectively at most 512,

256, and 128 customers, and due to hardware constraints the branching factor is restricted to 32.

Local Search for CRP

In [2] the authors propose a general constraint-based local search algorithm to tackle bounded minimum spanning tree with side constraints. Broadly speaking, the algorithm comprises two phases. First, in an intensification phase, the algorithm improves the current solution, little by little, by performing small changes. It employs a move operator in order to move from one solution to another in the hope of improving the value of the objective function. Second, in the diversification phase, the algorithm perturbs the incumbent solution in order to escape from difficult regions of the search. The algorithm switches from intensification to diversification when a local minimum is reached.

The subtree operator (Fig. 1) moves a given node e_i and the subtree emanating from e_i from the current location to another in the tree. As a result of this, the predecessor of e_i is not connected to e_i, and all successors of e_i are still directly connected to the node. e_i can be placed as a successor for another node or in the middle of an existing edge. In order to complete the intensification and diversification phases, the subtree operator requires four main functions: removing a subtree; checking whether a given solution is feasible or not; inserting a subtree; finding the best location of a subtree in the current subtree; and finding the best location of a subtree in the current network. The general idea of the LS algorithm in the intensification phase is described as follows:

1. Randomly select a node (e_i);
2. Delete the emanating subtree of e_i in the current solution;
3. Identify the best location, i.e., a new predecessor e_p and a potential successor e_s for e_i satisfying all constraints;
4. Insert e_i as a new successor of e_p, and if needed, add e_s as a new predecessor of e_p.

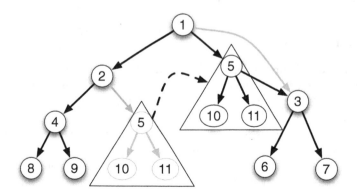

Fig. 1. Example of the subtree move-operator

During the diversification phase the third step performs a random selection of the new candidate location of e_i in the tree. It is worth noticing that the local search operator has been used in several network design applications such as [3,4].

A solution is represented by a tree whose root-node is the facility and the number of immediate successors of the root-node is the number of cables starting at the facility. Notice that the facility acts as the root-node of each cable tree network. Without loss of generality we add a set of dummy clients (or copies of the facility) to the original set of clients for the purpose of ease of representation to distinguish the cable tree-networks. More precisely the set of clients, $\{v_0, \ldots, v_n\}$, is modified to $\{v_0, v_1, \ldots, v_m, v_{1+m} \ldots, v_{n+m}\}$. Recall that n is the number of clients and m is the upper bound on the number of cables that can start at the facility. In the latter set v_0 is the original facility, each $v_i \in \{v_1 \ldots v_m\}$ denote the starting point of a cable tree network, and $\{v_{1+m}, \ldots, v_{n+m}\}$ denote the original set of clients. We further enforce that each dummy client is connected to v_0 and the distance between the dummy client and the facility is zero, i.e., $\forall 1 \leq i \leq m, d_{0i} = 0$.

Default Stopping Criterion. We use search stagnation as the default stopping criterion and compute a bound on the solution. In particular, we stop the algorithm whenever no improvement greater than 0.01% in the incumbent solution has been observed for 30 consecutive seconds.

3   Traveling Salesman Problem

The *Traveling Salesman Problem* (TSP) is a well-known combinatorial optimisation problem with applications in multiple areas ranging from transportation to VLSI circuit design, and bioinformatics. The TSP involves finding the shortest tour among a predefined list of n cities such that all cities are visited only once. We focus our attention on the 2D Euclidean TSP in which we are given a set of points in a plane, and the distance between two points is the Euclidean distance between their corresponding coordinates.

Local Search for TSP

The Lin-Kernighan heuristic (LKH) [5] is an efficient local search algorithm to solve TSPs with tens of thousands of cities. Although this is an incomplete algorithm it is known to provide near-optimal solutions within very short computational time.

The 2-opt operator [6] is one of the first local search move operators to solve TSP instances. Figure 2 shows an example of the operator by removing (black dotted lines) edges (a, b) and (c, d) from the current tour and reconstructing the tour, in the new solution, by connecting a with d and b with c (grey lines). Notice that after removing the edges in the 2-opt there is only one feasible way of reconnecting the tour. For instance, in our example adding edges (a, c) and

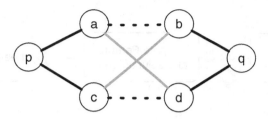

Fig. 2. Example of the 2-opt operator.

(b, d) will not result in a valid solution. In general, the k-opt move removes k edges from the current tour; k ranges from two to the number of cities. For $k > 2$ there are multiple ways of reconstructing the solution, in this case the algorithm typically uses the one with the shortest tour.

Currently, there are a large variety of heuristics to create the initial tour or solution such as Nearest Neighbour and Quick-Boruvka; see [7] for a complete description of the algorithms. However, LKH suggests the use of randomly generated tours as an initial solution. Other features of the LKH algorithm include the ability to restrict the use of certain edges within a given distance of a given city, and efficient data structures to check whether a solution is a valid tour or not.

Default Stopping Criterion. The algorithm stops when no improvement can be observed for a given solution, that is, when no neighbouring solution can improve the current tour. Additionally, we also use a certain time limit to stop the execution of the algorithm.

4   Machine Learning Methodology

Supervised Machine Learning exploits data labelled by the expert to automatically build hypotheses emulating the expert's decisions. Formally, a learning algorithm processes a training set $\mathcal{E} = \{(x_i, y_i), x_i \in \Omega, y_i \in \{1, -1\}, i = 1 \ldots n\}$ comprising n examples (x_i, y_i), where x_i is the example description (e.g. a vector of values, $\Omega = R^d$) and y_i is the associated output. The output can be numerical (i.e., regression) or a class label (i.e., classification). The learning algorithm outputs a hypothesis $f : \Omega \mapsto Y$ associating with each example description x the output $y = f(x)$. Among ML applications are pattern recognition, ranging from computer vision to fraud detection [8], predicting protein function [9], game playing [10], autonomic computing [11], etc.

In Fig. 3 we depict the general methodology to learn a stopping criterion for a given problem instance. Similar to other machine learning applications, the learning process takes place offline and involves computing features and the quality of the solution for each instance in the training set. Later on in the online testing phase, for each unseen instance, we compute the feature set and use the ML model to compute the expected outcome (i.e., solution quality) of the algorithm for the instance.

Offline

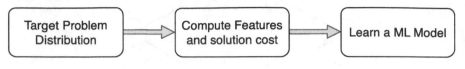

Online

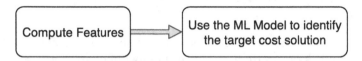

Fig. 3. ML methodology to learn a stopping criterion

In recent years, machine learning has been used extensively to design portfolios of algorithms. Informally, a portfolio of algorithms uses machine learning techniques to identify the most suitable algorithm to solve a given problem instance; see [12] for a recent survey. SATzilla [13] is probably the most popular portfolio of algorithms for SAT solving. Informally speaking, SATzilla uses a set of features to describe a given problem instance, the feature set encodes general information about the target instance, e.g., number of variables, clauses, fraction of Horn clauses, number of unit propagations, etc. Similar to our approach, SATzilla employs a training and testing phases. During the training phase the portfolio requires a set of target instances and a set of SAT solvers. During the testing phase, a set of pre-solvers are executed in a pre-defined manner for a short period and if no solution is obtained during the pre-solving time, the algorithm with minimal expected runtime is executed. SATzilla has shown impressive results during the past SAT competition winning several tracks in the annual SAT competitions.

Reinforcement learning, another branch of machine learning, has also been an effective alternative in the development of self-tuning algorithms. In contrast with the previous portfolio of algorithms approach, here the algorithms have the ability to adjust the parameters while solving a problem instance. An interesting application of these algorithms is the reactive search framework described and analysed in [14] where the authors proposed a mechanism for adjusting the size of the tabu list by increasing (resp. decreasing) its size according to the progress of the search.

CRP Instance Features

For each instance we compute two main type of features: basic and initial solution features. While a few of these features had already been used in the context of portfolio of algorithms for the TSP problem, many descriptors had to be added to consider properties of the CRP with distance, degree, and capacity constraints.

Basic Features. These encode general information of a given problem instance:

1. Number of nodes (1 feature);
2. Distance matrix (3 features): mean, median, standard deviation;
3. Distance to root-node (3 features): mean, median, standard deviation;
4. Pairwise distance (3 features): mean, median, standard deviation.

Initial Solution Features. For each instance we compute multiple initial solutions to extract the following features, and for each of the following five categories we compute the mean of 5 independent initial solutions:

1. Node out-degree (3 features): mean, median, standard deviation;
2. Root-node out-degree (3 features): mean, median, standard deviation;
3. Max. distance to root (3 features): mean, median, standard deviation;
4. Size of the cable tree networks (3 features): mean, median, standard deviation;
5. Number. of leaf nodes (3 features): mean, median, standard deviation.

TSP Instance Features

In [15] the authors describe 50 descriptors to characterise TSP instances. Computing the full set of features is computationally very expensive. Therefore we limit our attention to the following twelve features divided in three categories:

1. Problem size features (1 feature): number of cities;
2. Distance matrix features (3 features): mean, variation coefficient, skew;
3. Minimum spanning tree (8 features): after constructing the minimum spanning tree there are four features describing the distance (mean, variation coefficient, skew) and four features describing node degree statistics: mean, variation coefficient, and skew).

Stopping Criterion

The pseudo-code for a generic local search algorithm with the proposed stopping criterion is presented in Algorithm 1. The algorithm requires the target instance I, a regression model m, and a discrepancy $d \in (0, 1]$ indicating how far from

---

Algorithm 1. LocalSearchWithStoppingCriterion(*Problem-Instance I, ML-Model m, Discrepancy d*)

1: $v \leftarrow$ Compute-features(I)
2: $c \leftarrow$ Compute-expected-quality(m, v)
3: $target\_quality \leftarrow c + c \cdot d$
4: $s \leftarrow$ Initial-solution(I)
5: **while** default stopping criterion is not met and quality(s) > $target\_quality$ **do**
6:       $s \leftarrow step(I, s)$
7: **end while**
8: return s

---

the target solution should be from the expected solution. Certainly, the choice of a value close to zero for d leads to better solutions at a cost of using more time to reach the target solution.

The algorithm starts by computing the vector of features v for a given instance I (line 1), then we use the regression model m to compute the expected quality of the solution (line 2). Without loss of generality we assume a minimisation problem. Thus, we compute the target quality as the tolerance between the computed solution and the outcome of the regression model (line 3). Lines 4-7 sketch a general description of local search solvers, computing an initial solution (line 4), and iteratively moving from one solution to another using a given move operator, e.g., subtree or k-opt operators. We recall that we use the target solution in addition to the default stopping criterion of the algorithm. Thus, for instance, if the algorithm reaches a time limit we also stop the execution.

5   Empirical Evaluation

We consider a collection of 500 real-world CRP instances from a major broadband provider in Ireland. For TSP instances we randomly generated a set of 300 instances with the *portcgen* problem generator varying the number of cites from 500 to 5000.[1] We ran each local search algorithm 20 times on each instance (each time with a different random seed) with a 5-minute time cutoff and report the average time and solution quality for the instances. All experiments were performed on a 39-node cluster, each node features a Intel Xeon E5430 processors at 2.66 Ghz and 12 GB of RAM memory.

In order to validate our algorithms we used the traditional 10-fold cross-validation technique [16], that is, the entire dataset D is divided into 10 disjoint sets $\{D_1, D_2, \ldots, D_{10}\}$. For each dataset $D_{i \in 10}$, the regression model is learned with $D \setminus D_i$ and tested with D_i. In the following experiments we use the linear regression implementation available in the Weka toolbox (version 3.6.2) [17]. We use to state-of-the-art local search solvers [2,5] to tackle CRPs and TSPs, respectively.

We computed the baseline solutions using the default stopping criterion of each respective local search solver with a time limit of 300 seconds. We use this criterion as reference to compute the gap of the solutions as follows:

$$GAP(I) = \frac{ls\text{-}quality(I) - baseline\text{-}quality(I)}{baseline\text{-}quality(I)}$$

where *baseline-quality* and *ls-quality* indicate respectively the quality of the baseline solutions and the quality of the local search with the suggested stopping criterion for a given instance I.

We start our analysis with Table 1 in which we analyse the quality of the regression model to predict the quality of the baseline solutions. We report the correlation coefficient (CC) and the root mean-squared error (RMSE) between

---

[1] *portcgen* is available at http://dimacs.rutgers.edu/Challenges/TSP/codes.tar.

Table 1. Correlation coefficient and Root mean square error of estimations and runtime for the baseline computation

Problem	CC	RMSE	Runtime (s)	Feature time (s)
CRP	0.95	7.2	89.0	2.05
TSP	0.99	6.2	258.2	0.40

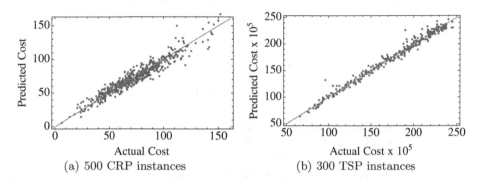

(a) 500 CRP instances

(b) 300 TSP instances

Fig. 4. Actual vs. Estimated quality for each CRP and TSP instance

the baseline solutions and our predictions, the average runtime and the average time for feature computation. As it can be observed the regression model greatly predicts the quality of the actual baseline solutions with a CC of 0.95 (CRP) and 0.95 (TSP), and RMSE of 7.2 (CRP) and 6.2 (TSP). We also remark that the average feature computation time is considerably lower than the actual average runtime of the algorithms. In Fig. 4 we provide a visual comparison of the actual baseline solution quality and the outcome of the regression model for the entire dataset. The diagonal line represents a perfect prediction, as it can be observed the predictions are highly correlated with the baseline results.

Fig. 5 shows the cumulative distribution of the feature computation time for all instances for both solvers. The feature computation time ranges from 0.6 to

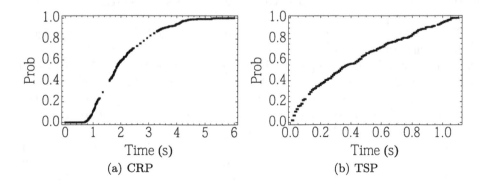

(a) CRP

(b) TSP

Fig. 5. Feature computation time

Table 2. CRP: statistics varying the discrepancy target to the expected solution

Discrepancy (%)	Time (s)	GAP(%)		KS(%)	WRS(%)
	avg	avg	std		
0	53.9	3.55	5.79	**55.9**	**20.4**
1	49.3	4.30	6.44	**36.9**	**9.63**
2	45.3	4.39	6.40	**32.9**	**8.46**
3	41.1	4.51	6.36	**25.7**	**7.45**
4	37.9	4.63	6.30	**22.6**	**6.44**
5	35.1	4.76	6.25	**22.6**	**5.56**
10	24.3	5.27	6.01	**9.5**	2.95
15	17.4	5.62	5.84	**6.9**	1.91

Table 3. TSP: statistics varying the discrepancy target to the expected solution, the last entry for the statistical tests are $5.4 \cdot 10^{-5}$ (KS) and $1.8 \cdot 10^{-5}$ (WRS)

Discrepancy (%)	Time (s)	GAP(%)		KS(%)	WRS(%)
	avg	avg	std		
0	153.4	0.76	1.72	**99.9**	**73.1**
1	121.9	1.09	1.91	**99.6**	**59.6**
2	84.61	1.54	2.11	**84.7**	**43.9**
3	58.7	2.10	2.32	**51.7**	**28.6**
4	39.0	2.73	2.48	**29.2**	**16.8**
5	22.8	3.43	2.62	**14.6**	**8.52**
10	5.5	7.50	3.11	0.01	0.02
15	3.7	11.51	3.62	0.00	0.00

5.7 seconds for CRP instances and from 0.01 to 1.1 seconds for TSP instances. We would like to recall that we use a subset of the complete feature set described in [15] for the TSP problem, the complete feature set required up to several minutes to compute the descriptors.

We now focus our attention on the evaluation of the local search algorithms with the suggested stopping criterion. To this end, we evaluate eight values for the minimum desired discrepancy of the target solution (i.e., 0–5%, 10% and 15%). Tables 2 and 3 depict the results showing: the average runtime, the gap with respect to the baseline solution, and the output of the Kolmogorov-Smirnov (KS) and Wilcoxon Signed-Rank (WSR) tests to check whether the solution with the new stopping criterion and the baseline solutions are statistically different or not. Bold entries indicate that the solution with the new stopping criterion is not statistically different than the baseline solution. Interestingly, in nearly all scenarios we observed that the results with the suggested stopping criterion are statistically the same as the baseline results, only 2 scenarios with KS and 4 scenarios with WSR failed the test with a 5% confidence level.

As expected, there is a trade-off between quality and speed. In our experiments we observed for CRP instances a reduction of up to 80% in the runtime to reach a solution 5.6% far from the baseline solution. Certainly, the runtime increases when enforcing better quality solutions, in particular we observed that for a discrepancy of 0% (i.e., setting as a target solution the predicted quality) the gap with respect to baseline solution is 3.5% and the runtime reduction is about 39%. A similar scenario can be observed for TSP instances where we observe a reduction of 97% (with discrepancy=5%) in the runtime to reach a solution with a gap of 3.4%, and enforcing the best possible quality we observe a solution with a gap of 0.7% with a reduction of 40% in the runtime with respect to the baseline solution.

We conclude our analysis with the cumulative distribution function (CDF) of the algorithms. Figure 6 describes the probability of a given algorithm to find a solution for a given instance with time (resp. quality) less than or equal to x. The x-axis denotes the time (resp. quality) and y-axis denotes the probability to reach a certain time (resp. quality). Figures 6(b) and (d) visually confirm the output of the KS and WRS tests, that is, the cost of the solutions are very close to the baseline for two discrepancy values of the CRP. For the TSP we

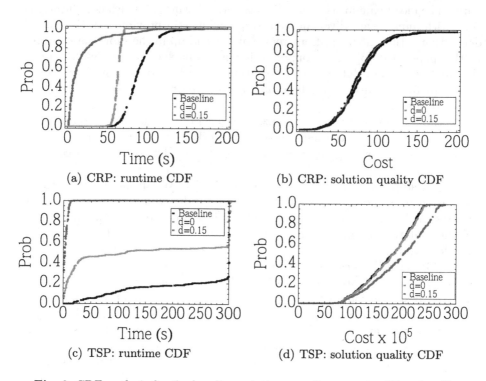

Fig. 6. CDF analysis for the baseline solutions vs. discrepancies 0% and 15%

observe that the baseline is very close to the baseline solutions for $d=0\%$ and slightly different for $d=15\%$. Figures 6(a) and (c) show that the baseline stopping criterion requires more time than the modified version of the algorithm with the suggested stopping criterion.

6   Related Work

Random restarts are often used to avoid getting trapped in local minima and search stagnation. Typically, the algorithm is stopped after completing a certain number of restarts or when a target solution (optimal or near-optimal) is obtained. However, in many real-world problems such a solution is not known. In [18] the authors propose a methodology to on-the-fly construct a probabilistic rule to estimate the probability of finding a solution at least as good as the current best solution in future restarts. The methodology starts by approximating the quality of the solution of the first k restarts with a Normal distribution. This distribution is then used to estimate the number of high quality solutions that will be observed in future restarts. The authors present empirical results with reliable predictions when k is large enough.

Alternatively, [19] uses the empirical distribution to identify the optimal number of diversification (or exploration) steps required to escape from a local minimum. The idea behind this approach is too avoid using too much computational time with unnecessary diversification steps, while still providing statistical evidence that future intensification steps will not end up in the same plateau area. In this paper we propose a stopping criterion to balance the trade-off between the quality of the solution and its runtime regardless of the methodology to escape plateaus and local minima, and without particular assumptions on the objective function.

In the field of continuous optimization several authors have proposed several stopping rules for the multistart framework. Here we briefly describe the basic ideas of a few stopping rules. We refer the reader to [20] for a complete description.

In [21] stopping rules are proposed based on Bayesian methods to stop as soon as there is enough evidence that the best value will not change in future restarts. A Bayesian stopping rule for GRASP is proposed in [22]. The authors assume that the distribution function is known beforehand and derive explicit rules for two particular cases.

In [23] the authors estimate the probability p that the best value observed in last restart is within certain threshold ϵ of the global optimum. p is approximated with the best solutions observed in previous restarts that at within ϵ of the incumbent solution. This approach was extended in [24] by counting the number of solutions that are within ϵ since the last update of the incumbent solution.

We remark that the above stopping rules for continuous optimization assume that the objective function satisfy certain properties, e.g., continuity, differentiable, Lipshitz condition, or that one should be able to find explicit formulas for the distribution function.

7   Conclusions

In this work, our objective was to study the application of machine learning techniques to carefully craft a stopping criterion for local search algorithms where the optimal solution is typically unknown beforehand. In particular, we use instance features to predict the cost of the solution for a given algorithm to solve a given problem instance. Interestingly, we have observed that machine learning can indeed provide very accurate estimations of the expected quality of two efficient local search solvers. We have observed that by using the estimation of the regression model we can reduce the average runtime by up to 80% for real-world CRP instances and by up to 97% for randomly generated instances with a minor impact in the quality of the solutions.

Further research will involve the use of the estimated target solution in the context of tree-base algorithms to tackle optimisation problems in two main directions: first reducing the computation time to reach the target solution, and second, pruning the search space by eliminating candidate solutions as soon as we are able to detect that no improvement is expected in the target solution.

Acknowledgments. This work was supported by DISCUS (FP7 Grant Agreement 318137) and Science Foundation Ireland (SFI) Grant No. 10/CE/I1853. The Insight Centre for Data Analytics is also supported by SFI under Grant Number SFI/12/RC/2289.

References

1. Davey, R., Grossman, D., Rasztovits-Wiech, M., Payne, D., Nesset, D., Kelly, A., Rafel, A., Appathurai, S., Yang, S.H.: Long-reach passive optical networks. J. Lightwave Technol. **27**(3), 273–291 (2009)
2. Arbelaez, A., Mehta, D., O'Sullivan, B., Quesada, L.: Constraint-based local search for the distance- and capacity-bounded network design problem. In: ICTAI 2014, Limassol, Cyprus, November 10–12, 2014, pp. 178–185 (2014)
3. Arbelaez, A., Mehta, D., O'Sullivan, B., Quesada, L.: Constraint-based local search for edge disjoint rooted distance-constrainted minimum spanning tree problem. In: CPAIOR 2015, pp. 31–46 (2015)
4. Arbelaez, A., Mehta, D., O'Sullivan, B.: Constraint-based local search for finding node-disjoint bounded-paths in optical access networks. In: CP 2015, pp. 499–507 (2015)
5. Helsgaun, K.: An effective implementation of the lin-kernighan traveling salesman heuristic. Eur. J. Oper. Res. **126**(1), 106–130 (2000)
6. Croes, G.A.: A method for solving traveling salesman problems. Oper. Res. **6**, 791–812 (1958)
7. Hoos, H., Stützle, T.: Stochastic Local Search: Foundations and Applications. Morgan Kaufmann, New York (2005)
8. Larochelle, H., Bengio, Y.: Classification using Discriminative Restricted Boltzmann Machines. In: ICML 2008, Helsinki, Finland, ACM 536–543., June 2008
9. Al-Shahib, A., Breitling, R., Gilbert, D.R.: Predicting protein function by machine learning on amino acid sequences - a critical evaluation. BMC Genomics **8**(2), 78 (2007)

10. Gelly, S., Silver, D.: Combining Online and Offline Knowledge in UCT. In: ICML 2007. vol. 227, pp. 273–280. ACM, Corvalis, Oregon, USA, June 2007
11. Rish, I., Brodie, M., Ma, S., et al.: Adaptive diagnosis in distributed dystems. IEEE Trans. Neural Netw. **16**, 1088–1109 (2005)
12. Kotthoff, L.: Algorithm selection for combinatorial search problems: a survey. AI Mag. **35**(3), 48–60 (2014)
13. Xu, L., Hutter, F., Hoos, H.H., Leyton-Brown, K.: Satzilla: portfolio-based algorithm selection for SAT. J. Artif. Intell. Res. **32**, 565–606 (2008)
14. Battiti, R., Tecchiolli, G.: The reactive tabu search. INFORMS J. Comput. **6**(2), 126–140 (1994)
15. Hutter, F., Xu, L., Hoos, H.H., Leyton-Brown, K.: Algorithm runtime prediction: methods & evaluation. Artif. Intell. **206**, 79–111 (2014)
16. Kahavi, R.: A study of cross-validation and bootstrap for accuracy estimation and model selection. In: IJCAI 1995, pp. 1137–1145 (1995)
17. Hall, M., Frank, E., Holmes, G., Pfahringer, B., Reutemann, P., Witten, I.H.: The weka data mining software: an update. SIGKDD Explor. **11**, 10–18 (2009)
18. Ribeiro, C.C., Rosseti, I., Souza, R.C.: Effective probabilistic stopping rules for randomized metaheuristics: GRASP Implementations. In: Coello, C.A.C. (ed.) LION 2011. LNCS, vol. 6683, pp. 146–160. Springer, Heidelberg (2011). doi:10.1007/978-3-642-25566-3_11
19. Bontempi, G.: An optimal stopping strategy for online calibration in local search. In: Coello, C.A.C. (ed.) LION 2011. LNCS, vol. 6683, pp. 106–115. Springer, Heidelberg (2011). doi:10.1007/978-3-642-25566-3_8
20. Pardalos, P.M., Romeijn, H.E.: Handbook of Global Optimization. Springer, Heidelberg (2002)
21. Boender, C.G.E., Kan, A.H.G.R.: Bayesian stopping rules for multistart global optimization methods. Math. Program. **37**(1), 59–80 (1987)
22. Orsenigo, C., Vercellis, C.: Bayesian stopping rules for greedy randomized procedures. J. Global Optim. **36**(3), 365–377 (2006)
23. Dorea, C.C.Y.: Stopping rules for a random optimization method. SIAM J. Control Optim. **4**, 841–850 (1990)
24. Hart, W.E.: Sequential stopping rules for random optimization methods with applications to multistart local search. SIAM J. Optim. **9**(1), 270–290 (1998)

Surrogate Assisted Feature Computation
for Continuous Problems

Nacim Belkhir[1,2](✉), Johann Dréo[1], Pierre Savéant[1], and Marc Schoenauer[2]

[1] Thales Research & Technology, Palaiseau, France
{nacim.belkhir,johann.dreo,pierre.saveant}@thalesgroup.com
[2] TAO, Inria Saclay Île-de-France, Orsay, France
marc.schoenauer@inria.fr

Abstract. A possible approach to Algorithm Selection and Configuration for continuous black box optimization problems relies on problem features, computed from a set of evaluated sample points. However, the computation of these features requires a rather large number of such samples, unlikely to be practical for expensive real-world problems. On the other hand, surrogate models have been proposed to tackle the optimization of expensive objective functions. This paper proposes to use surrogate models to approximate the values of the features at reasonable computational cost. Two experimental studies are conducted, using a continuous domain test bench. First, the effect of sub-sampling is analyzed. Then, a methodology to compute approximate values for the features using a surrogate model is proposed, and validated from the point of view of the classification of the test functions. It is shown that when only small computational budgets are available, using surrogate models as proxies to compute the features can be beneficial.

Keywords: Empirical study · Black-box continuous optimization · Surrogate modelling · Problem features

1   Introduction

Different optimization algorithms, or, equivalently, different parameterizations of the same algorithm, will in general perform non-uniformly on different classes of optimization problems. Today, it is widely acknowledged in the domain of optimization at large that the quest for a general optimization algorithm, that would solve all problems at best, is vain, as proved by different works around the *No Free Lunch theorem* [1,12]. Hence, tackling an unknown optimization problem amounts to choose the best algorithm among a given set of possible algorithms (*algorithm selection*), and/or the best parameter set for a given algorithm (*algorithm configuration*).

Such a choice can be considered as an optimization problem by itself, thus pertaining to the *Programming by Optimization* (PbO) paradigm proposed by [4]: given a new optimization problem instance (an objective function to minimize or maximize), a set of algorithms with domains for their parameters, and

© Springer International Publishing AG 2016
P. Festa et al. (Eds.): LION 2016, LNCS 10079, pp. 17–31, 2016.
DOI: 10.1007/978-3-319-50349-3_2

a performance measure, find the best[1] algorithm or parameter setting to solve this problem. However, such a meta-optimization problem is in general difficult to solve (hierarchical search space, multi-modal landscape, ...) and thus requires running different algorithms with different parameterizations, where each of these runs will in turn call the objective function a large number of times: in total, the number of calls to the objective function will be huge, making such an approach intractable in most real-world situations with expensive objective functions.

The PbO approach can however be applied to classes of objective functions: the best algorithm/parameter setting can be learned off-line, once and for all for a given class of functions, and the optimal setting (algorithm + parameters) applied to all members of that class. This is the case for instance in operational contexts, where the same type of problem, but with slightly different settings, has to be solved again and again. However, in the general case of black-box optimization problems, very little domain knowledge is known (the type of search space for sure, maybe some relevant parameters) and such characteristics are not sufficient to reliably choose an algorithm and its parameters.

Another approach is then to compute some characteristics of objective functions, aka *features*, without much domain knowledge, and, thanks to a large example base of algorithms performances on known objective functions, to learn a *performance model* of several algorithms and their parameters. A well-known success in that direction has been obtained in the SAT domain [13], but dozens of features had been proposed in the literature to describe SAT problems and try to understand what makes a SAT problem hard for this or that algorithm. This situation is quite unique, and is an appeal for research regarding the design and study of features in other domains.

In particular, in the domain of continuous optimization (the search space is a subspace of $\mathbb{R}^d$, for some d), several recent works [8,9,11] have proposed many different features to try to understand the landscape of continuous optimization and, ultimately, solve the Algorithm Selection and/or Configuration problem. Note that a large body of mathematical programming algorithms exist, and are proved to be optimal for specific classes of objective functions: Linear Programming should be applied if the function (and the constraints) are linear; gradient-based algorithms should be applied if the function is convex, twice differentiable and well-conditioned, etc. But these are rare exceptions in the real world, where the general problem remains open, and the feature-based approach seems worth investigating, in particular considering the promising initial results obtained by [9] (more in Sect. 2).

Unfortunately, the computation of all proposed features is based on many sample points, i.e., values of the objective function at given (generally random) points of the search space. In real-world situations, where the objective

---

[1] The performance measure generally involves time-to-solution (CPU time, or number of function evaluations) and precision/accuracy of the solution returned by the algorithm (precision of the solution for continuous optimization problems, number of constraints violated for Constraint Programming problems, etc.).

function is expensive and the computational budget limited, the computation of the features as proposed in [9] might simply be impossible. A first research question is hence to study how badly the features behave when the size of the sample,used to compute those features, decreases.

Yet, a prominent approach has already been proposed to handle expensive objective functions in optimization. It relies on *surrogate models*, i.e., models of regression or interpolation of the objective function built upon sample points gathered during the run of the algorithm, and used from time to time in the optimization algorithm in lieu of the actual objective function.

Building on this idea, the present work investigates the use of surrogate models to compute problem features: based on a small sample set, a surrogate model is developed. The features of the surrogate model can then be easily computed, since the cost of evaluating a surrogate model is negligible compared to the original objective function. The second issue investigated in this paper relates to the accuracy of the features of the surrogate model as approximations of the features of the original objective function, and will be studied here experimentally, using the well-known BBOB test set of functions [3]. Note that this paper will not directly tackle the algorithm selection or algorithm configuration problem, left for further work. The accuracy of feature sets will be assessed first by a direct comparison with the features computed using a very large sample set, as well as by their abilities to correctly recover the BBOB (hand-designed) classes, as in [9] (see Sect. 4 for more details on the BBOB testbench).

The paper is organized as follow, Sect. 2 surveys the features proposed in the literature. Section 3 will introduce the methodology proposed in this paper. Section 4 describes the experimental context, while Sect. 5 describes the experimental protocol in detail, including the performance measures used to validate the results presented in Sect. 6. Finally these results are summarized and hints for further works are given in Sect. 7.

2   Problem Features for Continuous Optimization

Let $\mathcal{F}$ be the objective function at hand, defined on a domain $\mathcal{D} \subset \mathbb{R}^d$ for some given dimensions d: d is the only high-level feature given a priori as domain knowledge. All (other) features considered here will be computed from a sample set $\mathcal{X} = (x_1, \ldots x_p)$ of points of $\mathcal{D}$, and the corresponding values of the objective function $\mathcal{Y} = (y_1, \ldots, y_p)$ (with $y_i = \mathcal{F}(x_i)$ for all i).

A first series of features used here is taken from [9], where low-level features are computed directly from $(\mathcal{X}, \mathcal{Y})$. High-level features,or fitness landscape properties(e.g. multi-modality,separability, convexity,etc), are then built from different statistics on these low-level features. A total of 62 low-level features is computed, giving in turn 8 high-level features. The low-level features are grouped in the following classes:

- *Distribution*: these features consider the distribution of objective values in $\mathcal{Y}$, computing the skewness and the kurtosis of the distribution, but also estimating the number of peaks of the distribution.

- *Level Set*: from a given threshold in the objective function values (e.g., some quantiles), a classification technique such as LDA, QDA or MDA is trained to predict the position w.r.t. the threshold. The distributions of the misclassification errors for the different classifiers are used as proxies to the multi-modality of the objective function.
- *Meta-Model*: First, linear and quadratic regression models for $(\mathcal{X}, \mathcal{Y})$ are computed. The resulting R^2 coefficients for the accuracy of the models, as well as some statistics on the relative sizes of the coefficients of the models give some indication on the shape of the landscape.
- *Convexity*: estimate the probability of convexity and linearity of the objective function by selecting two random points from the sample $\mathcal{X}$ and generating a new sample point in the segment and comparing the objective value of the new sampled point and the same convex combination of the two initial objective values. The probability of convexity is computed by averaging the number of trials where the computed difference is lower than a predefined negative threshold, while the probability of linearity is computed by considering all absolute differences that are smaller than the absolute value of the predefined threshold.
- *Curvature*: considers the numerical approximation of the gradient in each point of a sub-sample of $\mathcal{X}$ by the Richardson's extrapolation method, and the resulting features consider the basic statistics of the respective derivatives, the condition number of the similar numerical approximation of the Hessian.
- *Local Search*: a local search algorithm, e.g., Nelder-Mead, is run from starting points in a sub-sample of $\mathcal{X}$; the final solutions of all runs are clustered in order to identify the local optima of the objective function, the basin sizes are approximated by the number of local searches which terminate there, and other statistics gathered during the different runs give other indicators of the landscape properties.

Furthermore, a series of features termed *Dispersion* was originally proposed in [8]. Their computation analyzes the distance between candidate solutions for a percentile of the best solutions of the optimization problem by comparing them to the mean or median distance between solutions in the whole initial sample. Different percentiles are considered, giving in total 16 features.

Finally, features related to *Information Content* have been proposed in [11]. They are related to the number of the binary decisions needed to find the information, such that each candidate solution is binarized w.r.t. the fitness value of their nearest neighbors' fitness. These metrics give information about the smoothness, or the existence of a global structure of the objective function. A total of 5 dispersion features are considered in [11].

The recent works that defined those features [9,11] successfully demonstrated that these features could be used in order to classify the optimization problems w.r.t. their BBOB classes (see definition in Sect. 4). However, some of these features (local search, curvature, convexity) require additional samples. On the other hand, in [2,11], only features that can be computed with a fixed initial sample are considered; and the results demonstrate that such limited number of features (33 features) can nevertheless be used to correctly identify the BBOB classes. It should be emphasized, however, that all the above-mentioned works consider large samples of candidate solution (between $500 \times d$ and $5000 \times d$), which is not practical if the objective function is expensive.

3 Surrogate Assisted Feature Computation

3.1 Sub-sampling

Real world applications where numerical simulations are involved, usually result in very expensive objective functions, for which the computation of a single value might take a few hours. In such case, the features described in previous Sect. 2 might help choosing the right algorithm, and/or the right parameters of a given algorithm, thus saving computation time for its optimization process. The computation of these features should not cost more than the optimization – which might be the case if around $10^3 \times d$ evaluations are needed, as in [2,9,11]. Hence making the use of features no applicable in real world applications.

A first solution is of course to simply use a smaller sample set. But the price to pay will be a poor approximation of the actual features, which in turn might result in a wrong choice for the optimization algorithm or its parameters. As it can be expected (see Sect. 6.1), a too small sample set results in a poor approximation with a high variance w.r.t. the choice of the sample set.

3.2 Surrogate Modeling

Coming from another field, numerical engineers have tackled this problem using *Response Surface Methods* for many decades now: after few iterations of any optimization algorithm, many points of the search space have already been evaluated, and this sample set can be used to build a *surrogate model* of the objective function, that can in turn be used in the optimization algorithm as a proxy for the actual objective, being costless to compute (see e.g., [6,7] for surveys in the engineering domain and in in Evolutionary Computation domain, respectively). The most critical issue to be addressed when using surrogate modelling techniques is the choice of the function space where to look for a model. Several approaches have been proposed in the literature to solve such regression problem, from Neural Networks to Gaussian Processes (aka Kriging) to Support Vector Machines and Regression Random Forests, and it is beyond the scope of this paper to describe them in detail.

Although a surrogate model to be used within an optimization process should be accurate, taking into account the local characteristic of the problem. A surrogate model whose purpose is to compute features of the function at hand should globally approximate the objective function. Both goals are different, and this should be taken into account.

3.3   Accuracy vs Efficiency

The baseline of the approach proposed in this paper is to use small sample sets to build a surrogate model of the objective function at hand, to compute the features of that surrogate model using as many samples as needed, and to use these features in lieu of the unreachable actual features of the true objective function.

However, this approach must be validated – empirically – against the simple (without surrogate) sub-sampling approach (compute the features on small sample sets). Furthermore, some parameters of the approach (the model for the surrogate and its hyper-parameters, the number of samples to be used for evaluating the features of the surrogate model) must also be tuned. This raises the question of what measure should be used to assess the quality of the approximated features.

An obvious measure is simply the error made on the feature values: for each feature, the "exact" value can be computed using as large sample sets as needed (say, same sizes than in [2, 9, 11]). Then the L^2 norm of the error vector, difference between the approximated and "exact" values for all features gives a first idea of how good the approximation is.

However, the ultimate reason for computing the features is to solve the algorithm selection and/or configuration problem. And it might be the case that the error in solving the latter problem varies differently from the L^2 error on the feature values (e.g., some features that are poorly approximated are also not important for the algorithm selection and/or configuration).

Because there is not yet any standard algorithm selection or classification problem that would allow us to do such a validation, some obvious proxy classification problems that are used in [2, 9, 11], retrieving the known classes of problems manually defined on BBOB testbench (described in next Section). Next Section will give the technical details of these experiments, and their results will be presented in the following Section.

4   Experimental Settings

BBOB testbench

All the experiments presented here use test functions from the Black Box Optimization Benchmark (BBOB)[2] [3]. The BBOB benchmark contains 24 analytically defined continuous objective functions with known global optimum. All these functions are defined on the d-dimensional domain $[-5, 5]^d$, with $d \in \{2, 3, 5, 10, 20, 40\}$. In order to avoid possible biases, 15 variants of each function are considered, obtained from the original function by a translation of the position of the optimum, plus a rotation of the coordinate system for the non-separable functions.

These 24 functions have been manually classified in five classes of problem, with 5 separable functions, 4 uni-modal functions with low or moderate conditioning, 5 uni-modal functions with high conditioning, 5 multi-modal

---

[2] http://coco.gforge.inria.fr.

functions with regular global structure, 5 multi-modal functions with weak global structure.

Sample Sets
All sample sets are drawn uniformly on $[-5,5]^d$ for a BBOB function in dimension d (see above). In the remaining of the paper, all sample set sizes are normalized w.r.t. the dimension d of the definition domain of the objective function. For the sake of brevity, we will only mention the ratio between the sample set size and the dimension: "a sample of size k" will actually mean "a sample of size $k \times d$". In all experiments, $k \in \{30, 50, 100, 500, 1000, 2000\}$ (the largest value 2000 was decided after the first experiments, see Sect. 6.1).

Features
As discussed in Sect. 2, only part of the features described in [8,9,11] will be used here, namely: 3 Distribution features, 9 Meta-Model features, 16 Dispersion features, and 5 Information Contents features. All considered features are implemented as an R package publicly available at http://github.com/flacco, thanks to Pascal Kerschke.

Surrogate Models
As discussed in Sect. 3.2, several approaches to surrogate modelling will be used and compared: Gaussian Processes, Random Forests, Support Vector Machines with polynomial or RBF kernel, denoted respectively GP, RF, SVM_P and SVM_{RBF}. But learning a surrogate model requires some hyper-parameters to be tuned: a grid search is performed in the hyper-parameter space (4 parameters for GP, 5 for the other models), using a 5-folds cross-validation procedure — randomly re-sample (Bootstrap) the sample set, using 80% for training the surrogate and the remaining 20% for testing— for 300 iterations, optimizing the approximation accuracy on the test set. All surrogate modelling procedures are implemented w.r.t the python scikit-learn library[3].

5   Experimental Protocol and Validation

Experimental Protocol and Notations
For a given objective function $\mathcal{F}$ (one trial of one instance of one $d-$dimensional function from BBOB), the basic experiment goes as follows: one sample sets of given size is drawn from the definition domain of $\mathcal{F}$; the features are computed on this sample set, and a surrogate model using one of the chosen modelling techniques. The sample set is then completed with more samples, using the surrogate model in lieu of the original function. Approximate features are then computed using this extended sample set.

An immediate validation of such approximated feature values can be made by comparing them to the "exact" values: a proxy for these values will be the features computed with the largest initial sample set, of size 2000 (see Sect. 6.1). However, the global validation (see below) requires to compute such

---

[3] http://scikit-learn.org/.

approximated features for all BBOB functions, and for several different sample set sizes.

Let $\mathcal{X}^s, s = 1, \ldots, S$ be some sample sets from the domain of $\mathcal{F}^4$. Features $\Phi(\mathcal{F}^s)$ (vector of $\mathbb{R}^F$ if F is the number of features) are computed for $\mathcal{F}$ from sample set $(\mathcal{X}^s, \mathcal{Y}^s)$ (with $y_i = \mathcal{F}(x_i)$ for all $(x_i, y_i) \in (\mathcal{X}^s, \mathcal{Y}^s)$, see Sect. 2). Let us denote by $\Phi(\mathcal{F}^*)$ the features computed from the largest sample set (of size 2000).

Each sample set $\mathcal{X}^s$ is also used to learn some surrogate models $\widehat{\mathcal{F}_t^s}$ using different surrogate modelling techniques $t = T_1, \ldots, T_T$. For each (s, t), the set of features $\Phi(\widehat{\mathcal{F}_t^{s,s'}})$ is computed for $\widehat{\mathcal{F}_t^s}$, after completing the sample set $\mathcal{X}^s$ with s' new points using $\widehat{\mathcal{F}_t^s}$, resulting in sample set $(\widehat{\mathcal{X}_t^{s,s'}}, \widehat{\mathcal{Y}_t^{s,s'}})$ of size $s + s'$ (i.e., $\mathcal{X}^s \subset \widehat{\mathcal{X}_t^{s,s'}}$ and, for all $(x_i, y_i) \in (\widehat{\mathcal{X}_t^{s,s'}}, \widehat{\mathcal{Y}_t^{s,s'}})$, $y_i = \mathcal{F}(x_i)$ if $x_i \in \mathcal{X}^s$ and $y_i = \widehat{\mathcal{F}_t^s}(x_i)$ otherwise). All approximate features $\Phi(\mathcal{F}^s)$ and $\Phi(\widehat{\mathcal{F}_t^{s,s'}})$ can then be compared to the "exact" features $\Phi(\mathcal{F}^*)$, and their accuracy assessed using the L^2 norm in $\mathbb{R}^F$ of the errors on the feature values ($Err(\mathcal{F}^s) = ||\Phi(\mathcal{F}^s) - \Phi(\mathcal{F}^*)||_2$, $Err(\widehat{\mathcal{F}_t^s}) = ||\Phi(\widehat{\mathcal{F}_t^{s,s'}}) - \Phi(\mathcal{F}^*)||_2$).

However, as discussed in Sect. 3.3, another comparison is needed between the approximate features and the "exact" values, that relates to the ability of the approximate features to be sufficient to correctly classify the BBOB classes. Such validation requires the computation of all approximate features with same sizes of sample sets for all instances of functions of BBOB testbench.

Classification Efficiency

However, measuring the efficiency of a set of approximated features as a whole goes through using them as input for learning a classifier in order to discriminate the five BBOB classes. This is done using a 5-folds cross-validation, repeated 100 times, procedure as follows. Let us denote $Cl(\mathcal{F})$ the class (in 1..5) a given function $\mathcal{F}$ belongs to. For a given sample set size s, the example set for the classification task consists of $(\Phi(\mathcal{F}^s), Cl(\mathcal{F}))$ pairs when dealing with features computed on $\mathcal{F}$ and $(\Phi(\widehat{\mathcal{F}_t^{s,s'}}), Cl(\mathcal{F}))$ when dealing with surrogate model t built on $\mathcal{F}$ ($t \in \{GP, RF, SVM_P, SVM_{RBF}\}$). Such example set is made of 5 trials × 5 instances × 24 functions × 5 dimensions. Out of these 3000 examples, 80% are randomly chosen *without replacement and avoid any overlapping the training and test set*, equally distributed in the 5 classes, to build the training set, on which a Random Forest classifier is trained (with default hyper-parameters from scikit-learn).

The accuracy of the resulting classifier should then be assessed on the remaining 20% of the global example set. However, different scenarii are possible in real-world situations. A first scenario is when the training phase is done on easy functions, for which it is possible to compute the features with large enough sample sets, and the unknown functions on which to perform algorithm selection/configuration (the test phase) are all expensive. An "orthogonal"

---

[4] By abuse of notation, s will denote both the sample set and the (normalized) size of the sample set.

scenario is when the functions available for training are also expensive. In the latter case, only approximate features will be available for training, either computed on small samples, or computed using a surrogate of the functions used for training.

Two situations similar to the ones described above will be experimented with here, involving different sample sizes for the training of the classifier and its test.

When no surrogate model is involved (Sect. 6.1), an experiment studying the efficiency of the approximate features as a basis for classification (Sect. 6.1) is defined with only two parameters: the size s_{train} of the sample set used to learn the features for the training set, and the size s_{test} of the sample set used to learn the features for the test set. The classification accuracy of the resulting classifier will be denoted $Eff(s_{train}, s_{test})$.

But when analyzing the efficiency of surrogate assisted feature computation (Sect. 6.2), an experiment is defined with 5 parameters: the type T of surrogate model (in $\{GP, RF, SVM_P, SVM_{RBF}\}$), and, for both the training features and the test features, the sizes of the original sample sets used to learn the surrogate models (respectively s_{train}^{org} and s_{test}^{org}), and the additional number of points added to these original sample sets using the surrogate models (respectively s_{train}^{surr} and s_{test}^{surr}). The classification accuracy of the resulting classifier will be denoted $\widehat{Eff}(T, s_{train}^{org}, s_{train}^{surr}; s_{test}^{org}, s_{test}^{surr})$. Note that if one of the s_{train}^{surr} or s_{test}^{surr} is 0, only the true values of $\mathcal{F}$ are used in the corresponding step. In particular $\widehat{Eff}(T, s_{train}^{org}, 0; s_{test}^{org}, 0) = Eff(s_{train}^{org}, s_{test}^{org})$ (the surrogate model is never used).

6   Results

Two series of experiments will be presented here. The first one (Sect. 6.1) doesn't involve any surrogate model, and aims at studying how the features diverge from their "exact" baseline values when the size of the sample decreases. The goal of the second series (Sect. 6.2) is to check whether using a surrogate model built on the same available small sample set to complement it can help to cope with such divergence. In both series, the divergence with the baseline values will be assessed by the accuracy of the approximated values (individual comparison for a given feature and a given function, and their aggregation in the L^2 error – Sects. 6.1 and 6.2 respectively), and by the efficiency of the whole set of approximate features, using them to discriminate the 5 BBOB classes, as explained in previous Section – Sects. 6.1 and 6.2 respectively).Due to space constraints, only few typical figures are displayed[5].

6.1   Effects of Sub-sampling

Accuracy of Sub-Sampled Features. Five test functions (F1, F8, F13, F17, F23, one per BBOB class) are used here to assess the effect of sub-sampling

---

[5] Additional plots are available at https://drive.google.com/open?id=0B9GuQcCjvwt FM2VLeVEyMGtFQnM.

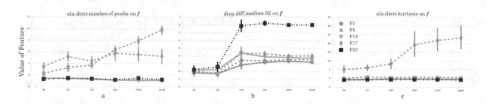

Fig. 1. Examples of effect of sub-sampling (x-axis) on feature values for the different test functions in dimension $d = 5$.

on the feature values. Figure 1 shows some typical feature behaviors on those 5 functions: whereas plot (a) display a smooth behavior, where feature values stabilize for $s \geq 100$, both other plots show that even $s = 2000$ might still be somehow too small for the multi-modal functions F17 and F23. However, most features on most functions exhibited a smooth behavior, and were stable for $s \geq 500$, justifying the decision to take $\Phi(\mathcal{F}^{2000})$ as $\Phi(\mathcal{F}^*)$. It is also clear from Fig. 1 that small sample sizes (e.g., 30) will provide a poor approximation of the feature values, and might not allow to discriminate among different functions.

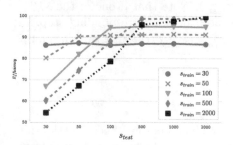

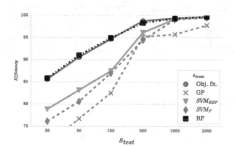

Fig. 2. $Eff(s_{train}, s_{test})$ vs s_{test} for different values of k_{train}.

Fig. 3. $Eff(s, s)$ (black circles) and $\widehat{Eff}(T, s, *; s, *)$ for different T.

Efficiency of Sub-Sampled Features. Figure 2 displays the efficiency of the approximated features to discriminate among BBOB classes (see Sect. 5). Each line corresponds to a sample size s_{train} used to train the classifier, and each point corresponds to a different sample size s_{test} used to compute the features of the test instance to be classified.

Two conclusions can be drawn from this figure. First, there is no reason to use a larger test sample size s_{test} than the size s_{train} that was used to train the classifier, as the efficiency does not increase after s_{test} has reached s_{train}. Second, if you know that only a limited budget will be available at test time (i.e., all new instances will be very expensive), then you should train the classifier with a small budget too: for a given s_{test}, the best efficiency is obtained by the classifier trained with $s_{train} = s_{test}$.

A possible explanation, to be investigated deeper in further work, is that sub-sampling does not only increase the variance of the feature values, but it also induces some bias that might be also tracked by using the same sample size for training than for testing.

6.2   Surrogate Assisted Feature Approximation

In this Section, we try to improve the accuracy and efficiency of the approximate features by adding samples computed using some surrogate model, as described in Sect. 3.

Accuracy of Surrogate-Assisted Features. Figure 4 displays feature values for 3 different features (columns) and 2 different surrogate models (top: Gaussian Processes, bottom: Random Forests), for the 5 test functions F1, F8, F13, F17, F23. The effects of small sample sizes are here rather similar to those on the function alone as displayed in Fig. 1. However, it should be noted that the different surrogate models can give different behaviors on the same feature.

An alternative point of view and a comparison with the approximated features directly computed using the initial small sample with exact objective values is given in Fig. 5. Here, the Random Forest surrogate model is used to add 2000 points to the initial sample set. It is clear (and results on other feature confirm this trend) that Random Forests give a much smaller error than Support Vector Machines (with both polynomial and RBF kernel), and even more so with Gaussian Processes (not shown here for space reason). More interestingly, using the Random Forest surrogate model often results in more accurate approximate features than computing their values only on the few available exact values (see the $s = 30$ histograms on Fig. 5).

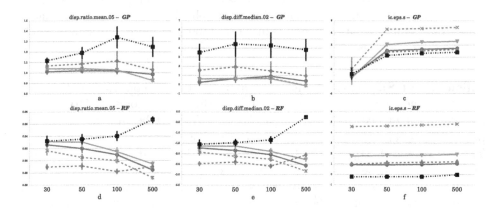

Fig. 4. Values of 3 features vs initial sample size s, for the 5 test functions ($d = 20$), using a surrogate model (GP for (a, b, c), RF for (d, e, f)) to add $s' = 2000$ points to the sample set.

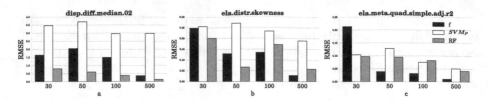

Fig. 5. Accuracy error for 3 features on F13 ($d = 5$), for different values of s (x-axis). For each s, $Err(s,s)$, $Err(\widehat{\mathcal{F}_{SVM_P}^{s,2000}})$ and $Err(\widehat{\mathcal{F}_{RF}^{s,2000}})$ are plotted.

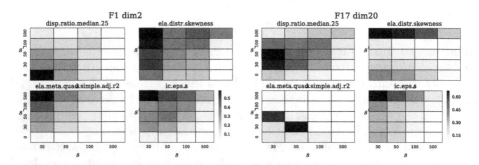

Fig. 6. Accuracy error of 4 features on F1, d=2 (left), and on F17, d=20 (right). For each (s, s'), the error $Err(\widehat{\mathcal{F}_{RF}^{s,s'}})$ is plotted (the darker the higher).

But whereas adding as many points as possible (2000 here) using the surrogate model seems a natural way to go, one can wonder if this is always the best thing to do. And the answer is no: Fig. 6 plots the error on the feature values obtained when adding s' (y-axis) sample points using a surrogate model (here, a Random Forest), to a sample set of size s (x-axis). In many cases, the error increases with s', or displays an unstructured behavior, in particular for small values of s. Only for large-enough values of s (500 and more, not shown here) does the error decrease.

Efficiency of Surrogate-Assisted Features. Let us now look at the efficiency of the approximated features to discriminate among BBOB classes (as explained in Sect. 5). Figure 3 displays $Eff(s,s)$ (the upper hull of the plots on adjacent Fig. 2) as well as the different $\widehat{Eff}(T,s,*;s,*)$, for $T \in [GP, SVM_P and SVM_{RBF}, RF]$. It is again obvious here that the Random Forest model outperforms all others, which is consistent with previous results (as well as with all other not presented results). From now on, only RF surrogate model will be considered. But another observation that can be made here is that there is no gain to be expected by using the surrogate model during the learning phase too, as both plots for $Eff(s,s)$ and $\widehat{Eff}(RF,s,*;s,*)$ are almost identical.

A final experiment will try to answer the main question that motivated this work: can the use of a surrogate model improve the efficiency of the approximated

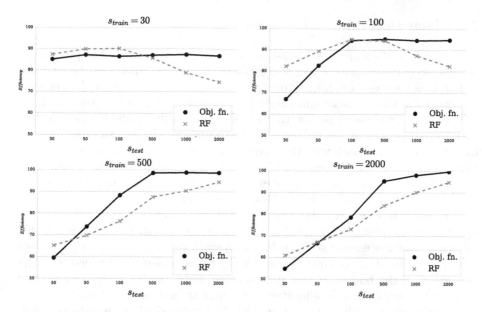

Fig. 7. Comparison, for different values of s_{train}, of the efficiency of sub-sampled features $Err(s_{train}, s_{test}$ (continuous black line) with that of surrogate assisted features $\widehat{Eff}(RF, s_{train}, 0; s_{test}, *)$ (grey dotted line).

Table 1. Mean and standard deviation of the efficiency of sub-sampled features $Eff(s_{train}, s_{test}$ (columns Obj. Fn.) and surrogate assisted features $\widehat{Eff}(RF, s_{train}, 0; s_{test}, *)$ (columns RF). Statistically significantly better results (Wilcoxon signed test with 95% confidence) are in bold.

$s_{train} = 30$			$s_{train} = 100$			$s_{train} = 500$			$s_{train} = 2000$		
s_{test}	Obj. fn.	RF	s_{test}	Obj. fn.	RF	s_{test}	Obj. fn.	RF	s_{test}	Obj. fn.	RF
30	85.1±1.6	**87.4±1.5**	30	66.9±3.7	**82.2±1.5**	30	59.5±3.7	**65.2±3.1**	30	54.8±3.5	**60.9±3.1**
50	87.3±1.4	**90.0±1.5**	50	82.6±3.0	**89.3±1.2**	50	**73.8±2.1**	69.8±3.1	50	66.6±3.0	**67.1±3.3**
100	86.5±1.5	**90.2±1.2**	100	94.6±0.9	**95.1±1.0**	100	**88.3±2.0**	76.3±2.8	100	**78.6±3.6**	73.1±3.0
500	**87.0±1.6**	85.6±2.0	500	**95.2±1.1**	94.3±1.0	500	**98.6±0.5**	87.6±1.8	500	**95.3±1.1**	84.0±2.4

features in the context of expensive objective functions? Fig. 7 displays 4 plots, corresponding to different values of s_{train}. On each plot, the black continuous line is $Eff(s_{train}, s_{test})$, i.e., the corresponding line of Fig. 2, and the dotted grey line shows $\widehat{Eff}(RF, s_{train}, 0; s_{test}, *)$, i.e., the efficiency obtained when using the RF surrogate model to augment the sample set with 2000 new samples. And indeed, there is some advantage in using the surrogate model during the test phase, the more so for small training budgets. Furthermore, this advantage of using the surrogate is statistically significant, as witnessed in Table 1 where the same data are given together with the standard deviations. Figures in bold are statistically better than the corresponding non-bold figures according to a Wilcoxon signed rank test with 95% confidence.

7   Discussion and Conclusion

In the context of continuous black-box optimization, this paper has proposed a methodology to compute features describing the characteristics of the problem at hand, as proposed in [2,9,11], using surrogate models to cope with expensive objective functions: to-date methods to compute such features rely on large sample sets of evaluated points, that are not practically available when dealing with expensive real-world problems. The performance of approximated features have been measured both with their *accuracy*, related to the error on their values when compared to values computed on very large sample sets, and with their *efficiency*, ability to train a classifier that can correctly discriminate the five classes of the test functions in the BBOB testbench [3].

The paper has first experimentally studied the loss of accuracy of the features due to sub-sampling, identifying a reasonable sample size beyond which the computed features exhibit stable values with small variance – 2000× the problem dimension. The study of the efficiency of the sub-sampled features led to a first conclusion: if the budget at test time is going to be small (expensive objective function), then the budget allocated to training should be small too.

Experiment involving surrogate models first surprisingly demonstrated that, in the context of the present work, only Random Forest surrogate models gave satisfactory results when used to augment the sample set used to compute the features, whereas for instance Gaussian Processes, very popular today when it comes to expensive optimization per se, gave the worst results of all.

But the most interesting observation is that when only few samples are available, using a surrogate model built on these small sample sets to augment the sample set can lead to better classification results (on BBOB classes) than the sub-sampled features directly computed on the small available sample sets. However, if some large training budgets are available, i.e., if there exists cheap functions that are representative enough of the expensive real-world objective functions that will be encountered later, then using a surrogate model on these expensive unknown function does not seem to be beneficial.

There are several directions for further work. First, several features that have been proposed in the literature have been discarded because they require additional function evaluations (e.g., the local search, curvature, and convexity features [9]). Surrogate models might also be useful in order to compute those features at low cost. Such progress would then make even more important to use some feature selection method rather than using all features for the classification task at hand.

Different paths can also be explored regarding the way the surrogate models are computed. First, as of today, only global surrogate models are considered. But it might be interesting to restrict the scope of the surrogate learning to better compute local search features for instance. Similarly, the performance measure used to construct the surrogate models is the standard approximation error on the known samples. But in the context of feature computation, some other measures could be considered – ultimately, surrogate models should be optimized for the quality of the approximate features they allow to compute.

Only global results on all features have been presented here. In particular, it could be the case that different surrogate models are the best choice to compute different features (e.g., SVM for convexity features, random forests for multi-modality features, etc.). A deeper insight on feature accuracy is needed here. Along the same line, some results presented here suggest that there might be a systematic bias induced by sub-sampling – something that requires deeper analysis too.

Last but not least, the link between those features and Algorithm selection and configuration remains to be established. Indeed, retrieving BBOB classes is a much easier problem than learning an empirical performance model, as in [5] for instance. Or those features might be used together with some latent features identification method, as in [10].

References

1. Auger, A., Teytaud, O.: Continuous lunches are free plus the design of optimal optimization algorithms. Algorithmica **57**, 121–146 (2009). https://hal.inria.fr/inria-00369788
2. Bischl, B., Mersmann, O., Trautmann, H., Preuß, M.: Algorithm selection based on exploratory landscape analysis and cost-sensitive learning. In: Proceedings of the 14th Annual Conference on Genetic and Evolutionary Computation, pp. 313–320. ACM (2012)
3. Hansen, N., Auger, A., Finck, S., Ros, R.: Real-parameter black-box optimization benchmarking 2010: experimental setup. Technical report RR-7215, INRIA (2010)
4. Hoos, H.H.: Programming by optimization. Commun. ACM **55**(2), 70–80 (2012)
5. Hutter, F., Xu, L., Hoos, H.H., Leyton-Brown, K.: Algorithm runtime prediction: methods and evaluation. Artif. Intell. **206**, 79–111 (2014)
6. Jin, R., Chen, W., Simpson, T.W.: Comparative studies of metamodeling techniques under multiple modeling criteria. Struct. Multidiscip. Optim. **23**, 1–13 (2000)
7. Jin, Y.: A comprehensive survey of fitness approximation in evolutionary computation. Soft Comput. **9**(1), 3–12 (2005)
8. Lunacek, M., Whitley, D.: The dispersion metric and the CMA evolution strategy. In: Proceedings of the 8th Annual Conference on Genetic and Evolutionary Computation, pp. 477–484. ACM (2006)
9. Mersmann, O., Bischl, B., Trautmann, H., Preuss, M., Weihs, C., Rudolph, G.: Exploratory landscape analysis. In: Proceedings of 13th GECCO, pp. 829–836. ACM (2011)
10. Mısır, M., Sebag, M.: Algorithm selection as a collaborative filtering problem. Technical report, INRIA-Saclay (2013). http://hal.inria.fr/hal-00922840
11. Munoz, M., Kirley, M., Halgamuge, S.K., et al.: Exploratory landscape analysis of continuous space optimization problems using information content. IEEE Trans. Evol. Comput. **19**(1), 74–87 (2015)
12. Wolpert, D., Macready, W.: No free lunch theorems for optimization. IEEE Trans. Evol. Comput. **1**(1), 67–82 (1997)
13. Xu, L., Hutter, F., Hoos, H.H., Leyton-Brown, K.: Satzilla: portfolio-based algorithm selection for SAT. J. Artif. Intell. Res. **33**, 565–606 (2008)

MO-ParamILS: A Multi-objective Automatic Algorithm Configuration Framework

Aymeric Blot[1,2,3]($\boxtimes$), Holger H. Hoos[3]($\boxtimes$), Laetitia Jourdan[1],
Marie-Éléonore Kessaci-Marmion[1], and Heike Trautmann[4]

[1] Université de Lille, Inria, CNRS, UMR 9189 – CRIStAL, Lille, France
`aymeric.blot@inria.fr`,
`{laetitia.jourdan,me.kessaci}@univ-lille1.fr`
[2] École Normale Supérieure de Rennes, Rennes, France
[3] University of British Columbia, Vancouver, BC, Canada
`hoos@cs.ubc.ca`
[4] University of Münster, Münster, Germany
`trautmann@wi.uni-munster.de`

Abstract. Automated algorithm configuration procedures play an increasingly important role in the development and application of algorithms for a wide range of computationally challenging problems. Until very recently, these configuration procedures were limited to optimising a single performance objective, such as the running time or solution quality achieved by the algorithm being configured. However, in many applications there is more than one performance objective of interest. This gives rise to the multi-objective automatic algorithm configuration problem, which involves finding a Pareto set of configurations of a given target algorithm that characterises trade-offs between multiple performance objectives. In this work, we introduce MO-ParamILS, a multi-objective extension of the state-of-the-art single-objective algorithm configuration framework ParamILS, and demonstrate that it produces good results on several challenging bi-objective algorithm configuration scenarios compared to a base-line obtained from using a state-of-the-art single-objective algorithm configurator.

Keywords: Algorithm configuration · Parameter tuning · Multi-objective optimisation · Local search algorithms

1 Introduction

The performance of many algorithms strongly depends on the setting of their parameters. In particular, state-of-the-art solvers for prominent $\mathcal{NP}$-hard combinatorial decision and optimisation problems, such as propositional satisfiability, scheduling, vehicle routing and mixed integer programming critically rely on configurable heuristics whose parameters have a strong impact on the overall performance achieved on a given class of problem instances. In many cases, these parameters interact with each other in complex and non-intuitive ways,

© Springer International Publishing AG 2016
P. Festa et al. (Eds.): LION 2016, LNCS 10079, pp. 32–47, 2016.
DOI: 10.1007/978-3-319-50349-3_3

and manually configuring them to optimise performance on a particular class of instances is a difficult and tedious task.

State-of-the-art automatic algorithm configuration procedures from the literature, such as ParamILS [6], SMAC [5] and irace [2], only consider a single performance objective when optimising the configuration of an algorithm; they are particularly frequently used for minimisation of running time or maximisation of solution quality. However, there are many situations in which multiple competing performance objectives matter when configuring a given algorithm, such as running time and memory consumption, and running time and solution quality, and there are well-known benefits in exploring the tradeoffs between such competing objectives.

To the best of our knowledge, automatic configuration for multiple performance objectives has been studied only recently, and only in the context of racing algorithms [13,14]. In this work, we introduce MO-ParamILS, an extension of the prominent ParamILS algorithm configuration framework [6,7] that allows us to deal with multiple performance objectives. Like the single-objective ParamILS framework that inspired it, MO-ParamILS implements two configuration procedures: MO-BasicILS and MO-FocusedILS. As we will demonstrate, both are effective in dealing with five bi-objective configuration scenarios, with MO-FocusedILS producing better results than MO-BasicILS.

The remainder of this paper is organised as follows. Section 2 describes the single- and multi-objective algorithm configuration problems and introduces definitions and notation required later. Section 3 presents the MO-ParamILS framework along with the MO-BasicILS and MO-FocusedILS multi-objective configuration procedures implemented within it. Section 4 presents an empirical evaluation of these configurators on five bi-objective configuration scenarios involving prominent solvers for MIP and SAT. Finally, Sect. 5 provides conclusions and an outlook on future work.

2   Automatic Algorithm Configuration

Automatic algorithm configuration deals with the optimisation of the performance of an algorithm through the automatic configuration of its parameters. In the following, we first describe the general context of algorithm configuration, before giving a formal definition of the algorithm configuration problem and its extension to a multi-objective setting.

2.1   General Context

Complex algorithms, especially ones for solving hard computational problems, often expose numerous parameters that can be optimised to achieve good performance in different application scenarios. General-purpose solvers for problems such as mixed integer programming (MIP) or propositional satisfiability (SAT) are usually designed to perform well across a broad range of instance types, but can be tuned manually for performance on particular sets of problem instances.

The algorithm configuration problem is an optimisation problem that aims at finding the best possible parameter configuration of a given algorithm w.r.t. its performance on a given set or distribution of problem instances. Note that when talking about methods for solving this algorithm configuration problem, there are two levels of algorithms involved: the *target algorithm* – a lower-level algorithm for some problem, such as MIP or SAT, whose performance is to be optimised, and the *configurator* – a higher-level algorithm used for optimising the performance of the target algorithm.

There is a sizeable literature on automatic algorithm configurators, including procedures based on sequential model-based optimisation, such as SMAC [5]; racing algorithms, such as irace [10,12]; and model-free search algorithms, such as CALIBRA [1] and ParamILS [6,7]. These all address the *single-objective* automatic algorithm configuration problem, where the performance of the target algorithm is assessed by a single scalar value, such as the running time or solution quality, or a fixed aggregation of multiple scalar values. Recently the idea of a more general *multi-objective* automatic algorithm configuration problem has begun to emerge, *e.g.*, in the work of Zhang *et al.* on multi-objective configurators based on racing [13,14].

On the other hand, good examples of target algorithms are metaheuristics for $\mathcal{NP}$-hard problems, or commercial solvers such as CPLEX with a broad range of parameters. During the development of such algorithms, automated configurators can be used to assess and optimise the performance of different design choices, as well as to find good default parameter settings.

Target algorithm parameters can be numerous and varied in their type and function. They can control low-level aspects of target algorithm behaviour, such as probabilities for certain types of operations, up to high-level aspects, such as computation strategies or problem representations. We distinguish three main types of parameters: *categorical parameters*, which have a finite number of unordered discrete values, often used to select between alternative mechanisms, *integer parameters*, which have discrete and ordered domains, and finally, *continuous parameters* that take numerical values on a continuous scale. In addition, *conditional parameters* exist that depend on the setting of other parameters, as well as *forbidden parameters*, which describe forbidden parameter combinations, to avoid known incorrect or undesirable behaviour of the target algorithm.

2.2 Problem Statement and Notations

The single-objective algorithm configuration problem is defined as a tuple $< \mathcal{A}, \Theta, \mathcal{D}, o, m >$, where

- $\mathcal{A}$ is the parameterised target algorithm,
- Θ is the search space of possible configurations of $\mathcal{A}$,
- $\mathcal{D}$ is a distribution of problem instances,
- o is a cost function, and
- m is a statistical population parameter.

A configuration $\theta \in \Theta$ is one possible setting of the parameters of $\mathcal{A}$. The cost function o is the performance objective for a single execution of algorithm $\mathcal{A}$ on an instance $\pi \in \mathcal{D}$, such as the final accuracy or the total running time. The statistical parameter m is used to aggregate the values of the cost function o over a set of instances, $e.g.$, the arithmetic mean or the median. The aggregated cost of one configuration $\theta \in \Theta$ of an algorithm $\mathcal{A}$ over all instances π from $\mathcal{D}$ is then defined as $c(\theta) := m(O(\theta))$, where $O(\theta)$ is the distribution of costs induced by the function o on $\mathcal{D}$.

The single-objective automatic algorithm configuration problem then consists of finding a configuration $\theta^* \in \Theta$ such that $c(\theta^*)$ is optimised. While in general, performance measures may be minimised or maximised, in the following, we assume (without loss of generality) that $c(\theta)$ is to be minimised.

Unfortunately, the cost $c(\theta)$ of a configuration θ often cannot be computed directly, as $\mathcal{D}$ is usually not finite or much too large to explore exhaustively. Usually, the cost is estimated based on a finite set of instances from $\mathcal{D}$. In this context, we use R to denote a list of runs of a given configurator and represent each run by a triple $< \theta, \pi, o >$, where

- $\theta \in \Theta$ is the configuration considered,
- π from $\mathcal{D}$ is the instance on which the target algorithm is evaluated,
- o is the observed cost of the run.

The *estimated cost* $\hat{c}(\theta)$ of a configuration θ given a sequence of runs R is then determined as the aggregate m over the cost o of all runs $< \theta, \pi, o >$ for some π from $\mathcal{D}$. If the target algorithm is stochastic, a configuration is combined with a specific random seed in order to ensure fair comparisons.

As a simple example, let us consider the configuration of a general SAT solver in an application scenario where we want to minimise the average running time for a certain type of SAT instances. By using a SAT solver that is specifically configured to achieve this, we can save time on instances to be solved in the future compared to using the solver in its default configuration, and eventually these time savings will exceed the effort required for configuring the solver. Here, the target algorithm $\mathcal{A}$ is the SAT solver, with configuration space Θ, and $\mathcal{D}$ is the distribution of SAT instances of interest. The cost function o reflects the solving time for a given SAT instance, and m is defined as the arithmetic mean.

However, in many cases, when optimising the performance of a given algorithm, there is more than one performance metric of interest, which gives rise to a multi-objective optimisation problem. To capture this, we consider an extension of the single-objective algorithm configuration problem that involves a vector of cost functions $o := (o_1, \ldots, o_n)$ where each o_i is a single-objective cost function, and a vector of statistical parameters $m := (m_1, \ldots, m_n)$. Theoretical cost vectors $c(\theta)$ and estimated cost vectors $\hat{c}(\theta)$ are defined based the component-wise scalar cost and estimated cost introduced earlier.

The multi-objective automatic algorithm configuration problem, given a dominance relation $\prec$ over configurations, then consists of finding a set of configurations $\Theta^* \subseteq \Theta$ such that no $\theta \in \Theta^*$ is dominated w.r.t. $\prec$ by any other $\theta' \in \Theta$.

In the following, the dominance relation $\prec$ we consider is Pareto dominance, i.e., for $u := (u_1, \ldots, u_n)$ and $v := (v_1, \ldots, v_n)$, u is said to *dominate* v (denoted by $u \prec v$) if, and only if

$$\forall i \in \{1, \ldots, n\}: u_i \leq v_i \ \wedge \ \exists i \in \{1, \ldots, n\}: u_i < v_i$$

This relation is transferred to configurations by their costs: a configuration θ_1 dominates a configuration θ_2 iff $c(\theta_1)$ dominates $c(\theta_2)$. We will refer to a set of mutually non-dominated configurations as an *archive*. Adding the dominance relation to the multi-objective automatic algorithm configuration problem reflects the overall aim of generating configurations with performance characteristics (according to the given objectives) that are not dominated by any other configuration available. Thus, we are interested in (ideally all) existing tradeoff solutions. We note that, just as in the single-objective case, practical configurators may only find suboptimal solutions to a given configuration problem.

3   From ParamILS to MO-ParamILS

In this section, we first outline the existing single-objective ParamILS framework and then describe our new multi-objective framework, along with the two MO-ParamILS variants we study in the following, MO-BasicILS and MO-FocusedILS.

3.1   Single-Objective ParamILS

ParamILS [6] is an automatic algorithm configuration framework that optimises a single performance metric using iterated local search, a well-known stochastic local search method [11]. The configuration process starts by evaluating a given default configuration along with r configurations chosen uniformly at random from the given configuration space Θ. The best of these $r + 1$ configurations is used as the starting point for the iterated local search process, which can be seen as a sequence of three phases that is repeated until a given time budget is exhausted. Throughout the search process, we keep track of the *incumbent configuration* θ^*, i.e., the best configuration seen so far. In the first phase, the current configuration θ is perturbed, by performing s random steps in the *one-exchange neighbourhood* (where two configurations are neighbours if, and only if, they differ by the value of a single parameter). In the second phase, randomised iterative first improvement local search is performed within the same neighbourhood, excluding all configurations that have been visited previously during the same local search phase. The local search process ends when all neighbours of a given configuration have been checked without achieving an improvement. If the configuration θ' thus obtained is better than the configuration θ from which the last perturbation phase was started, we set θ to θ' (and update the incumbent). To provide additional diversification to the search process and guarantee probabilistic approximate completeness, with a fixed probability $p_{restart}$, θ is reset to a configuration chosen uniformly at random from the entire space Θ.

We note that, in light of the usually high cost of evaluating configurations of the given target algorithm, ParamILS maintains a cache of the results from all target algorithm runs performed during the search process and only performs target algorithm runs after checking that the respective results are not available from that cache.

3.2   Multi-objective ParamILS

We now describe our multi-objective extension of the ParamILS framework. The main difference between ParamILS and MO-ParamILS (outlined in Algorithm 1) lies in the use of a multi-objective iterated local search process, in which an *archive* (*i.e.*, set of non-dominated configurations) is iteratively modified rather than a single configuration of the given target algorithm. Likewise, the incumbent is now an archive. Like ParamILS, MO-ParamILS exposes three parameters: the number r of initial random configurations, the number s of random search steps performed in each perturbation phase and the restart probability $p_{restart}$.

Algorithm 1. Multi-objective ParamILS

Data: Initial archive, algorithm parameters r, $p_{restart}$ and s
Result: Archive of incumbents, *i.e.*, overall best configurations found

current_arch ← initial archive;
for $i \leftarrow 1 \ldots r$ **do**
   | conf ← random configuration;
   | update(conf, current_arch);
   |__ archive(conf, current_arch);

until *termination criterion is met* **do**
   | **if** *first iteration* **then**
   | | arch ← current_arch;
   | **else**
   | | **if** *with probability $p_{restart}$* **then** // Restart
   | | | conf ← random configuration;
   | | | current_arch ← {conf};
   | | |__ arch ← current_arch;
   | | **else** // Random sampling and random walk
   | | | /* Incumbents are not forgotten between restarts */
   | | | conf ← random configuration of current_arch;
   | | | **for** $i \leftarrow 1 \ldots s$ **do**
   | | | |__ conf ← random neighbour of conf;
   | | |__ arch ← {conf};

   | arch ← local_search(arch);
   | **foreach** *conf* in *arch* **do**
   | | update(conf, current_arch);
   | |__ archive(conf, current_arch);

return the archive of incumbents;

Function 2. archive(new_conf, arch)

Data: Single configuration new_conf, archive arch
Result: Updated archive arch

foreach *conf* in *arch* do
   if *dominates(new_conf, conf)* then
      | arch ← arch \ {conf};
   else if *dominates(conf, new_conf)* then
      | return arch;

arch ← arch ∪ {new_conf};
return arch;

The initialisation of the search process does not change conceptually, except that an initial set of default configurations can be provided and is combined, with the r randomly chosen configurations, into an archive. We ensure that whenever we add a new configuration to an archive a, all Pareto-dominated configurations in a are discarded (see Function 2), so that an archive always contains only non-dominated configurations.

MO-ParamILS prominently uses the two following functions: dominates() and update(). The function dominates() compares two configurations using strict Pareto dominance on the respective cost (estimate) vectors. The function update() is, unless explicitly specified otherwise, the only function that runs the target algorithm and updates the cost vector of a configuration; it ensures that a given configuration can subsequently be compared to another configuration using dominates(). It also maintains a cache of all target algorithm runs performed throughout the multi-objective search process and ensures that the overall best configurations, the archive of incumbents, is kept up-to-date. We will discuss the instantiations of update() and dominates() for MO-BasicILS and MO-FocusedILS, the two MO-ParamILS variants we used in our experiments, later in this section.

We use a simple variant of the perturbation mechanism from ParamILS, in which a single configuration is selected uniformly at random from the current archive and modified by a sequence of s random search steps in the 1-exchange neighbourhood; the resulting configuration is then stored as a new archive, which forms the starting point of the subsequent local search phase [4]. The restart mechanism remains unchanged, except that it now replaces the current archive with one containing a single configuration chosen uniformly at random from the entire configuration space Θ. As in ParamILS, we use default values of $r := 10$, $p_{restart} := 0.01$, and $s := 3$ in our experiments.

The subsidiary local search process used in MO-ParamILS is outlined in Function 3. From a wide range of existing multi-objective local search procedures [9], we chose this one, because it is conceptually simple and resembles the subsidiary local search procedure used in ParamILS; it has also been shown to be very efficient [3]. At each step of the local search process, all configurations in the

---

Function 3. localSearch(init_arch)

Data: Initial archive of configurations `init_arch`
Result: Best archive of configurations found
Side effect: Change or update the incumbent if necessary

`current_arch ← init_arch;`
`tabu_set ← current_arch;`
repeat
   /* Selection */
   `current_set ← current_arch;`
   `candidate_set ← ∅;`
   foreach *current* **in** *current_set* **do**
      foreach *neighbour* **in** *randomised neighbourhood of* **current do**
         /* Exploration */
         if *neighbour ∈ tabu_set* **then**
            `next ;`
         `tabu_set ← tabu_set ∪ {neighbour};`
         `update(neighbour, current);`
         if *dominates(neighbour, current)* **then**
            `candidate_set ← candidate_set ∪ {neighbour};`
            break ;
         if not *dominates(current, neighbour)* **then**
            `candidate_set ← candidate_set ∪ {neighbour};`
   /* Archive */
   foreach *conf* **in** *candidate_set* **do**
      `archive(conf, current_arch);`
until *candidate_set = ∅;*
return `current_arch;`

---

current archive are explored individually. When exploring a configuration θ, its neighbours are evaluated in random order (excluding any configurations already visited earlier in the same local search phase), until one is found that strictly dominates θ or all neighbours have been visited. All non-dominated neighbours encountered during this process are added to the current archive, making sure that dominated solutions are removed. (Notice how this can be seen as a generalised version of the acceptance criterion used in the single-objective ParamILS framework.) The local search then stops when there is no more unvisited neighbour that can be added to the archive.

3.3   MO-BasicILS

The key idea behind BasicILS(n) is to evaluate configurations on a fixed set of n training instances, selected uniformly at random (without replacement) from the given training set $\mathcal{D}$ [7]. This can be easily carried over to the MO-ParamILS framework of Algorithm 1, by defining `update()` and `dominates()` the way that the latter always compares configurations based on their quality vectors on the same instances set, and the former ensures that all target algorithm runs required in this context are performed.

The disadvantage of the resulting MO-BasicILS procedure, as in the case of BasicILS, lies in the difficulty of choosing n: if n is too small, solution quality estimates can be inaccurate, leading to poor generalisation of the performance of the configurations obtained from MO-BasicILS to unseen test instances; if n is too large, much effort is wasted on evaluating poorly performing configuration, compromising the efficiency of the search process. In our experiments, we used a default setting of $n := 100$.

3.4   MO-FocusedILS

The key idea behind FocusedILS is to avoid the potential problems arising from the use of a fixed number of instances for evaluating configurations by starting comparisons between configurations on a small initial set of instances and then increasing the number of instances as better and better configurations are found [7]. Based on the same idea, MO-FocusedILS allows poor configurations to be dominated very soon, while promising configurations are evaluated increasingly more accurately as the search process progresses.

Towards this end, MO-FocusedILS uses a slightly weaker dominance relation in the function dominates(), which adds the condition that a configuration θ dominates a configuration θ' if, and only if, θ has been run on every instance θ' has been run and θ dominates θ' on those instances. Note that when θ and θ' have been run on the same instances, this corresponds to standard Pareto domination, and as the number of instances grows, it approximates Pareto domination on the true (theoretical) cost vectors arbitrarily accurately.

In practice, new runs are performed for the configuration that has been evaluated on fewer instances so far, until either of the two configurations being compared dominates another; the instances for these new runs are chosen according to a random permutation of the training instance set that has been determined when initialising MO-FocusedILS and then remains fixed. This ensures that for two configurations θ and θ' and their respective sequences of runs R_θ and $R_{\theta'}$, either $R_\theta \subseteq R_{\theta'}$ or $R_\theta \supseteq R_{\theta'}$, and that $R_\theta \cap R_{\theta'}$ is either equal to R_θ or to $R_{\theta'}$.

The update() function of MO-FocusedILS handles the comparison of a single configuration θ and an archive a by adding new runs of θ until there is at least one configuration $\theta' \in a$ for which $R_\theta \supseteq R_{\theta'}$ or θ' dominates θ.

Like Focused-ILS, MO-FocusedILS additionally requires an intensification mechanism that ensures that over the course of the search process, good configurations are evaluated on an increasing number of instances. This mechanism is outlined in Procedure 4: it simply performs new runs for a given configuration until its new cost vector Pareto dominates its cost vector before intensification. Procedure 4 is called at the beginning of every local search phase, to help start the local search process with a better cost estimate, after every local search phase, to further increase the accuracy of cost estimates, and each time the

---

Procedure 4. intensify(conf)

Data: Single configuration `conf`
Side effect: Updates the level of detail of `conf`

repeat
> `old_cost` ← cost(`conf`);
> perform a new run of `conf`;
> `new_cost` ← cost(`conf`);

until *pareto_dominates(old_cost, new_cost)*;

`update()` function compares two configurations with the same number of runs. (Alternative, but less efficient intensification mechanisms might perform a fixed number of new runs, or a number of runs given by a function of the time spent since intensification was last performed.)

4 Experiments

In this section, we present results for MO-ParamILS for two different multi-objective automatic algorithm configuration problems. First, we study the trade-off between running time and solution quality for an anytime optimisation algorithm. Our second example involves the simultaneous optimisation running time and memory usage. In both cases, we consider two optimisation objectives; however, MO-ParamILS is not restricted to such bi-objective algorithm configuration problems.

4.1 Experimental Protocol

To assess the performance of MO-ParamILS we consider five configuration scenarios, described by Table 1. These scenarios use three datasets and two target algorithms, which belongs to ACLib[1], a comprehensive algorithm configuration library, and are already known and have been studied in single-objective algorithm configuration.

Details of the two algorithms are precised by Table 2. Note that the neighbourhood relation of ParamILS considers all parameters as categorical; henceforth for integer or continuous parameters the set of values have been discretised before all experiments.

Our experimental protocol involves three consecutive steps, namely *training*, *validation* and *test*. In the training step, the configurator is run 25 times on 25 different permutations of the training set, resulting in 25 archives. Because the configurations produced at the end of the training phase have not necessarily been evaluated on precisely the same training instances, in the validation step, each final configuration of the 25 training runs is reassessed on the same subset of the training instances. At the end of this validation phase, all configurations

---

[1] http://aclib.net

Table 1. Configuration scenarios.

Dataset	Algorithm	Walltime	Performance objectives	Abbrv
Regions200	CPLEX	1 day	Quality, Cutoff	RCut
Regions200	CPLEX	1 day	Quality, Running Time	RRun
CORLAT	CPLEX	1 day	Quality, Cutoff	CCut
CORLAT	CPLEX	1 day	Quality, Running Time	CRun
QUEENS	CLASP	1 day	Memory usage, Running Time	QUEENS

Table 2. Target algorithm parameters (with number of possible values).

Algorithm	Categorical	Integer	Continuous	Total configurations
CPLEX	5 (2)	65 (2–7)	2 (5–6)	$2.26 \cdot 10^{46}$
CLASP	15 (2–5)	43 (2–16)	8 (6–14)	$9.96 \cdot 10^{48}$

have been assessed on the same set of problem instances and can therefore be meaningfully compared in order to identify the ones that are Pareto-optimal w.r.t. performance objectives on solved problem instances and percentage of unsolved problem instances. Then, in the test step, the configurations of the archive obtained from the validation step are reassessed on a disjoint set of testing instances. We use this protocol to compare the configurations obtained from MO-BasicILS, from MO-FocusedILS, from an approach only using SO-FocusedILS, as well as the default configuration. In each of these cases, we use the same 25 permutations of the 1000 training instances, the same subset of 100 training instances for the validation, and the same 1000 testing instances.

Regarding MO-BasicILS, its parameter n is set to 100, meaning that estimations of configuration performance use 100 training instances. Regarding the SO approach, we used SO-FocusedILS with every available improvement (*e.g.*, aggressive capping). For the four CPLEX scenarios, we ran SO-FocusedILS separately on the 5 different cutoff values chosen to obtain a total wall-clock time of one day (that is, for 1, 2, 3, 5 and 10 CPU seconds cutoffs, the walltime for the 1 CPU second cutoff is $^1/_{(1+2+3+5+10)} \times 24\,\mathrm{h}$). For the CLASP scenario, we ran SO-FocusedILS separately on each of the two objectives for 12 hours.

In the CLASP scenario, failure by CLASP to find a solution within 300 seconds in a particular instance is penalised by counting any such run as 10 times the cutoff time (*i.e.*, using the well-known PAR10 performance metric [6]). In the CPLEX scenarios, we penalised failure by CPLEX to return a MIP gap value by setting the MIP gap value to 10^{10} for such runs, thus making sure that such configurations tend to be avoided by our configuration approaches.

Performance assessment has been carried out using the PISA framework [8]. For the CPLEX scenarios, we used the data without timeout. For validation and test steps, the final fronts are compared using the hypervolume and ε indicators. First, all fronts for a given step and scenario are normalised so the values of every objective vector lie in the interval $[1, 2]$. Then, a reference front is computed by merging every front and applying Pareto dominance. The indicator values are then computed between each front and the reference front.

SPRINT-Race [14] is a recent multi-objective racing algorithm, and we originally considered including it in our performance comparison. However, both CPLEX and CLASP algorithms have very large configuration spaces (10^{46} and 10^{48} configurations, respectively), which implies that the only way to apply SPRINT-Race would be in combination with a sampling technique. Furthermore, the implementation of SPRINT available from its authors requires as input the exhaustive evaluation of all configurations on all instances, making it impractical to use for our configuration scenarios.

4.2   Results

Empirical results from the test phases are shown in Fig. 1, considering only instances solved before the given timeout. The corresponding number of unsuccessful runs are given in Table 3. Table 4 shows the performance assessment for test results for both indicators. For each scenario, the best value is highlighted.

As can be seen from Table 4, MO-FocusedILS finds considerably better Pareto fronts for the test sets of all our multi-objective configuration scenarios than our baseline single-objective approach in terms of hypervolume and ε indicator. In all but one case, MO-FocusedILS also produces better results than MO-BasicILS,

Table 3. Average percentages of timeouts for final CPLEX configurations.

Approach	Validation				Test			
	RCut	RRun	CCut	CRun	RCut	RRun	CCut	CRun
MO-FocusedILS	1.3	0.7	4.2	3.6	0	0	1.06	2.89
MO-BasicILS	0.1	0.6	3.6	2.9	0.04	0	0.47	3.78
SO approach	0.3	0.4	4.8	5.1	0.12	0	1.87	1.87
Default	0	0	2.2	2.2	0	0	0.14	0.14

Table 4. Hypervolume (top) and ε indicator values (bottom) for final test fronts.

Approach	RCut	RRun	CCut	CRun	Queens
MO-FocusedILS	9.02e−03	**2.07e−03**	**2.37e−02**	**7.63e−04**	**1.57e−02**
MO-BasicILS	**2.46e−03**	5.41e−02	5.53e−02	1.02e−01	5.49e−02
SO approach	3.82e−02	5.82e−02	3.35e−01	1.72e−01	3.04e−02
Default	2.43e−01	3.57e−01	2.70e−01	5.30e−01	1.08e+00
MO-FocusedILS	**1.44e−02**	**9.05e−03**	**9.00e−02**	**8.06e−04**	**2.64e−02**
MO-BasicILS	1.80e−02	1.71e−01	1.11e−01	1.48e−01	8.35e−02
SO approach	5.77e−02	1.38e−02	3.33e−01	1.42e−01	6.52e−02
Default	2.22e−01	2.69e−01	2.33e−01	3.90e−01	1.00e+00

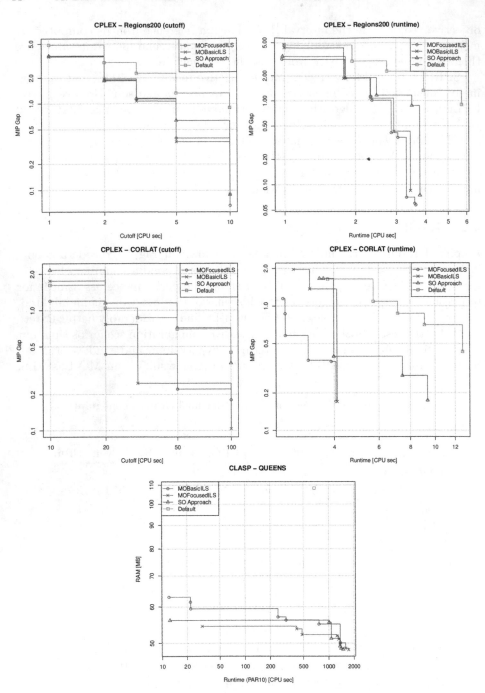

Fig. 1. Final test fronts for all five scenarios

which, in most cases, still produces better results than the single-objective approach, but with less of a margin. Figure 1 provides additional details by showing the Pareto fronts for all three multi-objective configuration approaches as well as the performance (trade-off) achieved by the default configuration; we note that the latter still produces a trade-off curve, because of the anytime nature of CPLEX.

When analysing these results, we noticed that MO-FocusedILS evaluates many more unique configurations than MO-BasicILS (4752 *vs* 166 on average, over all five scenarios). This clearly indicates the efficacy of the way in which MO-FocusedILS controls the number of runs per configuration performed and mirrors analogous findings for BasicILS *vs* FocusedILS in the single-objective case [6].

On all five scenarios, the default configuration of CPLEX or CLASP produce few unsuccessful runs on training or test instances. Our three approaches lead to configurations generating about as many timeouts as the default configuration. However, by also taking in account the configurations returned that have both more timeouts and better performances on successful instances, we were able to achieve even better results at the cost of a small loss of generality, as shown in Table 3. While our CLASP scenario uses PAR10 scores to take into account instances that could not be solved within the given cutoff time, as previously mentioned, the final Pareto fronts we produce for the CPLEX scenarios do not reflect a small number of instances for which no MIP gap was obtained within the allotted running time. The fraction of the validation and test sets on which this happened is shown in Table 3; as seen there, timeouts generally occur for a small fraction of instances, and while that fraction tends to increase as we configure CPLEX, it remains low enough in all cases to not raise serious concerns.

5   Conclusion

We have introduced MO-ParamILS, an extension of the prominent ParamILS automatic algorithm configuration framework for solving the multi-objective algorithm configuration problem. To the best of our knowledge, while MO-ParamILS is not the first multi-objective algorithm configurator, it is the first to be able to effectively deal with the highly-parameterised target algorithms usually considered in standard, single-objective algorithm configuration scenarios, as demonstrated in our experiments on five bi-objective configuration scenarios involving CPLEX and clasp, two prominent solvers for mixed integer programming (MIP) and propositional satisfiability (SAT) problems, respectively.

As is the case for their single-objective analogues, MO-FocusedILS typically performs better tha MO-BasicILS, but both approaches are able to produce sets of non-dominated configurations that cover an interesting range of trade-offs in all five scenarios we studied, and were considerably more effective in doing so than a base-line approach using a state-of-the-art single-objective configurator.

We believe that automatic multi-objective configurators, such as MO-FocusedILS, will be very useful in many application situations where there is no clear and obvious way to trade off multiple performance criteria for a given target algorithm. In future work, it might be interesting to apply multi-objective configuration to multi-objective optimisation procedures as target algorithms; these are notoriously difficult to configure, and we believe that doing so automatically, based on multiple performance objectives, could be quite attractive. It would also be interesting to exploit the potential for parallelisation inherent in the MO-ParamILS framework; while using standard configuration protocols, the current version of MO-ParamILS can exploit parallel computing resources (just like single-objective ParamILS), there is considerably more room for easy parallelisation in the multi-objective extension presented here. Furthermore, we believe that it might be interesting to explore advanced methods for ensuring effective coverage of the true tradeoff curves (or surfaces) of a given multi-objective configuration scenario within the MO-ParamILS framework.

Finally, we are interested in exploring multi-objective extensions of sequential model-based algorithm configuration methods, in particular SMAC [5]. We also see potential value in effective multi-objective extensions of configuration procedures such as irace [2].

References

1. Adenso-Díaz, B., Laguna, M.: Fine-tuning of algorithms using fractional experimental designs and local search. Oper. Res. **54**(1), 99–114 (2006)
2. Birattari, M., Yuan, Z., Balaprakash, P., Stützle, T.: F-Race and iterated F-Race: an overview. In: Bartz-Beielstein, T., Chiarandini, M., Paquete, L., Preuss, M. (eds.) Experimental Methods for the Analysis of Optimization Algorithms, pp. 311–336. Springer, Berlin (2010)
3. Blot, A., Aguirre, H., Dhaenens, C., Jourdan, L., Marmion, M.-E., Tanaka, K.: Neutral but a winner! How neutrality helps multiobjective local search algorithms. In: Gaspar-Cunha, A., Henggeler Antunes, C., Coello, C.C. (eds.) EMO 2015. LNCS, vol. 9018, pp. 34–47. Springer, Heidelberg (2015). doi:10.1007/978-3-319-15934-8_3
4. Geiger, M.J.: Foundations of the Pareto iterated local search metaheuristic. CoRR abs/0809.0406 (2008). http://arxiv.org/abs/0809.0406
5. Hutter, F., Hoos, H.H., Leyton-Brown, K.: Sequential model-based optimization for general algorithm configuration. In: Coello, C.A.C. (ed.) LION 2011. LNCS, vol. 6683, pp. 507–523. Springer, Heidelberg (2011). doi:10.1007/978-3-642-25566-3_40
6. Hutter, F., Hoos, H.H., Leyton-Brown, K., Stützle, T.: ParamILS: an automatic algorithm configuration framework. JAIR **36**, 267–306 (2009)
7. Hutter, F., Hoos, H.H., Stützle, T.: Automatic algorithm configuration based on local search. In: AAAI 2007, pp. 1152–1157 (2007)
8. Knowles, J., Thiele, L., Zitzler, E.: A tutorial on the performance assessment of stochastic multiobjective optimizers. Technical report 214, Computer Engineering and Networks Laboratory (TIK), ETH Zurich, Switzerland, revised version (2006)
9. Liefooghe, A., Humeau, J., Mesmoudi, S., Jourdan, L., Talbi, E.: On dominance-based multiobjective local search: design, implementation and experimental analysis on scheduling and traveling salesman problems. J. Heuristics **18**(2), 317–352 (2012)

10. López-Ibáñez, M., Dubois-Lacoste, J., Stützle, T., Birattari, M.: The irace package, iterated race for automatic algorithm configuration. Technical report, TR/IRIDIA/2011-004, IRIDIA, Université Libre de Bruxelles, Belgium (2011)
11. Lourenço, H., Martin, O., Stützle, T.: Iterated local search: framework and applications. In: Gendreau, M., Potvin, J.-Y. (eds.) Handbook of Metaheuristics, vol. 2, pp. 363–397. Springer, New York (2010)
12. Marmion, M.-E., Mascia, F., López-Ibáñez, M., Stützle, T.: Automatic design of hybrid stochastic local search algorithms. In: Blesa, M.J., Blum, C., Festa, P., Roli, A., Sampels, M. (eds.) HM 2013. LNCS, vol. 7919, pp. 144–158. Springer, Heidelberg (2013). doi:10.1007/978-3-642-38516-2_12
13. Zhang, T., Georgiopoulos, M., Anagnostopoulos, G.C.: S-Race: a multi-objective racing algorithm. In: GECCO 2013, pp. 1565–1572 (2013)
14. Zhang, T., Georgiopoulos, M., Anagnostopoulos, G.C.: SPRINT multi-objective model racing. In: GECCO 2015, pp. 1383–1390 (2015)

Evolving Instances for Maximizing Performance Differences of State-of-the-Art Inexact TSP Solvers

Jakob Bossek$^{(\boxtimes)}$ and Heike Trautmann

Department of Informations Systems, University of Münster, Münster, Germany
{bossek,trautmann}@wi.uni-muenster.de

Abstract. Despite the intrinsic hardness of the Traveling Salesperson Problem (TSP) heuristic solvers, e.g., LKH+restart and EAX+restart, are remarkably successful in generating satisfactory or even optimal solutions. However, the reasons for their success are not yet fully understood. Recent approaches take an analytical viewpoint and try to identify instance features, which make an instance hard or easy to solve. We contribute to this area by generating instance sets for couples of TSP algorithms A and B by maximizing/minimizing their performance difference in order to generate instances which are easier to solve for one solver and much harder to solve for the other. This instance set offers the potential to identify key features which allow to distinguish between the problem hardness classes of both algorithms.

Keywords: TSP · Instance hardness · Algorithm selection · Feature selection

1 Introduction

The traveling salesperson problem (TSP) is one of the most famous NP-hard combinatorial optimization problems of highly practical relevance (logistics, circuit boards assembly, etc.). Given a set of N cities and positive distances d_{ij} from city i to city $j, 1 \leq i, j \leq N$ with $i \neq j$, the task is to construct a roundtrip tour of minimal total distance that visits each city exactly once and returns to the origin.

We focus on the *2D Euclidean TSP* which refers to points in the Euclidean plane und thus results in a Euclidean distance matrix. Respective solvers can be distinguished into two classes. For exact algorithms like Concorde [1] optimality of the found solution after algorithm completion can be guaranteed. However, the required runtime might be quite high, especially for large instances. State of the art solvers in inexact TSP solving proved to be able to find solutions of very high quality and simultaneously much faster. Those are therefore of crucial interest, especially for efficient algorithm selection approaches [9].

In [8] we were able to show that per-instance algorithm selection between LKH [6] and the recently introduced evolutionary algorithm EAX [13] together

© Springer International Publishing AG 2016
P. Festa et al. (Eds.): LION 2016, LNCS 10079, pp. 48–59, 2016.
DOI: 10.1007/978-3-319-50349-3_4

with specific restart variants as presented in [8] is very promising w.r.t. to improving the state of the art in TSP solving.

Per-instance algorithm selection makes use of a comprehensive set of instance features from the literature[1] [7,12,15,16]. A crucial aspect is the identification of features which are useful for determining the instance hardness for different solvers. While much progress has been made in this respect, e.g. in [5] where LKH behavior is related to the transition from highly structured, polynomially solvable TSP instances to instances with increasingly random distributions of nodes, this issue is not yet fully understood. This paper specifically aims at improving the analysis of performance differences of the current state of the art inexact TSP solvers LKH+restart and EAX+restart as well as their orignial versions without restart mechanism. More specifically, we use an evolutionary approach to evolve instances on which the solvers exhibit maximum performance difference, i.e. which are easier to solve for one solver and harder for the other. The used evolutionary algorithm, inspired by [16], was introduced in the context of analysing problem hardness of the 2-opt heuristic in [11] and further adapted in [12,14]. However, we now refrain from focussing on a single solver but directly use the performance ratio of two solvers as the fitness function inside the EA.

Section 2 provides details of the TSP solvers and feature sets, while the evolutionary algorithm is presented in Sect. 3. The conducted experimental results are illustrated and discussed in Sect. 4 followed by summarizing remarks and an outlook on future research perspectives in Sect. 5.

2   Solvers and Features

TSP Solvers. Both LKH, a stochastic local search algorithm based on the Lin-Kernighan procedure [6], and EAX [13], a recently introduced evolutionary algorithm utilizing a specialized new edge assembly crossover procedure, are focussed. LKH is the state of the art in inexact TSP solving since its introduction in 2000. We used the reference implementation LKH 2.0.7 based on the former implementation 1.3 [10]. In [8], we introduced a dynamic restart mechanism as the underlying stochastic search process tends to stagnate too early quite frequently. This version is termed LKH+restart. The first indication that EAX could be competitive to LKH was given in [13], which was confirmed in [8]. In the latter paper, a restart strategy for EAX, EAX+restart, was implemented based on the original internal termination criterion. Once this is met, a restart is conducted and this procedure is repeated until a given accuracy or running time limit is reached.

It could be shown [8] that the respective restart variants outperform the orignial solver versions. Moreover, EAX+restart emerged as the single best solver over the set of considered representative instances. However, an algorithm selection model based on features that are quite cheaply computable could be learnt that managed to perform even better.

---

[1] All these feature sets have been recently made availabe in a single R-Package [4].

Features. Established feature sets for characterizing Euclidean TSP instances, i.e. the feature set described in [12] as well as in [7], are used for characterizing both the evolved as well as the baseline (random, TSPLIB) instances. We denote the former as *TSPMeta* and the latter as *UBC* features. The recently introduced additional set of features based on k-nearest neighbours [15] will be focussed in future studies. Both considered sets contain a comprehensive collection of features including e.g. features characterizing the distance structure, identifying possible clusters of nodes, statistics based on angles between cities and its two nearest neighbours as well as minimum spanning tree information. The R-package `salesperson` [4] containing all relevant feature sets is used for the feature computation task. Having already the algorithm selection task in mind for further studies, we restricted our feature set to cheaply computable features, i.e. we excluded the local search, branch and cut, and clustering distance features from the UBC feature set (UBC (cheap)) as motivated in [8].

3   On Evolving Instances

A simplified pseudocode is given in Algorithm 1: the initial population is generated by placing the desired number of nodes uniformly at random in the unit square $[0, 1]^2$. Subsequently, the next generation is obtained by selecting two parents from the mating pool, applying crossover as well as two mutation strategies in a row, namely uniform and gaussian mutation. Uniform mutation is applied with a very low probability. This operator replaces the node coordinates of selected nodes with new randomly chosen coordinates and thus may be

---

Algorithm 1. Evolving EA

```
 1: function EA(fitnessFun, popSize, instSize, generations, timeLimit)
 2:     poolSize = ⌊ popSize / 2 ⌋
 3:     for i = 1 → popSize do
 4:         population[i] = GENERATERANDOMINSTANCE(instSize)         ▷ in [0, 1]²
 5:     end for
 6:     while stopping condition not met do
 7:         for i = 1 → popSize do
 8:             fitness[i] = FITNESSFUN(population[i])
 9:         end for
10:         matingPool = CREATEMATINGPOOL                ▷ 2-tournament-selection
11:         offspring[1] = GETBESTFROMCURRENTPOPULATION           ▷ 1-elitism
12:         for i = 2 → popSize do
13:             Choose p1 and p2 randomly from the mating pool
14:             offspring[i] = APPLYVARIATIONS(p1, p2)
15:             Rescale offspring to [0, 1]²                      ▷ Algorithm 2
16:             Round to cell grid                               ▷ Algorithm 3
17:         end for
18:         population = offspring
19:     end while
20: end function
```

---

termed a global mutator. In contrast, gaussian mutation works locally by adding normally distributed noise to the point coordinates. The two sequential mutation strategies together enable small local as well as global structural changes of the offspring resulting from the crossover operation. Furthermore, a 1-elitist strategy is adopted to ensure survival of the current fittest individual.

A final rescaling of the evolved instances ensures the complete coverage of $[0, 1]^2$ in that the minimum and maximum coordinates are placed on the boundary of the instance space (see Algorithm 2). Therefore the area will be covered quite homogenously and instances become comparable in this regard. Afterwards the instance nodes are rounded to the nearest grid cell center after discretizing the plane using a grid with *cells* sections (see Algorithm 3). This relates to the aim of evolving practically relevant structures (e.g. in the design of circuit boards) and will furthermore affect some features which incorporate the proportion of distinct distances. Note that this strategy conceptually differs from rounding to a predefined number of digits. Figure 1 (taken from [12]) visualizes both rescaling and rounding.

Algorithm 2. Rescale Instance

1: **function** RESCALE(*instance*)
2:     $mins \leftarrow$ COLUMN_MINS(*instance*)
3:     $maxs \leftarrow$ COLUMN_MAXS(*instance*)
4:     $\delta \leftarrow maxs - mins$
5:     $scaled \leftarrow \emptyset$
6:     **for** $city \in instance$ **do**
7:         $scaled \leftarrow scaled \cup \{(city - mins)/\delta\}$
8:     **end for**
9:     **return** $scaled$
10: **end function**

Algorithm 3. Round instance

1: **function** ROUND(*instance, cells*)
2:     $gridRnd \leftarrow$ CREATEGRID(*resolution = cells*)
3:     $instRnd \leftarrow$ FLOOR(*instance * cells*)/*cells*
4:     **for** $i = 1$ to *instSize* **do**
5:         $instRnd[i,] \leftarrow SetToCellCenter(instRnd, gridRnd)$
6:     **end for**
7:     **return** $instRnd$
8: **end function**

Mersmann et al. [11] had chosen the *approximation ratio*, i.e., the arithmetic mean of the tour length computed by the considered stochastic algorithm divided by the length of the optimal tour computed by Concorde as the fitness function to be optimized. The first idea was to adopt this approach with slight modifications, since we aim to generate instances for pairs of algorithms A and B, i.e.,

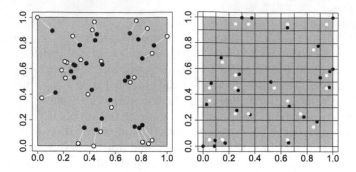

Fig. 1. Rescaling of an instance of size 25 (left). The original instance is reflected by black dots. Rounding of an instance of size 25 to grid cell centers (right). The rounded instance is visualized by white dots.

to focus on the ratio of the arithmetic means of the respective tour lengths [3]. However, we observed LKH and EAX and the restart variants respectively to perform extraordinary well even on large instance sizes up to 1000 in preliminary experiments. Hence, we experienced the approximation ratio to be unrewarding for our aims since we observed the "best" generated instance to have an approximation ratio of approximately 1 in our first series of experiments. In fact, the chosen state of the art solvers turned out to be too powerful for this scenario. Instead, our approach is slighty different. We use the *penalized average runtime* consumed by an algorithm to find the optimal tour (computed by Concorde in advance) as the performance of the algorithm. However, an algorithm reaching the cutoff time is not penalized with 10 times the latter within the EA as it is the standard procedure in par10. The cutoff time itself is used as otherwise the probability that such solutions would be removed at later stages of the EA run would be extremely low. However, for the final instance evaluation, the classsical par10 score is used. For ensuring integer values of the distance matrix inside the EA, the euclidean distance matrix is computed on the original coordinates, multiplied by a scaling factor of 100 and afterwards rounded to the nearest integer.

Let $R_A(I)$ denote the slightly modified penalized average runtime of algorithm A on instance I as explained above. Then the runtime proportion $R_{(A,B)}(I)$ for a pair of algorithms (A, B) is defined as

$$R_{(A,B)}(I) = \frac{R_A(I)}{R_B(I)}.$$

This *runtime ratio* serves as the fitness function in our investigations. We thus moved the focus to the time aspect instead of the solution quality. We minimize $R_{(A,B)}$ in order to generate instances which are easier to solve for algorithm A and harder to solve for algorithm B respectively.

4 Experiments

4.1 Experimental Setup

We generated each 25 instances of size 300 for each solver pairing (EAX, LKH), (LKH, EAX), (EAX_RESTART, LKH_RESTART) and (LKH_RESTART, EAX_RESTART) resulting in an evolved instance set of 100 TSP instances. Inside the evolutionary algorithm, each solver is replicated three times on each instance due to limited computational budget. The final evaluation of the instances, however, is based on 10 replications and the classical par10 score. All internal termination criteria besides cutoff time of two minutes were deactivated in all runs to get reasonable estimates of the performance measure.

Based on preliminary experiments and the experience of [11,12] the EA parameters were set to $popSize = 30$, $generations = 5000$, $uniformMutationRate = 0.001$, $normalMutationRate = 0.1$, $cells = 100$, and the standard deviation of the gaussian mutation operator $normalMutationSD = 0.025$.

As a baseline, 100 random instances were generated by placing the desired number of nodes uniformly at random in the unit square $[0,1]^2$. The Euclidean distance matrix was computed, multiplied by a scaling factor of 100 and subsequently rounded to the nearest integer. Moreover, Euclidean TSPLIB instances with node sizes between 200 and 400[2] were chosen in order to allow comparisons to practically relevant instances. For all considered algorithms the respective par10 scores were applied for comparison.

All experiments were run on the parallel linux computer cluster PALMA at University of Münster, consisting of 3528 processor cores in total. The utilized compute nodes are 2,6 GHz machines with 2 hexacore Intel Westmere processors, totally 12 cores per node and 2 GB main memory per core.

4.2 Results

Figure 2 visualizes the par10 scores of all instances for both solver pairs (EAX vs. LKH, EAX+restart vs. LKH+restart) in a scatterplot. The instances are marked w.r.t. instance type, i.e. either "random", "evolved" or "tsplib". It becomes obvious that the introduced evolutionary approach very successfully generates instances with high performance differences of the solvers. Even in the restart scenario where solver runtimes are quite homogenously clustered in the center of the plot, the evolved instances can clearly be distinguished and are located far away from the bisecting line. Both optimization directions work well (two clusters of instances) while evolving instances which are easier for EAX+restart is even more successful (upper cluster). Moreover, the TSPLIB instances are much easier to solve compared to the remaining ones in both scenarios. There is only one exception (rd400) which is extremely hard for both solvers.

Figure 3 presents boxplots of the par10 scores distribution for each solver categorized by instance type and confirms the discussed findings. In all cases but the

---

[2] TSPLIB-Instances: a280, gil262, kroA200, kroB200, lin318, pr226, pr264, pr299, rd400, ts225, tsp225.

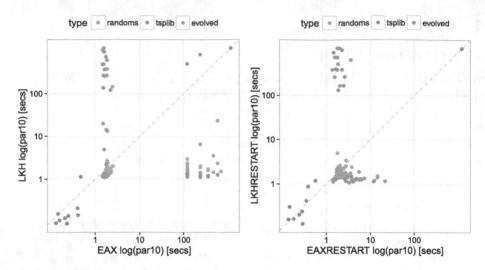

Fig. 2. Par10 scores (log-scale) of EAX vs. LKH (left) resp. EAX+restart vs. LKH+restart (right). Colors help to distinguish between the instance types. (Color figure online)

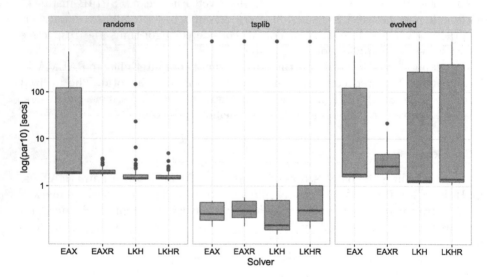

Fig. 3. Boxplots of the par10 scores (log-scale) for each solver categorized by instance type. (Color figure online)

original EAX the runtimes of the evolved instances are substantially higher compared to the random ones. Specifically, the variability of the runtimes increases reflected by the upper quartile, i.e. the upper border of the boxes. Frequently, even the runtime of Concorde was exceeded so that using an inexact solver was of no merit in retrospect. In general, as problem hardness tends to increase for

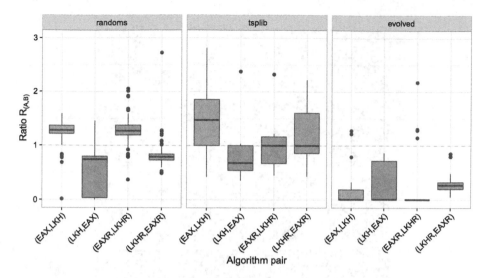

Fig. 4. Ratios of par10 scores for each solver pair, i.e. $R_{(A,B)}$. A value smaller than one means that the runtime of the first solver is smaller than the respective one of the second. This ratio is the fitness function to be minimized inside the evolutionary algorithm for evolving the respective instances. (Color figure online)

both solvers, local search in general becomes harder for the evolved instances. However, the extent varies by solver which results in low par10 score ratios. Par10 scores on the TSPLIB instances differ significantly from the remaining ones, the whole distribution is located far more to the left. Here, you can see the single hard outlier instance as well.

The actual par10 score ratios are presented by boxplots in Fig. 4. Values smaller than one reflect the superiority of the first solver from the pair (SolverA, SolverB), i.e. Solver A has a smaller par10 score than Solver B. As this ratio forms the objective function of the evolutionary algorithm, which has to be minimized, we expect these values to be substantially lower than one. Apart from very few outliers this is confirmed by the respective boxplots. Furthermore, significant differences are obvious compared to the random instances, even for the (LKH, EAX) pair where LKH already exhibits much lower runtimes than EAX on the random instances.

Representative instances are plotted in the Euclidean plane in Fig. 5. For each solver pair the four smallest instances regarding the par10 score ratio are displayed. Unfortunately, structural difference are very particular and cannot be clearly detected visually. Therefore, we additionally made use of machine learning techniques. A classification approach was conducted for each solver pairing with the aim of predicting the instance type (random, evolved) based on the TSP feature set comprising *TSPMeta* and UBC (cheap). To derive the most important features a nested feature selection with 10-fold crossvalidation was performed based on a simple classification tree. Deterministic forward and backward search

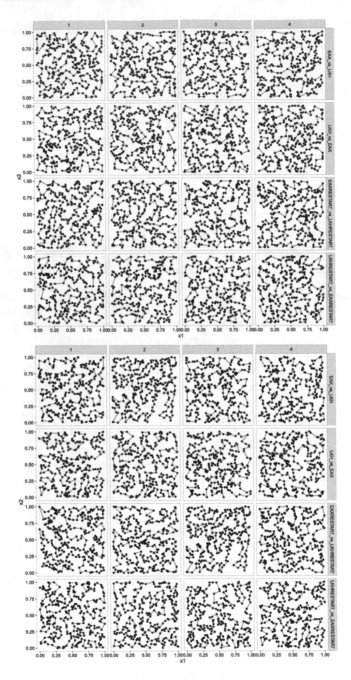

Fig. 5. Four lowest ranked instances regarding the par10 score ratio for all scenarios. Evolved instances are visualized on top, the random instances below.

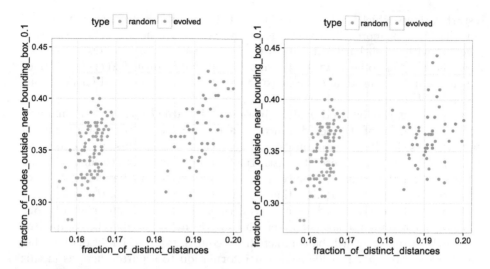

Fig. 6. Scatterplots of the two most important features selected during the feature selection step for predicting the instance type (random, evolved) based on instance features for the original versions (left) and the restart versions (right). (Color figure online)

served as the search strategy guiding the feature selection. Additionally an outer 10-fold crossvalidation was used in order to preserve that the selected features depend on the underlying fold. All computations were conducted with the R package mlr [2].

It turns out that the fraction of distinct distances is sufficient to separate random and evolved instances (see the scatterplots in Fig. 6) for both solver pairings $\{EAX, LKH\}$ and $\{EAX\_RESTART, LKH\_RESTART\}$. Misclassification errors vanish for both pairings. Thus, local search gets harder for higher fractions of distinct distances which was already hinted at in other studies. The observed characteristic of the evolved instances might be linked to the rounding strategy to grid cell centers applied to each individual in the evolutionary algorithm after mutation. However, the considered solvers face different levels of difficulties not solely depending on this fraction. The kind of classification probably will differ for instances evolved based on deactivated rounding to grid cell centers. We are going to investigate this in future work.

5   Conclusion

In this paper, we used an evolutionary approach to evolve TSP instances with maximal perfomance difference of LKH+restart vs. EAX+restart and of the pair of the respective original variants (without restart). For instances of size 300 a substantial decrease of solver performance ratios compared to the behaviour on random and TSPLIB instances could be obtained. This especially holds for the restart variants which are the state of the art inexact TSP solvers while it turned out that it was more effective to generate easier instances for EAX+restart together with much harder instances for LKH+restart than the opposite case. However, one has to be aware that a small performance ratio $R_A(I)/R_B(I)$ does not necessarily mean, that I is easy to solve for A. It may be hard for both, but easier for A.

Comparing random and evolved instances, it turned out that the number resp. the fraction of distinct distances is a central factor for separating both instance sets, i.e. local search in general gets harder given this situation. However, this feature is not suited for distinguishing between the performance of the solvers within the evolved instance sets. The next step will therefore consist of predicting the optimization direction for each solver pairing $\{A, B\}$ (easy for A or easy for B), i.e. a detailed analysis which features resp. feature combinations allow for identifying the kind of solver which performs worse than its competitor within the evolved set will be conducted. Of course, the fraction of distinct distances alone does not provide sufficient information to separate here as in this respect the set is quite homogenous. Preliminary studies indicate that features based on relating the node locations to the centroid of all nodes might play a role here.

In future work we will moreover work on adapting the EA to specifically focus on diversity of evolved instances to generate distinct structures as well as on assessing the influence of the internal rescaling and rounding steps. Moreover, larger instances will be addressed in a systematic way.

Acknowledgements. The authors acknowledge support from the European Center of Information Systems (ERCIS).

References

1. Applegate, D.L., Bixby, R.E., Chvatal, V., Cook, W.J.: The Traveling Salesman Problem: A Computational Study. Princeton University Press, Princeton (2007)
2. Bischl, B., Lang, M., Richter, J., Bossek, J., Judt, L., Kuehn, T., Studerus, E., Kotthoff, L.: mlr: Machine Learning in R. R package version 2.6 (2015). http://CRAN.R-project.org/package=mlr
3. Bossek, J.: Feature-basierte performance-analyse von Algorithmen für das travelling-salesperson-problem. Bachelor thesis, Technical University of Dortmund (2012)
4. Bossek, J.: salesperson: Computation of Instance Feature Sets and R Interface to the State-of-the-Art Solvers for the Traveling Salesperson Problem. R package version 1.0 (2015). https://github.com/wwu-wi/salesperson/
5. Fischer, T., Stützle, T., Hoos, H.H., Merz, P.: An analysis of the hardness of TSP instances for two high-performance algorithms. In: Proceedings of the 6th Metaheuristics International Conference, Vienna, Austria, pp. 361–367 (2005)
6. Helsgaun, K.: General k-opt submoves for the Lin-Kernighan TSP heuristic. Math Program. Comput. **1**(2–3), 119–163 (2009)
7. Hutter, F., Xu, L., Hoos, H.H., Leyton-Brown, K.: Algorithm runtime prediction: methods & evaluation. Artif. Intell. **206**, 79–111 (2014)

8. Kotthoff, L., Kerschke, P., Hoos, H., Trautmann, H.: Improving the state of the art in inexact TSP solving using per-instance algorithm selection. In: Dhaenens, C., Jourdan, L., Marmion, M.-E. (eds.) LION 2015. LNCS, vol. 8994, pp. 202–217. Springer, Heidelberg (2015). doi:10.1007/978-3-319-19084-6_18
9. Kotthoff, L.: Algorithm selection for combinatorial search problems: a survey. AI Mag. **35**(3), 48–60 (2014)
10. Lacoste, J.D., Hoos, H.H., Stützle, T.: On the empirical time complexity of state-of-the-art inexact TSP solvers, optimization Letters, to appear
11. Mersmann, O., Bischl, B., Bossek, J., Trautmann, H., Markus, W., Neumann, F.: Local search and the traveling salesman problem: a feature-based characterization of problem hardness. In: Hamadi, Y., Schoenauer, M. (eds.) LION 6. LNCS, vol. 7219, pp. 115–129. Springer, Heidelberg (2012). doi:10.1007/978-3-642-34413-8_9
12. Mersmann, O., Bischl, B., Trautmann, H., Wagner, M., Bossek, J., Neumann, F.: A novel feature-based approach to characterize algorithm performance for the traveling salesperson problem. Ann. Math. Artif. Intell. 1–32 (2013). http://dx.doi.org/10.1007/s10472-013-9341-2
13. Nagata, Y., Kobayashi, S.: A powerful genetic algorithm using edge assembly crossover for the traveling salesman problem. INFORMS J. Comput. **25**(2), 346–363 (2013)
14. Nallaperuma, S., Wagner, M., Neumann, F., Bischl, B., Mersmann, O., Trautmann, H.: A feature-based comparison of local search and the christofides algorithm for the travelling salesperson problem. In: Foundations of Genetic Algorithms (FOGA) (2013)
15. Pihera, J., Musliu, N.: Application of machine learning to algorithm selection for TSP. In: Fogel, D., et al. (ed.) Proceedings of the IEEE 26th International Conference on Tools with Artificial Intelligence (ICTAI). IEEE Press (2014)
16. Smith-Miles, K., van Hemert, J.: Discovering the suitability of optimisation algorithms by learning from evolved instances. Ann. Math. Artif. Intell. **61**(2), 87–104 (2011)

Extreme Reactive Portfolio (XRP): Tuning an Algorithm Population for Global Optimization

Mauro Brunato[✉] and Roberto Battiti

Department of Computer Science and Telecommunications, University of Trento,
Trento, Italy
{brunato,battiti}@disi.unitn.it

Abstract. Given the current glut of heuristic algorithms for the optimization of continuous functions, in some case characterized by complex schemes with parameters to be hand-tuned, it is an interesting research issue to assess whether competitive performance can be obtained by relying less on expert developers (whose intelligence can be a critical component of the success) and more on automated self-tuning schemes.

After a preliminary investigation about the applicability of record statistics, this paper proposes a fast reactive algorithm portfolio based on simple performance indicators: record value and iterations elapsed from the last record. The two indicators are used for a combined ranking and a stochastic replacement of the worst-performing members with a new searcher with random parameters or a perturbed version of a well-performing member.

The results on benchmark functions demonstrate a performance equivalent or better than that obtained by offline tuning schemes, which require a greater amount of CPU time and cannot take care of individual structural variations between different problem instances.

1 Introduction: Portfolios and Racing for Online Tuning

The design of methods with complicated relationships between the algorithmic building blocks suffers from a serious illness: it hides the relevance of the various blocks, and the abundance of hand-tuned parameters impedes an objective and scientific judgment. In many cases the role of a motivated researcher, with his brain in the loop, is critical to obtain positive results [5]. A kind of dangerous data-mining exists when using a fixed set of benchmark instances and experimenting ("interrogating the data") until the tuned algorithm "screams out" acceptable results [2]. Algorithmic self-tuning has been proposed as a method for mitigating this problem. A well-known case in which a form of "online" learning mechanism is active during the search assumes that the functions to minimize satisfies some statistical model. In this manner one can develop theoretically justified methods to generate new sample points based on information derived from previous samples [21]. Another notable example in the context of Lipschitz optimization algorithms is [17].

© Springer International Publishing AG 2016
P. Festa et al. (Eds.): LION 2016, LNCS 10079, pp. 60–74, 2016.
DOI: 10.1007/978-3-319-50349-3_5

The *algorithm portfolio* method [15] runs more algorithms concurrently, in a time-sharing manner, by allocating a fraction of the total resources to each of them. Dynamic strategies for controlling portfolios are considered in [9,10]. Statistical models of the quality of solutions generated by each algorithm are computed online and used as a control strategy, to determine how many cycles to allocate to each of the interleaved search strategies. A "life-long learning" approach for dynamic algorithm portfolios is considered in [13]. In [14,16] it is shown how to use features capturing the state of a solver during the initial phase of the run to predict the length of a run, to be used by dynamic restart policies.

A related strategy to optimize the allocation of time among a set of alternative algorithms for solving a specific instance is *racing*. Running algorithms are like horses: after the competition is started one gets more and more information about the relative performance and periodically updates the bets on the winning horses, which are assigned a growing fraction of the available future computing cycles. A racing strategy is characterized by two components: (i) the estimate of the future potential given the current state of the search, (ii) the subsequent allocation of hardware resources to speedup the overall minimization.

Racing is related to the *k-armed bandit problem*. One is faced with a slot machine with k arms which, when pulled, yield a payoff from a fixed but unknown distribution. One wants to maximize the expected total payoff over a sequence of n trials. If the distribution is known one would immediately pull only the best performing arm. What makes the problems intriguing is that one has to split the effort between *exploration* to learn the different distributions and *exploitation* to pull the best arm, once the winner becomes clear. One is reminded of the critical exploration-versus-exploitation dilemma observed in optimization heuristics, but there is an important difference: in optimization one is not interested in maximizing the total payoff but in maximizing *the best pull* (the maximum value obtained by a pull in the sequence). The paper [12] is dedicated to determining a sufficient number of pulls to select with a high probability a hypothesis whose *average* payoff is near-optimal. The max version of the bandit problem is considered in [10,11]. An asymptotically optimal algorithm is presented in [19], in the assumption of a generalized extreme value (GEV) payoff distribution for each arm.

When applied offline, racing algorithms search the parameter space by repeated executions of the underlying algorithm by mixing intensification and diversification phases [7]. Online racing algorithms, on the other hand, aim at performing parameter tuning during the optimization of a single instance; the goal is not to find the parametric configuration with the average best performance, but to single out good and bad runs, deciding on the spot when to replace any of them.

A way to estimate the potential of different algorithms is to put a threshold, and to estimate the probability that each algorithm produces a value above threshold by the corresponding empirical frequency. Unfortunately, the appropriate threshold is not known at the beginning, and one may end up with a trivial threshold - so that all algorithms become indistinguishable - or with an impossible threshold, so that no algorithm will reach it. The heuristic solution presented in [20] reactively learns the appropriate threshold. In spite of heuristics, the specific setting of the threshold is not clear and it looks like more work is needed.

The online racing and portfolios paradigms allow to implement a global optimization metaheuristic scheme by completely *decoupling* the underlying, fixed-meta-parameter searchers from the overall parameter-tuning heuristic. In the following, we assume that a limited amount of information is periodically provided by each portfolio member, and the portfolio can be dynamically tuned while running multiple members in time-sharing, or taking advantage of multi-core or multi-machine parallelism. Dynamic tuning consists of killing underperforming members and spawning new ones with given parameter values. For simplicity of implementation we do not consider here periodic dynamic re-allocation of hardware resources.

2   The XRP Reactive Portfolio

Consider a pool of searchers, with different parameters. Each searcher is periodically evaluated on the basis of its past performance. Since the runtime of each searcher may vary due to machine resource contention, decisions are based on performance with respect to the number of function evaluations (optimization steps). Given the online nature of the algorithm, a member which obtained good results might be preferred to (or run alongside) a more promising member whose results are still bad due to its later start. This form of exploration vs. exploitation balance must be carefully considered when planning a portfolio strategy aimed at making sound decisions about the future of each searcher.

Depending on the goal of the optimization strategy, both short- and long-term estimates may be necessary. If running on a bounded time budget, repeated checks of short-term predictions can be necessary in order to waste as few resources as possible on underperforming searchers. On the other hand, if resources are unlimited (i.e., an *anytime* scenario in which we want to achieve better and better optima while the search advances) then longer-term predictions can be more effective by giving to every searcher enough time to smooth out their short-term behavior.

In this paper we will focus on short-term strategies where the portfolio algorithm will decide about which member algorithm to continue according to the member's performance after a given time interval.

2.1   Prediction of "Time Before Next Record"

A fundamental building block of a portfolio selection procedure is the estimation of the time before the next improvement of a local searcher, so that runs for which improvements are not expected for a long time can be stopped in favor of more promising ones.

To evaluate whether such estimate can be applied, it is possible to resort to the extreme value statistic theory. Let $y_1 = f(\boldsymbol{x}_1), y_2 = f(\boldsymbol{x}_2), \ldots$ be the values of the target function along the search trajectory of an optimization algorithm. Let $L(n)$ be the iteration at which the nth record value is achieved. Clearly, $L(1) = 1$ (the first record is achieved at the first evaluation), while $L(n) = \min\{i | y_i < y_{L(n-1)}\}$ (the first evaluation at which the target value falls below the previous record).

A classical result [21] is that, if the y_i's are i.i.d. random variables (i.e., a searcher that evaluates a new random point at each iteration), and if N evaluations have already been performed, then the probability that the next record is achieved at iteration $N' > N$ does not depend on the past history or on the particular distribution the y_is are drawn from:

$$\Pr(L(n+1) = N' | L(n) \leq N) = \frac{N}{N'(N'-1)}. \tag{1}$$

Therefore, the probability that a new record is achieved *within* iteration N' (the corresponding c.d.f.) is also independent on the distribution and is given by

$$\Pr(L(n+1) \leq N' | L(n) \leq N) = \sum_{j=N+1}^{N'} \frac{N}{N'(N'-1)} - 1 - \frac{N}{N'} \tag{2}$$

The probability distribution is heavy-tailed, and the expected number of iterations before a new record is infinite. To define a viable criterion to model the time of the next record, it is possible to set a fixed probability value p and define the upper bound N'_p within which a new record will be achieved with probability p:

$$N'_p = \frac{N}{1-p}. \tag{3}$$

A simple experimental procedure can verify to what extent an actual search algorithm, whose y_is are not i.i.d., can be modeled by Eq. (2). Figure 1 (left) shows the comparison between the estimated and the actual probability of a record value being achieved in a specific interval by an optimization run. After fixing a target number of function evaluations $N = 100, 1000, 10000, 100000$ (see the legend) and a target probability value $0 \leq p < 1$ (horizontal axis), the upper bound N'_p is computed and the probability tested by running 1000 optimization runs and counting how many of them achieve a new record value between iterations N and N'_p. The estimated probability (vertical axis) is plotted against the target probability with a 95% confidence interval. We can observe that the i.i.d. hypothesis systematically underestimates the probability of achieving a record during a given interval with respect to the local searcher's behavior. This is expected, because subsequent values of the search are strongly dependent, and new records tend therefore to appear in bursts which are not considered in the i.i.d. hypothesis.

To mitigate the effect of record bursts caused by the dependence of subsequent evaluations, we may want to consider a whole record burst as a single

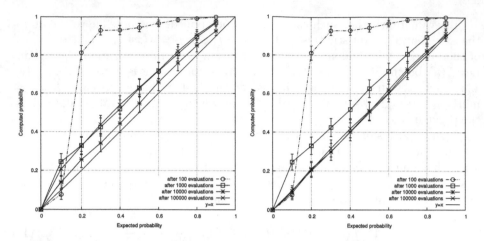

Fig. 1. Left: comparison between the expected probability of a new record event under the i.i.d. hypothesis and the actual probability computed on a sample of 1000 Inertial Shaker runs (see Sect. 3) on the 10-dimensional Rosenbrock function. Right: same comparison after collapsing record bursts.

record. To identify a burst, let us fix a small probability value $q = .01$; when a new record value y_N is achieved at iteration N, let us compute the target iteration N'_q; if a new record value is achieved within iteration N'_q, then we can say (with $1 - q = 99\%$ confidence) that it is not independent from y_N. By iterating this procedure, we identify a sequence of records, each being dependent on the previous one with high confidence. Let us define this sequence a *burst*, and only keep the initial (or final) iteration of a burst as the actual record. With this normalization, the correspondence between the target probability and its estimate, shown in Fig. 1 (right), is well within the 95% confidence interval for $N \geq 10000$.

This observation suggests that, while the i.i.d. hypothesis is too strong at the single function evaluation level, it holds with reasonable confidence at the burst level. In other words, the probability of a new record within a fixed number of iterations only depends on the number of iterations performed, and not on the past history of the search algorithm, *unless a record burst is currently taking place*, in which case a new record is more likely.

Based on the observations above, the evaluation of an algorithm's likelihood to produce a record in the near future will only be based on the number of iterations elapsed since the last record (as an indication of the run's likelihood to be in a dependence-fueled burst of records), without further consideration of the past history.

2.2   The Extreme Reactive Portfolio Algorithm

The Extreme Reactive Portfolio (XRP) procedure is outlined in Fig. 2. XRP works by maintaining and concurrently running a population $\mathcal{P}$ of searchers by removing underperforming ones and replacing them with new ones.

Function parameters	
f	Function to minimize
$\mathcal{A}$	Set of parametric search algorithm
$n_{\text{instances}}$	Number of concurrent instances
Δt	Time interval between population changes in seconds
p_{rnd}	Probability of replacement vs. perturbation
ϕ_{perturb}	Parameter perturbation amount
Local variables	
$\mathcal{P}$	Population of currently running search instances
$\mathcal{R}_{\text{combined}}$	Search instances sorted from worst to best combined performance
$\mathcal{R}_{\text{record}}$	Search instances sorted from best to worst record value

1. **function XRP** (f, $\mathcal{A}$, $n_{\text{instances}}$, Δt, p_{rnd}, ϕ_{perturb})
2. $\quad \mathcal{P} \leftarrow$ population of $n_{\text{instances}}$ random solvers from $\mathcal{A}$
3. $\quad$ **while** computational budget available
4. $\quad\quad$ Store progress information from running instances
5. $\quad\quad$ With Δt periodicity
6. $\quad\quad\quad \mathcal{R}_{\text{combined}} \leftarrow$ **worst_to_best**($\mathcal{P}$)
7. $\quad\quad\quad bad\_instance \leftarrow$ **select**($\mathcal{R}_{\text{combined}}$)
8. $\quad\quad\quad$ **with probability** p_{rnd}
9. $\quad\quad\quad\quad new\_instance \leftarrow$ new random instance from $\mathcal{A}$
10. $\quad\quad\quad$ **else**
11. $\quad\quad\quad\quad \mathcal{R}_{\text{record}} \leftarrow$ **best_to_worst**($\mathcal{P} \setminus \{bad\_instance\}$)
12. $\quad\quad\quad\quad good\_instance \leftarrow$ **select**($\mathcal{R}_{\text{record}}$)
13. $\quad\quad\quad\quad new\_instance \leftarrow$ **perturb**($good\_instance$, ϕ_{perturb})
14. $\quad\quad\quad \mathcal{P} \leftarrow \mathcal{P} \setminus \{bad\_instance\} \cup \{new\_instance\}$
15. $\quad$ **return** best configuration **in** $\mathcal{P}$

Fig. 2. The Extreme Reactive Portfolio (XRP) algorithm.

The procedure receives a set of n_{alg} parametric search algorithms $\mathcal{A} = \{a_1, \ldots, a_{n_{\text{alg}}}\}$. A generic search instance is given by $a_i(\alpha_{i,1}, \ldots, \alpha_{i,n_i})$, where algorithm a_i depends on n_i parameters. We will assume that the generic parameter α_{ij} (the j-th parameter of algorithm i) has a continuous distribution in a specified range of variability; however, given the nature of many search parameters, the distribution is not necessarily uniform. Therefore, each parameter will be defined by a generator function applied to a uniformly distributed parameter:

$$\alpha_{ij} = g_{ij}(u) \qquad \text{for} \quad u \in [\min_{ij}, \max_{ij}]. \qquad (4)$$

For instance, a parameter that needs to vary in $[10^{-5}, 10^{-1}]$ in logarithmic scale may be described by setting $\min_{ij} = -5$, $\max_i j = -1$ and $g_{ij}(u) = 10^u$.

XRP maintains a population $\mathcal{P}$ of running instances initialized at line 2, with each running instances being the instantiation of an algorithm randomly chosen from $\mathcal{A}$ with random parameters. The number of running instances is provided as a metaparameter; a possible choice can be given by the level of parallelism of the problem (e.g., the number of CPU cores in the machine if the evaluations do not depend on resources shared between cores), but in principle any number of instances is admissible.

In the main loop (lines 3–14) XRP maintains a history of each instance based on the instance reports, stored as a list of triplets (current record, number of evaluations performed, elapsed CPU time). Periodically, every Δt seconds, a change is performed in the running population by selecting a bad search instance for termination (lines 6–7) and replacing it with either a new random search instance (line 9) or the perturbed version of a well-performing instance (lines 11–13). The choice of the instance to be removed is driven by a combination of two factors: the record value, and the number of function evaluations since the last improvement. In particular:

– in line 6, the running instances are ordered from worst to best performing according to a performance index that depends on the record value and the number of evaluations elapsed since the last improvement. The performance index must be independent from the relative magnitudes of the two stated factors, therefore everything is computed on the basis of ranking. Let r_i^{record} be the ranking of instance i within $\mathcal{P}$ with respect to its record value (smallest first); let r_i^{fast} be the ranking of instance i with respect to the evaluations elapsed since the last improvement (smallest first). Then, function worst_to_best($\mathcal{P}$) sorts instances according to their combined rankings $r_i^{\text{record}} + r_i^{\text{fast}}$ in decreasing order, so that instances with a bad performance with respect to at least one criterion move up the final ranking.
– in line 7, the bad instance to be removed is selected. To improve differentiation, the worst instance (first in ranking) isn't always chosen; rather, the choice procedure is randomized by selecting the first instance in the given ranking with probability $1/2$, and every subsequent instance with half the residual probability, with the exception of the best instance (the last in the ranking) which is never selected, hence implementing a form of elitism.

After the selection, *bad_instance* will be replaced with a new instance, starting from a new random point, with two possible choices: the replication and perturbation of a well-performing instance, so as to increase the population density in proximity of good performers (intensification), or the creation of new random instance (diversification):

– With probability p_{rnd} a random instance is created (line 9).
– With probability $1 - p_{\text{rnd}}$, a running instance (different from the bad one) is selected with a probability that depends on its ranking with respect to the record value. In detail, function best_to_worst sorts all instances by increasing record value giving rank $\mathcal{R}_{\text{record}}$ which maps instance i to ranking r_i^{record} (line 11); an instance, called *good_selection*, is selected with the same random mechanism described above (function select, line 12) and a new instance is created by perturbing its parameters by a small amount (line 13).

Given a generic parameter $\alpha_{ij} = g_{ij}(u)$, in the notation of Eq. (4), the perturbed parameter is $\alpha'_{ij} = g_{ij}(u')$, where u' is a perturbed version of u obtained by using ϕ_{perturb} on u as a relative perturbation and clamping the result within the admissible range: $u' = \text{clamp}(u \cdot \text{rnd}(1 \pm \phi_{\text{perturb}}), [\min_{ij}, \max_{ij}])$.

3   The Competing Algorithms

The portfolio of competing algorithms is composed of two stochastic local search methods (RAS and IS), with a different adaptation of the dynamic local search area, and of a more complex global optimization scheme (CRTS) combining local search with prohibition-based diversification. No explicit restart is needed for the local search streams because of the online substitution of under-performing or stuck runs considered in our specific proposal.

The following sections briefly outline each technique and describe the parameters that they expose towards the portfolio algorithm.

Local Search with the Reactive Affine Shaker (RAS). The Reactive Affine Shaker Heuristic [8] (RAS) is a self-tuning local search algorithm which does not assume prior knowledge on the function to be minimized. The function is considered as a "black box" (oracle) which can be interrogated to get output values corresponding to input values. The RAS heuristic tries to rapidly move towards better objective values by maintaining and reshaping a bounded *search region* S around the current point x.

The search region is reshaped on the basis of success (or lack of success) during the last step: if a step in a certain direction yields a better objective value, then S is expanded along that direction; it is contracted otherwise. Therefore, once a promising direction is found, the probability that subsequent steps will follow the same direction is increased, and the search shall proceed more and more aggressively in that direction until bad results reduce its prevalence.

The RAS heuristic depends on two parameters, which are exposed to XRP: an initial box width coefficient $0 < \eta \le 1$, defining the size of the initial, isotropic search region with respect to the size of the function domain; and a contraction coefficient $0 < f_{\text{con}} \le 1$ governing the contraction of the search region along unsuccessful directions; a corresponding expansion coefficient for successful search directions is obtained automatically as $f_{\text{dil}} = f_{\text{con}}^{-1}$.

Local Search with the Inertial Shaker (IS). RAS requires matrix-vector multiplications to update the search region, and therefore slows down when the number of dimensions becomes very large. The simpler *Inertial Shaker* (IS) technique [3] can be a more effective choice in this case: the search box is always identified by vectors parallel to the coordinate axes (therefore the search box is defined by a single vector β and no matrix multiplications are needed) and a *trend direction* is identified by averaging the d previous displacements where d is the domain's dimensionality.

The IS heuristic exposes one parameter to XRP: the amplification coefficient, defined as $f_{\text{ampl}} > 0$, controlling the extent at which the trend direction mentioned above modifies the search box.

Continuous Reactive Tabu Search for Global Optimization. While the two previous heuristics do not contain an explicit diversification mechanism, and are therefore local search schemes, the *Continuous Reactive Tabu Search* heuristic [4] (CRTS) can use any of them as a basic searcher while incorporating elements to achieve global optimization.

The initial search region, specified by bounds on each independent variable, is recursively partitioned into a *tree of boxes* (with axes parallel to the coordinate axes). The tree is born with 2^d equally sized leaves, obtained by dividing in half the initial range on each variable. Each box is then subdivided into 2^d equally-sized children, as soon as two different local minima are found in it. The leaves of the tree partition the domain and are the admissible starting regions for the combinatorial component of CRTS.

While the underlying local search algorithm generates a search trajectory consisting of points $X^{(t)}$, CRTS maintains a trajectory consisting of *leaf-boxes* that are evaluated by sampling.

Moving from a leaf-box to another is done on the basis of their neighborhood, their sampled value and a prohibition list whose size is automatically adjusted depending on the search success.

The CRTS heuristic exposes one parameter to XRP: the tabu list size reduction factor $0 < f_{\text{red}} \leq 1$, controlling the extent by which the prohibition list is reduced in presence of successful results. The corresponding expansion factor is determined as $f_{\text{exp}} = f_{\text{red}}^{-1}$.

CRTS also exposes to XRP the f_{con} parameter if used in collaboration with RAS (CRTS+RAS) (η is automatically set by CRTS), and the f_{ampl} parameter if used in conjunction with IS (CRTS+IS).

4   Experimental Results

We tested XRP on various classical benchmark functions. The Rosenbrock, Sphere and Zakharov function families are unimodal, while Goldstein-Price, Hartmann, Rastrigin, and Shekel are multi-modal.

XRP was tested on an 8-core 2.33 GHz Intel Xeon server running a 64-bit Linux OS with kernel 3.13.0. The basic search algorithms (RAS, Inertial Shaker, CRTS), were implemented in C++, while the portfolio selection code was written in Python 2.7. The instances were executed as separate processes, reporting newly found minima via standard output.

Instances were created by randomly selecting a heuristic from the set $\mathcal{A} = \{a_1, \ldots, a_4\}$ of the four heuristics described in Sect. 3:

- a_1 is the Reactive Affine Shaker (RAS) with $n_1 = 2$ free parameters:
  (i) initial box width coefficient $\eta = \alpha_{11} = 10^{-u}$, u uniformly selected in $[\min_{11}, \max_{11}] = [1, 3]$;

(ii) contraction coefficient $f_{con} = \alpha_{12} = 1 - u^3$, u uniformly selected in $[min_{12}, max_{12}] = [.01, .7]$.

This choice of generator functions favors contraction coefficients close to 1, as small changes in the box size need to be more finely tuned with respect to large ones, and treats the initial box width as uniform in logarithmic values. The dilation coefficient is automatically determined as $f_{dil} = f_{con}^{-1}$.

- a_2 is the Inertial Shaker (IS) with $n_2 = 1$ free parameter, the amplification coefficient, defined as $f_{ampl} = \alpha_{21} = 1 - u^3$, with u uniformly selected in $[min_{21}, max_{21}] = [.01, .7]$.
  The rationale for the choice of the generator function is the same as in the RAS case.
- a_3 is CRTS with RAS as local searcher, with $n_3 = 2$ free parameters:
  (i) the prohibition reduction factor $f_{red} = \alpha_{31}$ is chosen uniformly in $[min_{31}, max_{31}] = [.5, 1]$;
  (ii) RAS's contraction coefficient $f_{con} = \alpha_{32}$ is defined as in a_1.
- a_4 is CRTS with IS as local searcher, with $n_4 = 2$ free parameters:
  (i) the prohibition reduction factor $f_{red} = \alpha_{41}$ is defined as in a_3;
  (ii) the amplification coefficient $f_{ampl} = \alpha_{42}$ is defined as in a_2.

Metaparameters Determination. As described in Sect. 2.2, XRP depends on a number of metaparameters. The number of concurrent instances has been set to the number of cores in the experimental server, $n_{instances} = 8$. The remaining three metaparameters have been set by experimenting on 60-s optimizations of the 80-dimensional Rastrigin function. The function was chosen because it is barely solvable within the allocated time limit, with many search instances being unable to find the minimum, and it wasn't used in the following assessment (only lower-dimensional instances will).

- Time interval Δt between two consecutive instance replacement events. Experiments show that a small time interval ($\Delta t \leq 2\,s$) imposes a significant startup overhead on the operating system, so that some processes failed to start within the next period. Tests using $\Delta t = 2.5\,s, 5\,s, 10\,s, 20\,s$ appear to place the best value in the interval $[2.5\,s, 10\,s]$; setting $\Delta t = 20\,s$ shows a performance deterioration because of the small instance turnover in the allotted minute. The selected value for subsequent tests is $\Delta t = 5\,s$.
- Probability of random replacement p_{rnd} governing the choice between a new random instance and a perturbed version of a running instance. Values of $p_{rnd} = 0, .25, .5, .75, 1$ have been tested. Tests on the extreme values ($p_{rnd} = 0$ and $p_{rnd} = 1$) show that many runs (20 of 30) do not converge to the minimum (due to excessive focus on some instances or, conversely, to excessive randomness). The central values seem to be approximately equivalent, so $p_{rnd} = .5$ has been selected.
- Relative parameter perturbation factor $\phi_{perturb}$. Values of 0, .1 and .2 have been tested; as expected, for $\phi_{perturb} = 0$ a fraction of the runs (17 of 30) do not converge to the minimum, probably because of too little differentiation between good instances. We chose $\phi_{perturb} = .1$.

In general, we observe that XRP is quite robust with respect to metaparameter changes, provided that they fall within a reasonable range, therefore no further optimization has been attempted beyond this preliminary investigation.

4.1 Evaluation with a Fixed Budget

For each of the benchmark functions listed in Table 1, 30 runs of XRP were performed with a total budget of 10^7 evaluations. The median and inter-quartile range of all record results are reported in the "XRP" column. Results are reported as the difference from the actual global minimum value (Goldstein-Price, Hartmann and Shekel families have non-zero global minima). The "XRP" column also shows the distribution of the winner algorithm over the 30 runs of XRP for each benchmark function (the search instance that achieved the record value over the portfolio selection run).

It is possible to observe that the majority of winners involves the simpler IS technique, with the RAS algorithm only appearing on two runs, both times in conjunction with CRTS. As expected, many low-dimensional instances ($d \leq 10$) benefit from the diversification capabilities of CRTS; the exceptions are the unimodal Rosenbrock and Zakharov instances, where a single local search run is sufficient, and the four-dimensional Shekel instances, whose few

Table 1. Median record values for 30 runs of XRP, best algorithm and iRace-determined algorithm for 10^7 evaluations of the target function. The gray bars show the distribution of the winner algorithm over the 30 runs.

		= IS	= CRTS+IS		= CRTS+AS		
		XRP		Winner		iRace winner	
Function	Dim	Median	IQR	Median	IQR	Median	IQR
Goldstein -Price	2	$5.62 \cdot 10^{-12}$	$1.85 \cdot 10^{-11}$	$1.77 \cdot 10^{-12}$	$1.47 \cdot 10^{-11}$	$7.89 \cdot 10^{-05}$	$1.27 \cdot 10^{-04}$
Hartmann	3	$1.24 \cdot 10^{-14}$	$5.51 \cdot 10^{-14}$	$6.66 \cdot 10^{-15}$	$2.89 \cdot 10^{-14}$	$3.52 \cdot 10^{-06}$	$3.67 \cdot 10^{-06}$
Hartmann	6	$1.33 \cdot 10^{-12}$	$2.59 \cdot 10^{-12}$	$5.44 \cdot 10^{-13}$	$1.99 \cdot 10^{-12}$	$5.05 \cdot 10^{-05}$	$4.53 \cdot 10^{-06}$
Rastrigin	10	$1.10 \cdot 10^{-08}$	$2.39 \cdot 10^{-03}$	$4.19 \cdot 10^{-09}$	$6.28 \cdot 10^{-04}$	$2.63 \cdot 10^{-02}$	$2.02 \cdot 10^{-02}$
Rastrigin	30	$2.01 \cdot 10^{-13}$	$1.55 \cdot 10^{-12}$	$8.53 \cdot 10^{-14}$	$2.39 \cdot 10^{-13}$	$6.06 \cdot 10^{-13}$	$2.14 \cdot 10^{-09}$
Rosenbrock	10	$3.35 \cdot 10^{-13}$	$3.29 \cdot 10^{-10}$	$5.38 \cdot 10^{-20}$	$1.40 \cdot 10^{-16}$	$7.82 \cdot 10^{-19}$	$4.86 \cdot 10^{-16}$
Rosenbrock	30	$1.72 \cdot 10^{-14}$	$3.99 \cdot 10^{-14}$	$6.73 \cdot 10^{-15}$	$7.91 \cdot 10^{-15}$	$6.99 \cdot 10^{-15}$	$3.75 \cdot 10^{-14}$
Shekel 5	4	$2.53 \cdot 10^{-10}$	$8.16 \cdot 10^{-10}$	$6.49 \cdot 10^{-11}$	$2.08 \cdot 10^{-10}$	$2.17 \cdot 10^{-02}$	$1.84 \cdot 10^{-02}$
Shekel 7	4	$1.87 \cdot 10^{-10}$	$5.98 \cdot 10^{-10}$	$4.49 \cdot 10^{-11}$	$1.53 \cdot 10^{-10}$	$2.19 \cdot 10^{-02}$	$1.58 \cdot 10^{-02}$
Shekel 10	4	$1.16 \cdot 10^{-10}$	$4.53 \cdot 10^{-10}$	$3.86 \cdot 10^{-11}$	$3.72 \cdot 10^{-10}$	$3.06 \cdot 10^{-02}$	$2.03 \cdot 10^{-02}$
Sphere	30	$3.78 \cdot 10^{-79}$	$7.31 \cdot 10^{-77}$	$1.47 \cdot 10^{-90}$	$1.55 \cdot 10^{-83}$	$2.35 \cdot 10^{-87}$	$4.11 \cdot 10^{-80}$
Zakharov	10	$8.50 \cdot 10^{-69}$	$1.55 \cdot 10^{-62}$	$4.74 \cdot 10^{-82}$	$9.85 \cdot 10^{-74}$	$1.72 \cdot 10^{-82}$	$1.31 \cdot 10^{-72}$
Zakharov	30	$3.95 \cdot 10^{-81}$	$7.96 \cdot 10^{-70}$	$3.38 \cdot 10^{-319}$	$2.39 \cdot 10^{-319}$	$2.98 \cdot 10^{-319}$	$1.67 \cdot 10^{-319}$

foxhole-shaped local minima trick the CRTS district evaluation mechanism, but are few (resp. 5, 7, and 10) and narrow enough for a well-tuned local searcher to jump in and out of them without the need to restart. High-dimensional instances ($d = 30$), on the other hand, the CRTS technique doesn't have the time to start evaluating all first-level districts (2^{30}), while the local searchers don't suffer from that disadvantage.

After each run, the search algorithm instance that achieved the record result was allocated the whole 10^7 evaluations budget for a separate run with the same random seed, so that its execution would exactly mimic its run within XRP, but for a longer time. The 30-test median and IQR are reported in column "Winner".

Results in column "iRace" will be discussed in Sect. 4.3. Observe that in most cases the XRP outcome is quite close to the "Winner" result, within the same order of magnitude. Notable exceptions are the simplest unimodal Sphere function, quadratic with spherical symmetry, and Zakharov's very flat global attractor, whose small final improvements tend to require quite long runs. Even in these cases, however, the XRP outcome is well within reasonable target limits.

The outcomes for the 10- and 30-dimensional Rosenbrock and Rastrigin functions are also reported in Fig. 3 for ease of comparison. Observe that, in some cases, the 30-dimensional problem is solved more efficiently than the 10-dimensional one, in particular for the Rastrigin function. When evaluating the 10-dimensional function, bad optimization instances can consume a large part of the budget before XRP can check and eventually stop them. This effect is mitigated by the longer time needed to evaluate the 30-dimensional version. Finer control over the instance performance, overcoming the strict periodicity of the current algorithm, will possibly remove this kind of artifact.

The "Random" box refers to a series of 30 runs of random heuristics with randomly determined parameters. Note that, in all cases, the choice of random instances achieves much worse results; thus motivating the use of online portfolio selection when no prior information about the optimal algorithm and parameter settings is given.

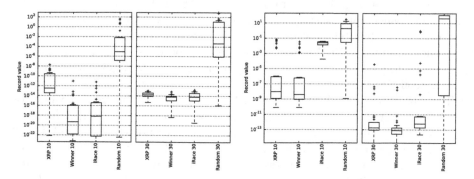

Fig. 3. Distribution of record values after 10^6 function evaluations for XRP, Winner, iRace-determined and random heuristic for the 10- and 30-dimensional Rosenbrock (left) and Rastrigin (right) functions.

4.2   Evolution of Record Values in Time

Figure 4 shows the evolution of the record value over 60 s of XRP, comparing it with the best instance running for the same time and with the same seed (see Sect. 4.1) and with a random instance. Curves represent median values of 30 runs, with bars representing the interval between the 1st and 3rd quartiles.

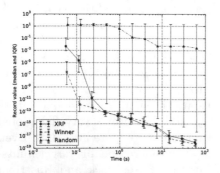 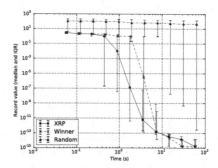

Fig. 4. Median and IQR of record values for 30 60-s runs of XRP, Winner heuristic and Random heuristic, for 30-dimensional Rosenbrock (left) and 30-dimensional Rastrigin (right). Note: inter-quartile bars have been slightly offset to tell them apart in spite of overlaps.

The overlap between the "XRP" and "Winner" IQR bars shows that many runs fall in the same range; while in the end the Winner instance always wins, for a short initial time (consider that Fig. 4 is a log-log plot) the overall XRP outcome can outperform the Winner alone due to the concurrent execution of many instances, in particular when a more difficult function such as Rastrigin is chosen (with its many local minima). Again, a random choice is almost always penalized.

4.3   Comparison with Offline Racing

To identify a good algorithm/parameter combination that would work over the whole set of benchmark functions, we trained the offline racing package iRace [18] on a budget of 1000 60-s optimization runs over the whole function set, with the same parameter distribution as the online procedure. The Inertial Shaker (IS) algorithm with amplification coefficient $f_{\mathrm{ampl}} = .7033$ was identified as the best tradeoff.

The last column of Table 1 shows the distribution of the record value for 30 10^7-iteration runs of the optimal algorithm on each benchmark function. We can observe that the resulting parameter set was not equally good for all functions; in particular, it performs very well on Zakharov, Sphere, and the 30-dimensional Rosenbrock and Rastrigin functions (also see Fig. 3 for a visual comparison of result distributions), while it seems to work less than optimally for other instances. This result confirms the heuristic observation that parameter tuning is needed at the instance level, and supports our rationale for online tuning (when the anytime nature of the solution method allows for it).

5   Conclusions

A novel online dynamic portfolio, XRP, for anytime optimization heuristics has been presented and tested on classical function minimization benchmarks. XRP can be applied to any collection of optimization heuristics with the only requirement to be able to report their record values as soon as they are achieved. While XRP itself depends on few simple metaparameters, preliminary investigation indicates that results are robust with respect to variations from their natural central values.

Further investigation will consider the use of XRP for more challenging very large-scale optimization problems, both continuous and discrete, and the implementation of more sophisticated models for the performance of the competing instances, in order to reliably identify unpromising executions and their most suitable replacements as optimization proceeds. An evaluation of schemes based on reinforcement learning [1,6] is also on the stack.

Acknowledgments. The research of Roberto Battiti was supported by the Russian Science Foundation, project no. 15–11–30022 "Global optimization, supercomputing computations, and applications."

References

1. Battiti, R., Brunato, M., Campigotto, P.: Learning while optimizing an unknown fitness surface. In: Maniezzo, V., Battiti, R., Watson, J.-P. (eds.) LION 2007. LNCS, vol. 5313, pp. 25–40. Springer, Heidelberg (2008). doi:10.1007/978-3-540-92695-5_3
2. Battiti, R., Brunato, M., Mascia, F.: Reactive Search and Intelligent Optimization. Operations research/Computer Science Interfaces, vol. 45. Springer, Heidelberg (2008)
3. Battiti, R., Tecchiolli, G.: Learning with first, second, and no derivatives: a case study in high energy physics. Neurocomputing **6**, 181–206 (1994)
4. Battiti, R., Tecchiolli, G.: The continuous reactive tabu search: blending combinatorial optimization and stochastic search for global optimization. Ann. Oper. Res. - Metaheuristics Comb. Optim. **63**, 153–188 (1996)
5. Battiti, R., Brunato, M.: The LION way. Machine Learning plus Intelligent Optimization. LIONlab, University of Trento, Italy (2014)
6. Battiti, R., Campigotto, P.: Reinforcement learning and reactive search: an adaptive max-sat solver. In: Ghallab, M., Spyropoulos, C.D., N.F., Avouris, N. (eds.) Proceedings ECAI 08: 18th European Conference on Artificial Intelligence, Patras, Greece, 21–25 July 2008. IOS Press, Amsterdam (2008)
7. Birattari, M., Stützle, T., Paquete, L., Varrentrapp, K.: A racing algorithm for configuring metaheuristics. In: Langdon, W.B., et al. (eds.) Proceedings of the Genetic and Evolutionary Computation Conference, pp. 11–18. Morgan Kaufmann Publishers, San Francisco (2002). AIDA-2002-01 Technical report of Intellektik, Technische Universität Darmstadt, Darmstadt, Germany
8. Brunato, M., Battiti, R., Pasupuleti, S.: A memory-based rash optimizer. In: Geffner, A.F.R.H.H. (ed.) Proceedings of AAAI-06 Workshop on Heuristic Search, Memory Based Heuristics and Their Applications, pp. 45–51, Boston, Mass. (2006). ISBN 978-1-57735-290-7

9. Cicirello, V.: Boosting stochastic problem solvers through online self-analysis of performance. Ph.D. thesis, Carnegie Mellon University (2003). Technical report CMU-RI-TR-03-27

10. Cicirello, V., Smith, S.: The max k-armed bandit: a new model for exploration applied to search heuristic selection. In: 20th National Conference on Artificial Intelligence (AAAI 2005), July 2005

11. Cicirello, V.A., Smith, S.F.: Heuristic selection for stochastic search optimization: modeling solution quality by extreme value theory. In: Wallace, M. (ed.) CP 2004. LNCS, vol. 3258, pp. 197–211. Springer, Heidelberg (2004). doi:10.1007/978-3-540-30201-8_17

12. Fong, P.W.L.: A quantitative study of hypothesis selection. In: International Conference on Machine Learning, pp. 226–234 (1995). http://citeseer.ist.psu.edu/fong95quantitative.html

13. Gagliolo, M., Schmidhuber, J.: Learning dynamic algorithm portfolios. In: Proceedings AI and MATH 2006, Ninth International Symposium on Artificial Intelligence and Mathematics, Fort Lauderdale, Florida, January 2006

14. Horvitz, E., Ruan, Y., Gomes, C., Kautz, H., Selman, B., Chickering, D.M.: A bayesian approach to tackling hard computational problems. In: Seventeenth Conference on Uncertainty in Artificial Intelligence, pp. 235–244, Seattle, USA, August 2001

15. Huberman, B.A., Lukose, R.M., Hogg, T.: An economics approach to hard computational problems. Science **275**, 51–54 (1997)

16. Kautz, H., Horvitz, E., Ruan, Y., Gomes, C., Selman, B.: Dynamic restart policies. In: Eighteenth National Conference on Artificial Intelligence, pp. 674–681. American Association for Artificial Intelligence, Menlo Park (2002)

17. Lera, D., Sergeyev, Y.D.: Acceleration of univariate global optimization algorithms working with lipschitz functions and lipschitz first derivatives. SIAM J. Optim. **23**(1), 508–529 (2013)

18. López-Ibánez, M., Dubois-Lacoste, J., Stützle, T., Birattari, M.: The irace package, iterated race for automatic algorithm configuration. Technical report TR/IRIDIA/2011-004, IRIDIA, Université Libre de Bruxelles, Belgium (2011)

19. Streeter, M.J., Smith, S.F.: An asymptotically optimal algorithm for the max k-armed bandit problem. In: Proceedings of the National Conference on Artificial Intelligence, vol. 21, pp. 135–142. AAAI Press, MIT Press/Menlo Park, Cambridge, London 1999 (2006)

20. Streeter, M.J., Smith, S.F.: A simple distribution-free approach to the max k-armed bandit problem. In: Benhamou, F. (ed.) CP 2006. LNCS, vol. 4204, pp. 560–574. Springer, Heidelberg (2006). doi:10.1007/11889205_40

21. Zhigljavsky, A., Žilinskas, A.: Stochastic Global Optimization, vol. 9. Springer Science & Business Media, Heidelberg (2007)

Bounding the Search Space of the Population Harvest Cutting Problem with Multiple Size Stock Selection

Laura Climent[✉], Barry O'Sullivan, and Steven D. Prestwich

Insight Centre for Data Analytics, Department of Computer Science,
University College Cork, Cork, Ireland
{laura.climent,barry.osullivan,
steven.prestwich}@insight-centre.org

Abstract. In this paper we deal with a variant of the Multiple Stock Size Cutting Stock Problem (MSSCSP) arising from population harvesting, in which some sets of large pieces of raw material (of different shapes) must be cut following certain *patterns* to meet customer demands of certain product types. The main extra difficulty of this variant of the MSSCSP lies in the fact that the available patterns are not known *a priori*. Instead, a given complex algorithm maps a vector of continuous variables called a *values vector* into a vector of total amounts of products, which we call a *global products pattern*. Modeling and solving this MSSCSP is not straightforward since the number of value vectors is infinite and the mapping algorithm consumes a significant amount of time, which precludes complete pattern enumeration. For this reason a representative sample of global products patterns must be selected. We propose an approach to bounding the search space of the values vector and an algorithm for performing an exhaustive sampling using such bounds. Our approach has been evaluated with real data provided by an industry partner.

1 Introduction

The *Cutting Stock Problem* (CSP) [6] is a well-known NP-hard optimization problem in operations research. This problem involves deciding which pattern should be applied to raw material stock in order to obtain sufficient amount of products to meet the demands while minimizing cost. The CSP can be modeled and solved as an Integer Linear Program (ILP). In this paper we deal with a variant of CSP introduced in [7] that we classify as $*/V/D/R$ using Dyckoff's typology [3], where $*$ means any dimensionality, V means that the raw material stock is sufficient to accommodate all the demanded products (hence only some selected stock pieces have to be cut), D means that all large pieces are different (in terms of shape) and R indicates many products demanded of few different types. The feature V (any demand can be fulfilled) entails that the raw material stock to be cut has to be selected. Each large piece has an associated "value"

© Springer International Publishing AG 2016
P. Festa et al. (Eds.): LION 2016, LNCS 10079, pp. 75–90, 2016.
DOI: 10.1007/978-3-319-50349-3_6

(e.g. typically values are proportional to their sizes) and the objective function is to minimize the total value of the raw material stock selected to fulfill the demand.

According to a later typology presented in [8] we are dealing with a variant of the *Multiple Stock Size Cutting Stock Problem* (MSSCSP). In [8] it is noticed that research on cutting and packing problems still is rather traditionally oriented. For instance, few recent papers consider heterogeneous assortments of large pieces. In addition, the variant analyzed in this paper has certain peculiarities that make it harder and consequently more challenging. This variant might emerge in real-life applications in which the number of raw material pieces available for cutting and their dimensions is uncertain because: (i) only a sample of the whole set of pieces is known, and/or (ii) the pieces might change dynamically with time. Population harvesting (e.g. plants, fish and animals [4]) are examples of both types of uncertainty: the dimensions of their raw elements might change due to their growth (case ii); and only some samples of the dimensions of the raw elements are taken by the industry (case i). Note that measuring all of them (there are possibly several hundred/thousand of elements) would be too costly.

Due to the above mentioned uncertainties associated with the raw material pieces of this type of MSSCSP, it is impossible to know all the patterns associated with each piece of stock (many dimensions of the stock are unknown). For this reason, in the literature and in industry, an algorithm that simulates the cutting of a whole set of raw material samples according to certain values vector has been generally used. The objective is to obtain similar results when such vectors and algorithms are used for cutting the real stock from which the sample data was acquired. The values vector is composed of continuous variables and each of them is associated with a product type. Note that we do not have access to the set of patterns to be cut in a direct manner, only via the application of this complex algorithm which is denoted throughout the paper as $\mathcal{A}$. The $\mathcal{A}$ algorithm selects the optimal cutting for each raw material sample based on the values of the products. The optimality criterion of such algorithm is to maximize the total value, which is the sum of the products of value and units cut of each product type. Then each combination of total amounts of products that can be cut from a set of raw material pieces represents a global products pattern for this set. By providing several different values vectors as input to the $\mathcal{A}$ algorithm, different global product patterns associated with a set of raw material pieces can be obtained. Once the best values vector has been selected in this cutting simulation process over the sample data, it is used as input to the $\mathcal{A}$ algorithm that cuts the real raw material (which is installed in the cutting machines). If the sample is representative of the whole population, the results of the cutting will be similar to that predicted in the simulation phase.

We would like to highlight that it is not straightforward to model and solve this variant of MSSCSP because the number of values vectors is infinite (continuous variables) and therefore, for realistic instances we can not enumerate all the global products patterns in a reasonable amount of time. For this reason a representative sample of them must be selected. This is a complicated task because the algorithm that generates the global products patterns (a) is complex and requires a great amount of time for realistic instances and (b) it matches many different values vectors to similar global products patterns. As an example of (b) consider a values vector whose associated global products pattern has the maximum possible amount of each type of product. Then, even if its associated value in the vector is increased (and the other values of the vector remain the same), the same global products pattern will be obtained. The latter fact, and the necessity of finding a representative set of global products patterns in a reasonable amount of time (case a), has motivated the work presented here.

Our main objective is to reduce the search space for the values vector by bounding it in such a way that areas that produce the same global products pattern are excluded (because we want to maximize how scattered the global products patterns produced are). To illustrate this we show a graph in Fig. 1 that represents the amount of certain products obtained after applying the $\mathcal{A}$ algorithm with different values vectors (other types of products can also be cut from the raw material). On the horizontal axis is the value associated with each type of product (normalized to the interval [0,1], and on the vertical axis are the product amounts. Note that the minimum amount that it is possible to obtain is zero units and the maximum amount is 211.74 units. The dashed rectangle includes different amounts of such products, so it is necessary to sample in this area in order to obtain a wide range of different global products patterns. Note that values less than or equal to the minimum value in the rectangle (lower bound), the $\mathcal{A}$ algorithm produces the same amount of product: zero. The opposite occurs with values that are at least the maximum value in the rectangle (upper bound): the amount obtained is the maximum. For this reason, using values that are outside the interval delimited by the lower and upper bound is a waste of time as no new global products patterns will be obtained. As mentioned, reducing the computational time is vital, especially in on-line problems such as CSP real-life applications with uncertainties.

In order to reduce the computational time for generating a representative sample of global products patterns, we present definitions and equations for

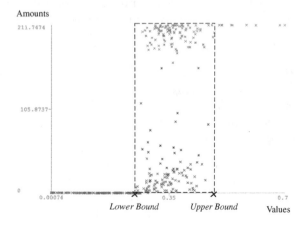

Fig. 1. Amounts obtained with $\mathcal{A}$ for several values associated with a product.

calculating the lower and upper values bounds of each product. These bounds ensure that the minimum/maximum amount of a type of product is obtained. As far as we know, there is no other reported technique that attempts to compute the lower and upper bounds, nor to reduce the values search space. The lower and upper bounds that we introduce can be used with any sampling technique (e.g. random sampling as the most naive approach) and consequently by bounding the search space, a representative set of global products patterns tends to be obtained more quickly. In addition, without loss of generality, in this paper we also propose a method for exhaustive sampling using such bounds.

The paper is structured as follows. First, the variant of the MSSCSP is formalized. An explanation of the $\mathcal{A}$ algorithm is also provided. Afterwards, the calculation of the upper and lower bounds is explained, and we also introduce an exhaustive sampling algorithm. The effectiveness of our method is shown with an evaluation with real-life instances. Finally we present our conclusions.

2   Problem Formalization

In this section we explain the new features of the variant of the MSSCSP, with respect to the traditional CSP formulation [6]. (Parts of the following explanations have been extracted from [7]). MSSCSPs have raw material pieces of different dimensions which we can cut at will. In the variant that we are dealing with, we have a fixed number of raw material pieces (possibly hundreds or thousands) each with its own dimensions σ_r. There are K subsets of raw material pieces and each subset has R pieces (R might be different for each subset but this is ignored to simplify the description). Either a subset is fully cut with a unique values vector or none of its pieces is cut (as previously motivated in Sect. 1, due to the uncertainty of the environment). Then, each subset of raw material pieces has its own associated global products patterns (which are not given and therefore we must sample them), which are the combinations of amounts of products that can be cut from it. A global products pattern $\boldsymbol{p} \in \mathbb{Q}_+^{|\mathcal{M}|}$ is noted as $\boldsymbol{p} = \langle a_1, \ldots, a_{|\mathcal{M}|} \rangle$, where $\mathcal{M}$ is the set of product types and a_j represents the amount of units of product $m_j \in \mathcal{M}$ cut from certain set of raw material pieces.

Definition 1. *We represent a type of product as a tuple $m_j = \langle s_j, z_j \rangle$, where:*

- *$s_j \in \mathbb{R}$ is the size of a piece of m_j. Depending on the number of dimensions analyzed, s_j can represent: lengths for 1-dimension (e.g. cm), areas for 2-dimensions (e.g. cm^2) or volumes for 3-dimensions (e.g. cm^3).*
- *z_j is the dimensions of m_j. For instance, if m_j has the shape of a rectangle, z_j would be the required length and width for m_j.*

As previously mentioned, in the variant of MSSCSP analyzed, the patterns are not known *a priori* and it is only possible to have indirect control over them via a list of continuous variables called a values vector. A values vector $\boldsymbol{v} \in \mathbb{Q}_+^{|\mathcal{M}|}$ is a vector of $|\mathcal{M}|$ continuous variables. Each v_j represents the value associated with the type of product $m_j \in \mathcal{M}$ per unit of s_j. For instance v_j could represent monetary units: €, \$, etc. per each unit of s_j, (e.g. €/m^3).

A set of products types $\mathcal{M}$ and a vector of dimensions σ of $|R|$ raw material pieces can be passed to an algorithm $\mathcal{A}$ which uses a values vector for calculating the corresponding p. Then, $\mathcal{A}$ can be represented as the following mapping function:

$$\mathcal{A}(\mathcal{M}, \langle \sigma_1, \ldots, \sigma_{|R|} \rangle, v) \rightarrow p \tag{1}$$

To make this variant of the MSSCSP amenable to an ILP approach, in [7] the infinite set of possible values vector was reduced to a finite set of n values vectors (which should be sufficiently representative) u_{ik} $(i = 1 \ldots n)$ for each subset of raw material pieces $k \in K$ (the same n is assumed for each subset to simplify the notation). Then global products patterns for each subset k are precomputed by using algorithm $\mathcal{A}$, storing the results in vectors of constants $u_{ik} = \mathcal{A} : (\mathcal{M}, \langle \sigma_1, \ldots, \sigma_{|R|} \rangle_k, v_{ik})$ $(\forall i, k)$. The ILP model is as follows:

$$\min \sum_{i=1}^{n} \sum_{k=1}^{K} c_k x_{ik} \qquad \forall x_{ik} \in \{0, 1\}$$

$$\text{s.t} \quad \sum_{i=1}^{n} \sum_{k=1}^{K} u_{ikj} x_{ik} \geq d_j \qquad \forall j \in \mathcal{M}$$

$$\sum_{i=1}^{n} x_{ik} \leq 1 \qquad \forall k \in K$$

where d_j is the targeted demand for each type of product, c_k is the value associated with the stock subset k and x_{ik} are the decision variables that indicate if the subset k is cut with the global product pattern i. The objective function is to minimize the total value of the sets of raw materials used for satisfying the demands. Note that if a subset of raw material pieces k is not used for satisfying the demands (and therefore it is not cut) then all its decision variables $(x_{ik} \forall i \in n_k)$ are zero. Note also that the first constraint ensures that the demands are fulfilled, and the second constraint prevents the use of more than one global products pattern in a set of raw material pieces. As mentioned, this set of representative global products patterns $(u_{ik} \forall i \forall k)$ must be generated in order to be able to solve this ILP with standard optimisation software; and it must adequately cover all possible global products patterns for each set of raw material pieces k. The main contribution of this paper resides in this task. By bounding the search space of values of the vector, we are reducing the likelihood of generating global products patterns that are not new (they are equal to a previous generated global products pattern). Hence, the global products patterns generated in a fixed amount of time tend to be more scattered and therefore more significant for the analyzed problem.

3 Algorithm $\mathcal{A}$

In [7], the $\mathcal{A}$ algorithm is treated as a black box. Here we analyze and use its properties, which allows us to reduce the values search space of this variant of the MSSCSP. For this reason, in this section we briefly explain the $\mathcal{A}$ algorithm.

This algorithm simulates the cutting of a set of raw material pieces R into certain products types $\mathcal{M}$. Each type of product m_i has an associated value v_i, representing how valuable each unit of product is (these values compose the values vector v, which is provided as an input). This algorithm selects the optimal cutting for each raw material sample, where the optimality criterion is to maximize the total value, which is the sum of the products of value and units cut of each product type.

Definition 2. *The total value of a piece of product $m_j \in \mathcal{M}$ is calculated as:*

$$t(m_j) = s_j v_j \tag{2}$$

Note that the input values vector has a direct impact on the amounts of each type of product that will be obtained from a certain raw material. The other factor that has an influence on the amounts is the dimensions of the raw material pieces and the dimensions of the product types. The greater a value v_i where the other values of v are fixed, the greater the number of products m_i its associated global products pattern p will have after running algorithm $\mathcal{A}$ (they will be equal only in the case of reaching a saturation point, see Fig. 1 as an example). In the same way, the opposite situation (at most number of pieces) holds when the values of v_i are decreased. In the next section, we use these properties of the $\mathcal{A}$ algorithm for bounding the values search space.

Typically, in the literature, the $\mathcal{A}$ algorithm has been implemented with Dynamic Programing (DP) [1,2]. DP is an approach that allows us to solve complex problems by dividing them into a collection of simpler subproblems. For such purpose, the sub-problems must be overlapping. The problem of cutting a raw material piece satisfies these properties, since it is a recursive one (i.e. maximize by cutting the first product and then maximizing the remainder).

4   Computing Lower and Upper Bounds

In this section we present definitions and equations of the lower and upper values vector that is provided as an input of the $\mathcal{A}$ algorithm. Each value v_i of such a vector (v) is associated with a type of product m_i with particular shape characteristics z_i. The main idea behind the lower and upper bounds is that when we apply the $\mathcal{A}$ algorithm (Eq. 1) to a certain raw material, when the value associated with a type of product is:

(i) at most the lower bound, it is ensured that the amount obtained of such product is the minimum.
(ii) at least the upper bound, it is ensured that the amount obtained of such product is the maximum.

A toy example is described for further explanation of the bounds.

Example 1. We consider a 2-dimensional space with two product types (m_1 and m_2) with a rectangle shape. Then z_1 and z_2 is represented with the height (h) and the width (w). The size of the product types, is determined by the equation of the area of a rectangle. Thus, $s_1 = l_1 w_1$ and $s_2 = l_2 w_2$. Figure 2 shows such products. Note that s_1 is greater than s_2. Without loss of generality, for this example, $w_1 = w_2$.

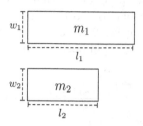

Fig. 2. 2- dimensional example of two product types.

4.1 Computing Lower Bounds

Given $\mathcal{M} = \{m_i, m_j\}$ where $s_j < s_i$ (in Example 1, $m_i = m_1$ and $m_j = m_2$), we want to calculate the lower value bound, which is the greatest value that we can assign to the associated variable of m_i in a values vector (v) in order to ensure that the minimum amount of m_i is obtained for any raw material. For such a purpose we present the following definition.

Definition 3. *The maximum number of pieces of m_j that can fit in one piece of m_i, according to their shape specifications (z_i and z_j), is $k(z_i, z_j)$. To simplify the notation, we will use k rather than $k(z_i, z_j)$ when the variables are obvious from context.*

We consider a subpart of the raw material from which we can only cut either one piece of m_i or k pieces of m_j. Note that the waste produced when cutting m_i is *smaller or equal* than when cutting k pieces of m_j. Figure 3 shows such situation for Example 1, where the size of the subpart of the raw material analyzed is ks_1. Note that for this example $k(z_1, z_2) = 1$. If the cut off products from this subpart is k pieces of m_2, there is an associated waste which is the size analyzed minus

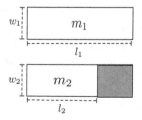

Fig. 3. Example for explaining the lower bound.

ks_2 (grey area in Fig. 3). Thus, if $v_1 = v_2$, cutting the type of product m_1 is the best option because the total value of a piece of m_1 is greater than the total value of k pieces of m_2: $t(m_1) > kt(m_2)$ due to $s_1 \geq ks_2$ (see Eq. 2).

For computing the lower value bound, we would like to know how much smaller the value v_i should be, in order to change the priority order where cutting k pieces of m_j is the most profitable. In order to compute this value, we compute for which v_i the priorities are equal. This situation occurs when the total profit of cutting a piece of m_i is equal to the total profit of cutting k pieces of m_j ($t(m_i) = kt(m_j)$). By applying Eq. 2 we obtain for which v_i the priorities are equal. Then, using an arbitrarily small positive number (denoted as ϵ), we obtain the lower bound:

$$v_i^{lb}(m_i, m_j) = \frac{k(z_i, z_j)s_j v_j}{s_i} - \epsilon, \text{ for } s_i \geq s_j. \tag{3}$$

From the above we can state that for any value $v_i \leq v_i^{lb}$ the total benefit of cutting k pieces of m_j is greater than the total benefit of cutting a piece of m_i. Therefore, when applying the $\mathcal{A}$ algorithm with v_i^{lb} to a complete raw material, the obtained number of pieces of m_i is the minimum and the obtained number of pieces of m_j is the maximum. Note that if at least one piece of m_j fits in a piece of m_i (e.g. Example 1 represented in Fig. 2) then the minimum amount of m_i is zero (e.g. Fig. 1). Otherwise there might exist subparts of the raw material in which m_j does not fit but m_i does, in which case the amount of m_i could be greater than zero.

4.2   Computing Upper Bounds

In this section, given the same $\mathcal{M} = \{m_i, m_j\}$ where $s_j < s_i$ (in Example 1, $m_i = m_1$ and $m_j = m_2$) we want to calculate the upper value bound, which is the lowest value that we can assign to the associated variable of m_i in a values vector (v) in order to ensure that the maximum amount of m_i is obtained for any raw material. For this purpose we define:

Definition 4. *The minimum number of pieces of m_j that are required to fit one piece of m_i into their global shape, according to their shape specifications (z_i and z_j), is $h(z_i, z_j)$. To simplify the notation, we will use h rather than $h(z_i, z_j)$ when the variables are obvious from context.*

We consider a subpart of the raw material from which we can only cut either one piece of m_i or h pieces of m_j. Note that the waste produced when cutting m_i is *at least* that when cutting h pieces of m_j. Figure 4 shows such situation for Example 1, where the size of the subpart of the raw material analyzed is hs_2. Note that for this example $h(z_1, z_2) = 2$. If the cut off product from this subpart is m_1, there is an associated waste which is the size analyzed minus s_1 (grey area in Fig. 4). Thus if $v_1 = v_2$ then cutting the type of product m_1 is the worst option because the sum of the total value of the h pieces of m_2 is greater than the total value of a piece of m_1: $t(m_1) < ht(m_2)$ because $s_1 \leq hs_2$ (see Eq. 2).

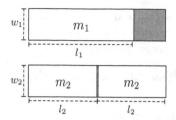

Fig. 4. Example for explaining the upper bound.

For computing the upper value bound we would like to know how much greater the value v_i should be, in order to change such priority order (a piece of m_i the most profitable). In order to compute such a value, as we did previously, we compute for which v_i the priorities are equal. This situation occurs when the total profit of cutting a piece of m_i is equal to the total profit of cutting h pieces of m_j ($t(m_i) = ht(m_j)$). By applying Eq. 2 we find for which v_i the priorities are equal. Then adding an arbitrarily small positive number (denoted as ϵ) we obtain the upper bound:

$$v_i^{ub}(m_i, m_j) = \frac{h(z_i, z_j)s_j v_j}{s_i} + \epsilon, \text{ for } s_i > s_j. \tag{4}$$

From the above we can state that, for any value $v_i \geq v_i^{ub}$, the total benefit of cutting h pieces of m_j is lower than the total benefit of cutting a piece of m_i. Therefore when applying the $\mathcal{A}$ algorithm with $v_i \geq v_i^{ub}$ to a complete raw material, the obtained number of pieces of m_i is the maximum and the obtained number of pieces of m_j is the minimum. This is because, when it is possible to cut from a certain subpart of the material h pieces of m_j, a piece of m_i will be cut instead. Note that the minimum amount of m_j does not have necessarily (and probably will not) to be zero. This is due to the fact that $s_j < s_i$ so there will probably be parts of the raw material in which m_i does not fit but m_j does.

4.3   Generalizing the Bounds

Previously we introduced the equations of the lower and upper bounds that ensure that we obtain the minimum and maximum amounts of a type of product in comparison with a smaller type of product. Here we extend these concepts to the case in which there are more than two types of product to be cut. First we denote the smaller subset of a product type m_i as:

Definition 5. $\mathcal{M}_i^< \subset \mathcal{M} : s_j < s_i, \forall m_j \in \mathcal{M}.$

We present two propositions:

(i) If v_i is equal to the minimum value of all the lower bounds associated with each smaller type of product, we can ensure that the minimum amount of m_i

will be obtained, since the total value of cutting k pieces (see Definition 3) of any of the smaller products is greater than the total value of cutting a piece of m_i. This is denoted as follows:

$$v_i^{lb}(m_i, \mathcal{M}_i^<) = \min_{m_z \in \mathcal{M}_i^<} v_i^{lb}(m_i, m_z). \tag{5}$$

(ii) If v_i is equal to the maximum value of all the upper bounds associated with each smaller type of product, we can ensure that the maximum amount of m_i will be obtained in the homogeneous case (combinations of products of the same type only), since the total value of cutting h homogeneous pieces of any other smaller products is lower than the total value of cutting a piece of m_i (see Definition 4). This is denoted as follows:

$$v_i^{ub}(m_i, \mathcal{M}_i^<) = \max_{m_z \in \mathcal{M}_i^<} v_i^{ub}(m_i, m_z). \tag{6}$$

For the case of combinations of h heterogeneous pieces (combinations of products of different types) of smaller products, unfortunately it is possible that some heterogeneous combinations have a *slightly* greater total value than a unit of product m_i. Then, we cannot assume Proposition (ii) for all the heterogeneous combinations. However, it is very unlikely that such proposition does not hold for real-life instances. This is because when we compute the bounds, we consider that the space left in the area analyzed is waste (worst scenario). However, in reality this space could be used to fit another piece of any product (generally several pieces fit in every big raw material piece). Then the total value of h heterogeneous pieces (excluding m_i) would be compared against the total value of n heterogeneous pieces (including at least a piece of m_i). (However in our proposition only a unique piece of m_i is considered, which has lower total value than combining such piece with another product). For this reason, generally for real instances the minimum/maximum amounts of products are obtained with greater/lower values (respectively) than the theoretical bounds presented in this paper.

We now define the interval of values between the lower and upper bounds of a type of product in relation to smaller types of products. This allows us a reduction of the values search space while ensuring that global products patterns with amounts between the minimum and maximum possible amounts (inclusive) are selected (according to Propositions (i) and (ii)). Such a set of values is defined as follows:

$$\mathcal{V}_i(m_i, \mathcal{M}_i^<) = [v_i^{lb}(m_i, \mathcal{M}_i^<), v_i^{ub}(m_i, \mathcal{M}_i^<)] \tag{7}$$

In [7] the authors generate a set of global products patterns with Monte Carlo simulation over a fixed interval for all the types of products (e.g. $[1, 1000]$). Instead, this simulation can be performed over $\mathcal{V}_i(m_i, \mathcal{M}_i^<)$ by fixing a basis value for the smallest product and computing the $\mathcal{V}$ interval for the bigger products (in increasing size order) for each sampling. As mentioned, by calculating the specific interval delimited by the lower and upper bounds for each type of product, we are reducing the likelihood of generating equal global products patterns, which implies a greater likelihood of obtaining scattered global products patterns. In the following we also introduce an exhaustive global products pattern generation algorithm that uses such intervals.

5   Exhaustive Global Products Patterns Generation Based on $\mathcal{V}$

In this section we explain how to generate all possible global products patterns for a set of product types with respect to some fidelity (denoted as f) over the values vector. First the $\mathcal{V}$ interval is discretized based on f (see Eq. 7), then we present an algorithm that exhaustively generates global products patterns.

5.1   A Fidelity-Based Discretization of the Interval $\mathcal{V}$

In the following, the interval $\mathcal{V}$ (Eq. 7) is discretized according to a fidelity variable f, which represents a value increment:

$$\mathcal{V}_i(m_i, \mathcal{M}_i^<, f) = \{v_i^{lb}(m_i, \mathcal{M}_i^<) + nf, \forall n \in \{0, \ldots, q\}\}$$
$$\text{where } q = \min \mathbb{N} : v_i^{lb}(m_i, \mathcal{M}_i^<) + qf \geq v_i^{ub}(m_i, \mathcal{M}_i^<) \quad (8)$$

The above set of values is expressed as a minimum value and a series of increments of value f over it. The minimum value of the set is the lower bound. The next values are obtained by incrementing f units in every step. The maximum number of such increments is denoted as q and it is the minimum natural number of increments of value f that are necessary in order to reach or exceed the upper bound. Note that the lower the fidelity is, the greater the set $\mathcal{V}_i$ is (with the exception of rare situations in which the lower and upper bounds are equal).

5.2   Algorithm for an Exhaustive Generation of Global Products Patterns

We introduce an algorithm that generates all the values vectors for a set of product types $\mathcal{M}$ by computing the discretized set $\mathcal{V}$ (see Eq. 8) for a given fidelity f. The corresponding global products patterns (denoted as C) of the values vectors are also computed by using the $\mathcal{A}$ algorithm over a given subset of raw material pieces with characteristics $\langle \sigma_1, \ldots, \sigma_{|R|} \rangle$. First, Algorithm 1 initializes an empty values vector. Then, it assigns a basis number (denoted as b) to the smallest type of product, where b can be randomly generated or it can be specified by the user. Note that this value remains fixed during the complete execution of the algorithm. Algorithm 1 also initializes the subset $\mathcal{M}_u \subset \mathcal{M}$, which contains the products whose values have not yet been assigned (at this stage, all the products except the smallest one).

Finally, Algorithm 1 calls the recursive procedure fixValue which, given the set of unassigned product types $(\mathcal{M}_u)$, selects the smallest one and computes its set of values ($\mathcal{V}$, see Eq. 8). Subsequently, each of the values of such a set is assigned iteratively to the analyzed type of product. In addition, the

---

Algorithm 1: Exhaustive Generation of Global Products Patterns

---

$\quad$ **Data**: $\langle \sigma_1, \ldots, \sigma_{|R|} \rangle, \mathcal{M}, f, b$

$\quad$ **Result**: C

$\quad m_i \in \mathcal{M} : s_i = \min_{z \in |\mathcal{M}|} s_z$;

$\quad v \leftarrow \emptyset$; // empty values vector

$\quad v_i \leftarrow b$;

$\quad \mathcal{M}_u \leftarrow \mathcal{M} \backslash \{m_i\}$; // Unassigned set of products

$\quad C \leftarrow \texttt{fixValue} (v, \sigma, \emptyset, \mathcal{M}, \mathcal{M}_u, f)$;

$\quad$ **return** C

---

already assigned type of product is deleted from $\mathcal{M}_u$. This process is repeated recursively until all the product types have already an assigned value. Note that the procedure $\texttt{fixValue}$ is recursively called with the updated set $\mathcal{M}_u$. Once all the product types have already an assigned value, the $\mathcal{A}$ algorithm is used for obtaining the global products pattern associated with the vector of values v. If such global products pattern is new, it is added into C. Finally, when all the values in the set $\mathcal{V}$ have been assigned, the procedure adds the type of product to the unassigned set of products ($\mathcal{M}_u$) and it returns the set of global products patterns computed C. Once all the runs of the procedure $\texttt{fixValue}$ have finished, the total set of global products patterns generated is returned to the Algorithm 1. (Note that all the patterns obtained by each call to the procedure are merged into C).

Procedure . $\texttt{fixValue} (v, \sigma, C, \mathcal{M}, \mathcal{M}_u, f) : C$

$\quad$ select $m_i \in \mathcal{M}_u : s_i = \min_{z \in |\mathcal{M}_u|} s_z$;

$\quad \mathcal{M}_u \leftarrow \mathcal{M}_u \backslash \{m_i\}$;

$\quad V \leftarrow \mathcal{V}_i(m_i, \mathcal{M}_i^<, f)$; // See Eq. 8

$\quad$ **for** $e \in V$ **do**

$\quad\quad | \quad v_i \leftarrow e$;

$\quad\quad | \quad$ **if** $\mathcal{M}_u = \emptyset$ **then**

$\quad\quad | \quad\quad | \quad \langle a_1, \ldots, a_{|M|} \rangle \leftarrow \mathcal{A}(\mathcal{M}, \langle \sigma_1, \ldots, \sigma_{|R|} \rangle, v)$;

$\quad\quad | \quad\quad | \quad$ **if** $\langle a_1, \ldots, a_{|M|} \rangle \notin C$ **then**

$\quad\quad | \quad\quad | \quad\quad \lfloor \quad C \leftarrow C \cup \{\langle a_1, \ldots, a_{|M|} \rangle\}$;

$\quad\quad | \quad$ **else**

$\quad\quad | \quad\quad \lfloor \quad C \leftarrow C \cup \texttt{fixValue} (v, \sigma, C, \mathcal{M}, \mathcal{M}_u, f)$;

$\quad \mathcal{M}_u \leftarrow \mathcal{M}_u \cup \{m_i\}$;

$\quad$ **return** C

---

6 Evaluation

In this section we first describe a case study of a real-world population harvesting problem [4]: forestry harvesting. Subsequently we evaluate several instances of this type with our approach. We do not compare our solution with other techniques because, as far as we know, there is no other approach that attempts to compute such bounds, nor to reduce the values search space. In the forestry harvesting problem the logs of the trees have to be cut into smaller log-pieces by harvesting machines in order to satisfy the demands of the customers. The products have shape specifications (minimum diameter, length, etc.) and their volumes are measured typically in cubic meters (m^3). As mentioned, in this variant of the MSSCSP there are several subsets of stock. For the forestry problem, we call each subset of trees a *block* and each of them has a given specific value. The objective function is to minimize the total value associated with the blocks harvested. (Recall that for the MSSCSP only some stock is selected for satisfying the demands and therefore only this selection is harvested).

In this section we evaluate the approach presented in this paper, which reduces the values search space by computing upper and lower values bounds associated with all the products. For such purpose, we used the exhaustive sampling approach introduced in this paper (Algorithm 1) for generating a set of global products patterns for each block. Lastly, by solving the ILP of the MSS-CSP variant analyzed (see Sect. 2) we obtain the optimal solution for such patterns. A solution of this variant of the MSSCSP consists in a selected vector of values (and its corresponding global products pattern) for each of the selected blocks to be cut. Note that there is no guarantee that this solution is the optimal based on all the possible global products patterns. Due to the impossibility of enumerating all the possible global products patterns for realistic instances, optimality can not be ensured for the analyzed variant of the MSSCSP. Hence the importance of obtaining a sparse (and therefore significant) sample of global products patterns.

We have performed the evaluation with a sample of real data from our industrial partner. The total volume of the sampled logs of the trees is $1191.3m^3$ and it is composed by eight blocks and four product types. We computed the global products patterns of each block by using Algorithm 1 with $b = 10$ and fidelities: $0.9, 0.7$ and 0.5. Then, we solved 50 randomly generated and satisfiable demands instances by solving their corresponding ILPs with CPLEX solver with a time cut-off of 1 h (but the average of the solving time was 16 min). The experiments were run on a 2.3 GHz Intel Core $i7$ processor. Our industrial partner also provided us with the software that carries out the DP-based simulation that implements $\mathcal{A}$ (see Eq. 1). Since DP is a complete algorithm, other DP implementations, such as the ones mentioned in Sect. 3, could have been used for $\mathcal{A}$, with equivalent results.

Figures 5 and 6 show the results obtained from the experiments performed. Specifically, Fig. 5 shows the number of global products patterns that Algorithm 1 computed for the tested fidelities. As mentioned, the lower the fidelity the higher the number of global products patterns, in general. For instance in Block 4 and Block 7 there is a difference of almost 400 global products patterns between fidelity 0.5 and 0.9. This is reflected in the computation times of the global products patterns (see the times between brackets in Fig. 5). Note that computing the global products patterns for all the blocks for fidelity 0.9 required one hour and a half, which is less than five times less the time required

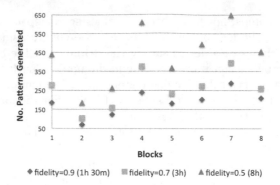

Fig. 5. Number of global products patterns generated for each block.

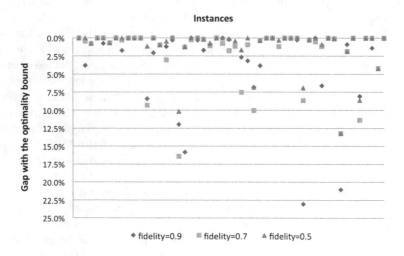

Fig. 6. Quality of the solutions obtained.

for fidelity 0.5 (eight hours). However, our algorithm allows the selection of the global products patterns granularity according to the available time.

The differences in the number of global products patterns computed among the different blocks depend on the characteristics of the blocks (such as number of pieces of raw material and their sizes). Note that it is more likely to generate equal global products patterns (which are rejected by Algorithm 1) for smaller block sizes. For our analyzed instance, Block 2 is the smallest block (in terms of total m^3) and it has the lowest number of generated global products patterns (see Fig. 5).

We compute an optimality bound independently of the global products patterns. Then, we relax the problem by assuming that any combination of amounts of products can be cut from each block (within its size). (It can be obtained by making such variations over the ILP of Sect. 2). Note that it is a bound on the optimality because at least such a total value must be expended for satisfying certain demands. However, it might occur that the global products patterns of the optimality bound do not exist in reality. In such a case, the optimal solution has a greater total value than the optimality bound.

Figure 6 shows the quality of the solutions obtained after solving the ILPs of the 50 random demands instances, which is expressed as a gap with the optimality bound (left axis). Note that for 41 instances we obtained a gap lower than 5%, which can be considered as near-optimal. In addition, for many of them the gap is 0%, which means that they are optimal. As expected, with better (lower) fidelities (which very often implies more global products patterns) the quality of the solutions are equal or better. Needless to mention the economical impact that the quality of these solutions has in the real-life applications (for real instances, the value of a single block could possible be several thousand €). In addition to the sampling performed with our approach, a complementary subsequent clustering algorithm (such as the one presented in [7]) could be applied in order to reduce the set of input global products patterns provided to the ILP (with the objective of speed up its solving time).

7   Conclusions

In this paper we have contributed to the literature by introducing an approach that bounds the search space of a variant of the MSSCSP that arises from population harvesting. For such a problem it is not possible to enumerate all the patterns in a reasonable amount of time, and therefore it is necessary to find a sparse set of global product patterns. In this paper we provide definitions and equations of the lower and upper values bounds of the products. By sampling in their interval, the likelihood of generating equal global products patterns is lower (and therefore patterns tend to be more scattered). Furthermore, we also have introduced an algorithm that exhaustively generates global products patterns according to their bounds and a fidelity parameter that fixes their granularity.

The evaluation performed with a harvesting problem from our industrial partner showed that the better the fidelity, the more global products patterns are generated, and the better the quality of the solutions tends to be. Most of the solutions obtained were near-optimal or optimal, specially for the best fidelity analyzed. We would like to highlight that the number of global products patterns, and how representative they are, affects the quality of the solutions, so it has an economic impact for the stock owners.

As future work, we will focus on applying this approach to other types of real-life MSSCSPs that fit into the population harvesting framework (e.g. we found similarities with a problem from the clothing industry [5]).

Acknowledgments. This research was supported in part by Science Foundation Ireland (SFI) under Grant Number SFI/12/RC/2289.

References

1. Anderson, D., Sweeney, D., Williams, T., Camm, J., Cochran, J.: An Introduction to Management Science: Quantitative Approaches to Decision Making. Cengage Learning, Boston (2015)
2. Bellman, R., Dreyfus, S.: Applied Dynamic Programming. Princeton University Press, Princeton (1962)
3. Dyckhoff, H.: A typology of cutting and packing problems. Eur. J. Oper. Res. **44**(2), 145–159 (1990)
4. Getz, W.M., Haight, R.G.: Population Harvesting: Demographic Models of Fish, Forest, and Animal Resources, vol. 27. Princeton University Press, Princeton (1989)
5. Gradišar, M., Kljacić, M., Resinovič, G., Jesenko, J.: A sequential heuristic procedure for one-dimensional cutting. Eur. J. Oper. Res. **114**, 557–568 (1999)
6. Kantorovich, L.V.: Mathematical methods of organizing and planning production. Manag. Sci. **6**(4), 366–422 (1960)
7. Prestwich, S.D., Fajemisin, A.O., Climent, L., O'Sullivan, B.: Solving a hard cutting stock problem by machine learning and optimisation. In: Appice, A., Rodrigues, P.P., Santos Costa, V., Soares, C., Gama, J., Jorge, A. (eds.) ECML PKDD 2015. LNCS (LNAI), vol. 9284, pp. 335–347. Springer, Heidelberg (2015). doi:10.1007/978-3-319-23528-8_21
8. Wäscher, G., Haußner, H., Schumann, H.: An improved typology of cutting and packing problems. Eur. J. Oper. Res. **183**(3), 1109–1130 (2007)

Designing and Comparing Multiple Portfolios of Parameter Configurations for Online Algorithm Selection

Aldy Gunawan[1]([✉]), Hoong Chuin Lau[1], and Mustafa Mısır[2,3]

[1] School of Information Systems, Singapore Management University,
Singapore, Singapore
{aldygunawan,hclau}@smu.edu.sg
[2] Department of Computer Science, University of Freiburg,
Freiburg im Breisgau, Germany
[3] College of Computer Science and Technology,
Nanjing University of Aeronautics and Astronautics, Nanjing, China
mmisir@nuaa.edu.cn

Abstract. Algorithm portfolios seek to determine an effective set of algorithms that can be used within an algorithm selection framework to solve problems. A limited number of these portfolio studies focus on generating different versions of a target algorithm using different parameter configurations. In this paper, we employ a Design of Experiments (DOE) approach to determine a promising range of values for each parameter of an algorithm. These ranges are further processed to determine a portfolio of parameter configurations, which would be used within two online Algorithm Selection approaches for solving different instances of a given combinatorial optimization problem effectively. We apply our approach on a Simulated Annealing-Tabu Search (SA-TS) hybrid algorithm for solving the Quadratic Assignment Problem (QAP) as well as an Iterated Local Search (ILS) on the Travelling Salesman Problem (TSP). We also generate a portfolio of parameter configurations using best-of-breed parameter tuning approaches directly for the comparison purpose. Experimental results show that our approach lead to improvements over best-of-breed parameter tuning approaches.

1   Introduction

Algorithm Selection [1] concentrates on choosing the best algorithm(s) from a set of algorithms for a given problem instance. The key idea is to build a model that provides a mapping between instance features and performance of a group of algorithms on a set of instances. The resulting model is used to make performance predictions for new problem instances. In relation to algorithm selection, i.e. Algorithm Portfolios [2] primarily focus on determining a set of algorithms for an algorithm selection process. The goal is to choose these algorithms in a way that their strengths complement each other or provide algorithmic diversity that hedge against heterogeneity in problem instances in pretty much the same spirit

© Springer International Publishing AG 2016
P. Festa et al. (Eds.): LION 2016, LNCS 10079, pp. 91–106, 2016.
DOI: 10.1007/978-3-319-50349-3_7

as investment portfolios to reduce risks in economics and finance [3]. Meta-learning [4] has also been proposed as a unified framework for considering the algorithm selection problem as a machine learning problem.

The idea of algorithm selection has also been investigated in the context of parameter tuning or configuration [5]. The goal is to determine a configuration for a target algorithm that will (hopefully) work well for given instances. Hyper-parameter tuning [6] has the same tuning objective but only for machine learning algorithms. The evolutionary algorithm and meta-heuristic community categorises such methods as parameter tuning and parameter control [7]. Parameter tuning occurs offline, while parameter control is concerned with the strategies for adapting parameters in an online manner via some rules or learning algorithms.

SATZilla [8] is a successful example of applying algorithm portfolios to solve the SAT problem, which has consistently ranked top in the various SAT competitions. Its success lies in its ability to derive an accurate runtime prediction model makes effective use of the problem-specific features of SAT. Hydra [9] is another portfolio-based algorithm selection method that combines with automatic configuration to solve combinatorial problems such as SAT effectively.

Several configurators have been proposed for optimisation algorithms. CALIBRA [10] combines Taguchi fractional experimental design and local search. ParamILS [5], as explained above, applies iterated local search to find a single parameter configuration. Racing algorithms like F-Race [11] look for effective parameter values by performing a race between different configurations. Instance-Specific Algorithm Configuration (ISAC) [12] incorporates a G-means clustering algorithm for clustering instances with respect to the features with an existing parameter tuning method, i.e. GGA. GGA is used to configure an algorithm for each instance cluster and works like other case-based reasoning related algorithm selection approaches. [13] proposed Randomized Convex Search (RCS) with the underlying assumption that the parameter configurations (points) lie inside the convex hull of a certain number of the best points. FocusedILS, derived from ParamILS, has been used to provide a number of different parameter configurations for a given single algorithm. It has also been applied for designing multiple parameter configurations for a planner called Fast Downward, in [14]. The results are used for seven portfolio generation methods to build sequential portfolios. [15] proposed a model-based approach, namely SMAC, that can be used to handle categorical parameters. AutoFolio [16] was developed for automated configuration at a higher level by applying SMAC to algorithm selectors. ADVISER [17] was introduced as a web-based platform for algorithm portfolio generation.

This paper seeks to extend the literature on automatic algorithm configuration. The experiments were conducted for combinatorial optimization problems. Rather than providing a single parameter configuration that works well in general or a pre-set schedule of algorithms, we work in the space of online algorithm selection and feeds this process with a portfolio of parameter configurations derived from Design of Experiments (DOE). Our aim is to develop a generic approach for designing algorithm portfolio of parameter configurations for use within an online algorithm selection process to solve combinatorial optimization problems. In particular, we consider generating different parameter configurations for a given target algorithm as the algorithm portfolio.

Our contributions are listed as follows:

- We apply DOE to build algorithm portfolios of parameter configurations for a given target algorithm. Unlike configurators like ParamILS, F-Race or CALIBRA that provide single value for each parameter, DOE provides a subregion of values for each parameter (that are statistically important compared to other regions).
- We propose a random sampling approach to determine a portfolio of parameter configurations from the subregions. Even though methods like ISAC [12] and Hydra [9] already deliver portfolios of configurations, these techniques run a tuner for multiple times, resulting in huge computational overheads. In our case, DOE and sampling is done once, which reduces computational overheads tremendously.
- We employ two online algorithm selection methods, namely Simple Random and Learning Automata. Again, the aforementioned portfolio-based methods that make use of parameter configurations are usually performed offline without any solution sharing, while our approach combines the strengths of multiple configurations by selecting them online and operating them on the same solution, which is different from standard algorithm configuration scenarios. Although dynamic portfolio methods [18] perform online selection, they also ignore solution sharing. The empirical results on two NP-hard problems, namely the Quadratic Assignment Problem (QAP) and Travelling Salesman Problem (TSP), show the advantage of using multiple configurations and solution sharing in algorithm selection.

The remainder of this paper is organized as follows. Section 2 summarizes the algorithm configuration problem. Section 3 details the DOE process. Section 4 explains how the proposed online algorithm portfolio approach works. Section 5 presents an empirical analysis on two problem domains: QAP and TSP. The paper is finalized by some concluding perspectives and future research ideas in Sect. 6.

2   Algorithm Configuration Problem

The algorithm configuration problem (ACP) [19] is about configuring a given target algorithm TA to perform well on a set of problem instances. The goal is to configure k parameters to set, $P = \{pr_1, \ldots, pr_k\}$, where each parameter has a range of values to be set, $pr_i \in D_i$. The configuration space involves $C = D_1 \times \ldots \times D_k$ many possible configurations. The objective is to come up with a configuration from such a, usually, large set to provide the best performance on an instance set, I. Thus, the ACP can be considered an optimisation problem where a solution is a configuration c_i of the algorithm TA on I. One issue with this idea is on solution evaluation. For assessing the quality of a c_i, TA with c_i

should run on I. Although the required computational time for this task varies w.r.t. TA, I and the problem domain of I, it is computationally expensive in general. Heuristic-based search and optimisation techniques such as GGA [20] and ParamILS [5] have been employed in order to overcome this issue.

Such tuning methods are eligible to deliver an effective configuration for a given algorithm. The idea of algorithm portfolios [2] have been used to take advantage of such techniques for building strong algorithm sets including algorithms with different configurations. Existing tuning based portfolio approaches like ISAC [12] and Hydra [9] were designed to address the offline algorithm selection problem. They pursue to the goal of specifying the best single algorithm configuration for solving a particular problem instance. These systems require a set of features representing instances to select algorithms after delivering a set of configurations derived from a computationally expensive training phase. For instance, Hydra mentions that it took 70 CPU days to construct a portfolio of configurations. A similar tool used for a SAT solver, i.e. SATenstein [21], spent 240 CPU days.

Unlike these cases, the aim of this study is to build a portfolio of configurations that can be used in an online setting. The online nature of our approach can allow changing configurations while a selected configuration is fixed for the offline ones. Our system performs like a parameter tuning tool where any domain specific features are not needed. Besides that, the tuning process is faster since the tuning operation is performed once while the tuners used in the aforementioned portfolio approaches run for multiple times. Although our approach is not directly comparable with these offline portfolio methods due to its distinct design, a state-of-the art parameter tuning approach, i.e. ParamILS which is also used in Hydra, is experimented for comparison.

3   Design of Experiments (DOE)

DOE is a well-studied statistical technique used in scientific/engineering decision-making to select and determine the key parameters of a particular process [22]. Some typical applications of DOE include (1) evaluation and comparison of basic design configurations, (2) evaluation of different materials, and (3) selection of design parameters.

Let us consider, in order to solve a particular problem, an algorithm (called the target algorithm) that requires a set of parameters to be set prior to the execution of the algorithm, a DOE-based framework was proposed in [23] to find ranges of parameters values which serve as input to existing configurators such as ParamILS [5]. The main goal is to find a parameter setting that performs best over a set of training instances and subsequently verifies the quality of this setting on a set of testing instances. In this paper, we utilize the first two phases of the framework, namely screening and exploitation phases, to provide promising sub-regions for the parameter configurations, as shown in Fig. 1. These phases are briefly explained in the following subsections.

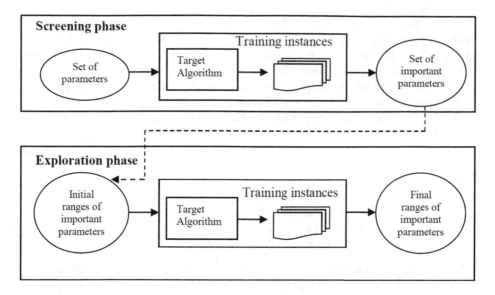

Fig. 1. DOE framework

3.1 Screening Phase

Suppose we have k parameters of a target algorithm to be tuned, where each parameter p_i (discrete or continuous) lies within a numeric interval. In the screening phase, a complete 2^k factorial design is applied to identify m parameters ($m \leq k$) which have significant effects to the performance of the target algorithm (the "important" parameters). This requires $n \times 2^k$ observations where n represents the number of replicates. Experiments are replicated to help identify the sources of variability and to better estimate the true effects of treatments.

In a 2^k factorial design, we examine the magnitude and direction of the effects to determine which parameters are likely to be important. The importance of a particular parameter p_i can be defined by conducting the test of significance on the main effect of the parameter with a significance level, e.g. $\alpha = 10\%$. Furthermore, the ranking of the critical parameters is determined by the absolute values of the main effects of those parameters. The direction of the parameter effects are determined by the sign of the values of the main effects. For instance, if the objective function of the target algorithm is a minimizing function, the value of a particular parameter should be set to a low value if its coefficient of the main effect is positive. The output of this phase consists of a reduced range for each important parameter and all unimportant parameters will be set to a constant value.

3.2 Exploration Phase

In the exploration phase, we treat the m important parameters determined from the screening phase, with the aim to find a promising range for them. We start exploring a larger space where the linear relationship is held and apply the standard approach for linear model checking and diagnosis [22].

The target algorithm is run with respect to the parameter configuration space θ which contains $(2^m + 1)$ possible parameter configurations with an additional setting defined by the centre points of the m important parameters. By adding centre points, protection against curvature is provided.

The form of the relationship between the objective function and the parameters is initially unknown. Thus, the first step is to assume a first-order (planar) model of the response surface. The planar model is given by the following approximating function:

$$Y = X\beta + \varepsilon \tag{1}$$

where:

Y is the vector of $(n \times 2^k)$ responses/objective function values
X is the $((n \times 2^k) \times m)$ matrix
β is a vector of size $(m \times 1)$
ε is the $((n \times 2^k) \times 1)$ error vector

Model adequacy checking involves two statistical tests, namely the interaction and curvature tests, are required. The planar model can still be applied as long as either one of them is not statistically significant. Otherwise the region of planar local optimality has been reached and the promising region has been found. We then continue the process by applying the steepest descent, in order to bring the parameter to the vicinity of the optimum values. Once the region of the optimum has been found (e.g. one of two statistical tests is statistically significant), the planar model is invalid and we can assume that we are in the promising range for each important parameter. The details of this framework can be referred to [23].

4 Solution Approach

Algorithm portfolios are often used with offline algorithm selection strategies. This work was inspired by hyper-heuristic and operator selection studies which encourage to use online algorithm selection [24]. Our approach is basically in two parts - portfolio generation and online selection. The portfolio generation part involves finding varying instantiations, as configurations, of the same algorithm with diverse problem solving capabilities. A resulting portfolio involving configurations of a particular algorithm is then used by online algorithm selection.

Algorithm 1. Online Portfolio-Based Algorithm Selection

TA: target algorithm with k parameters, θ: configuration space defined by the initial ranges of each parameter p_i ($\forall i = 1, 2, \ldots, k$), I: instances, z: portfolio size

Portfolio Design
1 Run DOE to obtain a promising range $[l_i, u_i]$ for each parameter p_i
2 Generate a portfolio of z parameter configurations from the promising ranges
Online Algorithm Selection
3 Apply a selection method (Simple Random (SR) or Learning Automata (LA)

4.1 Portfolio Design

Recall that DOE provides a promising range for each parameter p_i based on steepest descent of a linear response surface. More precisely, we have the promising interval $[l_i, u_i]$ for each parameter p_i. Given these intervals, we propose three different methods to generate a portfolio of parameter configurations. The first is to simply use a constant step size, namely, $[l_i, l_i + \delta, l_i + 2\delta, \ldots, u_i]$ where δ is a constant step size.

The second method is to perform intensification on the promising configuration space via random sampling. First, we generate n random samples of parameter configurations from the promising space. A contour plot is then generated where all sampled points (parameter configurations) having the same response are connected to produce the contour lines of the surface. This contour plot provides an approximate fitness landscape from which we can sample z points randomly, with a probability that decreases from one contour line to the next. More precisely, if the contour plot is divided into y contour lines, then we draw z_i samples from the region bounded by contour lines i and $i+1$, where $z = z_1 + z_2 + \ldots + z_y$ and $z_i > z_{i+1}$ for all $1 \leq i \leq y - 1$.

As a more informed strategy compared to using the contour plots, a clustering approach is utilized as the final approach. The k-means clustering is employed while k, i.e. the number of clusters, is determined by the Silhouette score [25]. Inspired from OSCAR [26], the idea is to cluster configurations w.r.t. their performance, i.e. solution quality, on the training instances. The performance measure used in our study is the percentage of deviation of the objective function value obtained by a particular parameter configuration from the best known solution. Normalized performance values are used as features characterizing configurations, similar to landmarking [27]. Finally, the configuration with the highest average rank from each cluster is then included in the portfolio.

4.2 Online Algorithm Selection

Even though the No Free Lunch theorem [28] states that there is no one algorithm performs well on all possible problem instances that are closed under permutation, it is usually the case that the search spaces of target problem instances do not have the property of 'closure under permutation'. Using multiple mutation operators in an evolutionary algorithm setting is theoretically shown to be effective by [29]. The advantage of using more than one algorithm in a hyperheuristic environment is theoretically explained in [30]. As a consequence, both experimental and theoretical studies suggest that online algorithm selection is useful for better performance.

In this section, we propose two approaches to perform online algorithm selection. First, Simple Random (SR) [31] randomly chooses a parameter configuration at each iteration. Although this is very naive approach, it can effectively manage a small-sized algorithm set. Second, Learning Automata (LA), a.k.a. stateless reinforcement learning, has been used to perform heuristic selection in [32] due to its nice convergence property to a Nash equilibrium. Formally, a learning automaton is described by a quadruple $\{\tilde{A}, \beta, p, U\}$. $\tilde{A} = \{\tilde{a}_1, \ldots, \tilde{a}_n\}$ is the set of actions available. p maintains the probabilities of choosing each of these actions. $\beta(t)$ is a random variable between 0 and 1 for the environmental response. U is a learning scheme used to update p [33].

A learning automaton operates iteratively by evaluating the feedback provided as the result of a selected action. The feedback from the environment is stated as the environmental response ($\beta(t)$) referring whether a selected action is favorable ($\beta(t) = 1$) or unfavorable ($\beta(t) = 0$). This feedback is then used to update the corresponding action probabilities. The Boolean feedback is only used in this online algorithm selection. The update operation will depend on the choice of the update scheme (U), namely *linear reward-penalty*, *linear reward-inaction* and *linear reward-ε-penalty* are the common update schemes which vary in the degree of rewarding or penalising a selected action with respect to the environmental response. All these update schemes use Eqs. 2 and 3. In these equations, the λ_1 and λ_2 values are the learning rates used to update the selection probabilities. The first one is used to reward an action while the latter parameter is to penalise an unfavorable action. The aforementioned three update schemes are determined based on how these two learning rates are set. If they are equal, the update scheme is described as linear reward-penalty (L_{R-P}). When the second rate is set to zero, the system is defined as linear reward-inaction (L_{R-I}). In the case of $\lambda_2 < \lambda_1$, it is defined as linear reward-ε-penalty ($L_{R-\varepsilon P}$).

$$p_i(t+1) = p_i(t) + \lambda_1\ \beta(t)(1 - p_i(t))$$
$$- \lambda_2(1 - \beta(t))p_i(t) \tag{2}$$
$$\text{if } \tilde{a}_i \text{ is the action taken at time step } t$$

$$p_j(t+1) = p_j(t) - \lambda_1\ \beta(t)p_j(t)$$
$$+ \lambda_2(1 - \beta(t))[(r - 1)^{-1} - p_j(t)] \tag{3}$$
$$\text{if } \tilde{a}_j \neq \tilde{a}_i$$

where r is the number of actions in $\tilde{A}$.

Applying the above to online algorithm selection, the actions are associated with the choice of algorithms during run time. In this paper, we have chosen the (L_{R-I}) update scheme, which means that λ_2 is set to 0. Moreover, the probability update process is performed based on two feedbacks, which are finding a new best solution and delivering an improved solution respectively. This means that two values are chosen for λ_1 depending on which of two feedbacks are received.

5   Experimental Results

In order to empirically show the performance of our proposed algorithm selection methods, we perform experiments on two classical combinatorial optimization problems - the Quadratic Assignment Problem (QAP) and Traveling Salesman Problem (TSP).

For each experiment, we perform six different selection methods: (1) SR with constant step size (SR-Constant), (2) SR with random sampling from the contour plot (SR-Contour), (3) LA with constant step-size (LA-Constant), (4) LA with random sampling from the contour plot (LA-Contour), (5) SR with clustering (SR-Cluster), and (6) LA with clustering (LA-Cluster). The learning rate (λ_1) for LA is set to 0.1 and 0.01 respectively for the feedbacks of finding a new best solution or improving the current solution.

5.1   Quadratic Assignment Problem

The QAP is interested in the minimum cost allocation of facilities to locations, taking the costs as the sum of all distance-flow products. A Simulated Annealing - Tabu Search (SA-TS) hybrid meta-heuristic [34] is used as the target algorithm with four parameters. Table 1 gives the details about the parameter configurations from [23].

The QAP benchmark instances are from QAPLIB [35]. The instances are grouped into four classes: unstructured instances (Group I), grid-based distance matrix (Group II), real-life instances (Group III) and real-life-like instances (Group IV). Due to the limitation of the target algorithm that can only handle symmetrical distance matrix, we only focus on instances from the first

Table 1. The parameter space of the QAP

Parameters	Initial range	DOE range		Step size	Contour plot	Clustering
Initial temperature (*Temp*)	[100, 7000]	Group I	[4378, 6378]	250	5 values	2 values
		Group II	[4238, 6238]	250	5 values	2 values
		Group III	[4000, 6000]	250	5 values	6 values
Cooling factor (*Alpha*)	[0.5, 0.95]	Group I	[0.935, 0.945]	0.005	5 values	2 values
		Group II	[0.935, 0.945]	0.005	5 values	2 values
		Group III	[0.85, 0.95]	0.05	5 values	6 values
Tabu list length (*TabuLngth*)	[5, 10]	Group I	5	-	-	-
		Group II	6	-	-	-
		Group III	6	-	-	-
Diversification factor (*Limit*)	[0.01, 0.1]	Group I	0.01	-	-	-
		Group II	0.1	-	-	-
		Group III	0.1	-	-	-

Group 3: Real-life instances

Factorial Fit: Dev versus Temp, Alpha, Tabu Lngth, Limit

Estimated Effects and Coefficients for Dev (coded units)

Term	Effect	Coef	SE Coef	T	P
Constant		13.834	0.3103	44.58	0.000
Temp	-9.207	-4.603	0.3103	-14.83	0.000
Alpha	-1.374	-0.687	0.3103	-2.21	0.028
Tabu Lngth	0.991	0.495	0.3103	1.60	0.113
Limit	0.137	0.069	0.3103	0.22	0.825
Temp*Alpha	-0.798	-0.399	0.3103	-1.29	0.200
Temp*Tabu Lngth	0.368	0.184	0.3103	0.59	0.554
Temp*Limit	0.096	0.048	0.3103	0.15	0.877
Alpha*Tabu Lngth	0.303	0.151	0.3103	0.49	0.626
Alpha*Limit	0.700	0.350	0.3103	1.13	0.262
Tabu Lngth*Limit	0.730	0.365	0.3103	1.18	0.242
Temp*Alpha*Tabu Lngth	0.360	0.180	0.3103	0.58	0.563
Temp*Alpha*Limit	-0.496	-0.248	0.3103	-0.80	0.426
Temp*Tabu Lngth*Limit	-0.011	-0.006	0.3103	-0.02	0.986
Alpha*Tabu Lngth*Limit	1.903	0.951	0.3103	3.07	0.003
Temp*Alpha*Tabu Lngth*Limit	-1.605	-0.803	0.3103	-2.59	0.011

......

Fig. 2. Screening phase (Group III)

three classes. By referring to [23] for classifying instances into training and test-ing instances, we conduct the experiments for two different set of instances: (1) testing instances and (2) all instances (training + testing instances). Instance classes consist of 11, 24, 14 training and 5, 11, 7 testing instances, respectively. Those instances are selected randomly.

The application of DOE screening phase yields the following result for Group III (Fig. 2). It reveals that two parameters (*Temp* and *Alpha*) are sta-tistically significant (with p-value $\leq 5\%$), while the effect of other parameters: *TabuLngth* and *Limit* are insignificant. Based on the coefficient value obtained, we determine the constant value for each insignificant parameter, e.g. the effect of parameter *Limit* is 0.137, so the value of this parameter is set to its lower bound value, which is 0.01 (Table 1).

Using this information in the DOE exploration phase, we find the promising planar region for both parameters (*Temp* and *Alpha*). The final ranges for both parameters are summarized in Table 1. The contour plots generated from random sampling for instances are as shown in Fig. 3. From the plots, we pick three and

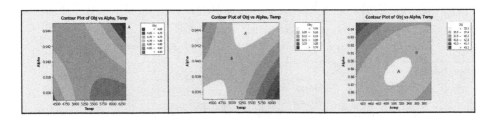

Fig. 3. Contour plot (Groups I, II and III, respectively)

Table 2. The performance of the tested approaches on the QAP instances with respect to the best known solutions (P: ParamILS, Cs: Constant, Ct: Contour, Cl: Cluster)

Instances	Metric	Methods									
		RCS	P	SR-P	LA-P	SR-Cs	LA-Cs	SR-Ct	LA-Ct	SR-Cl	LA-Cl
Group I	% Dev Avg (Test)	0.606	0.692	0.779	0.509	0.535	0.492	0.473	**0.471**	0.512	0.534
	% Dev Best (Test)	0.314	0.345	0.416	0.350	0.378	0.340	**0.301**	0.325	0.331	0.359
	% Dev Avg (All)	0.880	0.973	1.011	0.756	0.737	0.734	0.716	**0.700**	0.709	0.748
	% Dev Best (All)	0.505	0.581	0.618	0.470	0.449	**0.444**	**0.444**	0.476	0.556	0.465
Group II	% Dev Avg (Test)	0.214	0.210	0.394	0.139	0.168	**0.134**	0.149	0.151	0.157	0.136
	% Dev Best (Test)	0.030	0.024	0.061	0.025	0.022	0.018	**0.012**	0.032	0.028	0.029
	% Dev Avg (All)	0.262	0.247	0.417	**0.183**	0.192	0.189	0.189	**0.183**	0.188	0.195
	% Dev Best (All)	0.068	0.060	0.103	0.038	0.045	0.031	**0.030**	0.031	0.033	0.043
Group III	% Dev Avg (Test)	1.231	1.196	1.990	0.667	0.744	0.767	**0.636**	0.935	0.704	0.866
	% Dev Best (Test)	**0.000**	0.191	**0.000**	**0.000**	**0.000**	**0.000**	**0.000**	**0.000**	**0.000**	**0.000**
	% Dev Avg (All)	2.921	2.848	3.414	1.620	2.563	1.824	2.516	1.700	1.737	**1.590**
	% Dev Best (All)	1.114	0.873	0.278	0.252	1.170	0.241	0.266	**0.231**	0.239	0.251

two different parameter configurations randomly from two promising regions, A and B, respectively.

Unlike the contour plots case, in the clustering based portfolio generation approach, the number of configurations used in the resulting portfolio is automatically determined. The last three columns show details on step sizes, number of random samples used to generate the portfolio and number of parameter settings generated by the clustering method.

In order to compare the performance of our proposed approach, we also run the target algorithm with constant parameter values generated by RCS and ParamILS. Both configurators also use the same inputs from DOE range (Table 1). The parameter values are obtained from [23].

For each instance within a particular group, we perform 10 runs and compare the percentage deviations of the average objective function value of the solutions obtained and the best objective function value obtained against the best known/optimal solutions. In order to ensure the fairness among approaches, we use the same computational budget for each one. For example, ParamILS uses z time units so others also use z time units.

The results are summarized in Table 2. In general, we see that we can obtain better results by generating a portfolio of algorithms with different parameter configurations, either by applying Simple Random (SR) or Learning Automata (LA), compared against constant parameter values (RCS or ParamILS). The best performers are SR-Contour and LA-Constant. SR and LA with clustering (SR-Cluster and LA-Cluster) are also comparable with others. Those constant parameter values (RCS and ParamILS) do not perform well.

We also run ParamILS five times in order to generate five parameter configurations. Both selections methods, simple random and learning automata (SR-ParamILS and LA-ParamILS), are used to compare with others. The purpose of this comparison is to show how generating a portfolio of algorithms with different parameter values generated from the contour plot and the clustering method

Table 3. Wilcoxon Signed Rank Test of Table 2 results

Methods	Group I	Group II	Group III
RCS	5	5	5
ParamILS	6	5	5
SR-ParamILS	7	6	6
LA-ParamILS	3	1	1
SR-Constant	4	4	2
LA-Constant	2	1	2
SR-Contour	1	2	1
LA-Contour	1	2	4
SR-Cluster	3	3	1
LA-Cluster	4	1	3

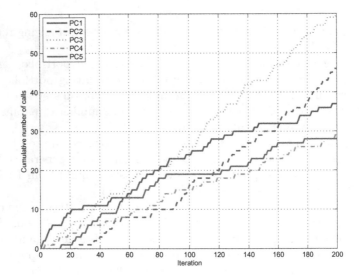

Fig. 4. The effect of learning automata on parameter configuration selection while solving a QAP instance, tho150

outperforms a portfolio of algorithms using constant step-size and best-of-breed parameter tuning approaches (e.g. ParamILS). The results are summarized in SR-ParamILS and LA-ParamILS columns of Table 2.

For further analysis, Wilcoxon Signed Rank Test is used to test all pairwise differences between each algorithm selection approach in terms of %Dev Avg (Test) values. The ranks are summarized in Table 3. Some methods have the same rank values, meaning that those methods are statistically indifferent. We observe that algorithm selection with LA methods outperforms other methods in all groups of testing instances. SR also performs well in Group III instances in terms of the percentage of average deviations for testing instances. In general, using the contour plot to generate a portfolio of promising parameter values outperforms other approaches.

Lastly, we provide a glimpse of the effectiveness of the generated portfolio by examining the frequency distribution of selection, as shown in Fig. 4. As shown in the figure, the cumulative frequency of choosing each parameter configuration vary over iterations, suggesting that different configurations are effectively used throughout the online selection process. And considering that LA outperformed both ParamILS and SR, we can conclude that the LA's learning process pays off.

5.2 Travelling Salesman Problem

The Travelling Salesman problem (TSP) requires finding a tour that visits all cities exactly once that minimises the total distance travelled. In our experiment, Iterated Local Search (ILS) with a 4-Opt perturbation [36] is used as the target algorithm.

Table 4 summarizes the list of the parameters to be tuned, the initial and final ranges for each parameters after applying DOE. Similar to QAP, the last column provides how we generate the algorithm portfolio. We only compare with ParamILS since RCS does not perform well in solving the QAP (Sect. 5.1). 47 TSP instances out of 70 instances from TSPLIB are used as the training instances while the rest (23 instances) are treated as testing instances.

The experiment result is presented in Table 5. We observe that our approach works well compared to existing configurators. In particular, the selection method using LA-Contour outperforms others. The performance of algorithm selection methods are ranked based on Wilcoxon Signed Rank Test as follows: LA-Contour $\approx$ LA-Constant $\approx$ SR-Cluster $\succ$ LA-ParamILS $\approx$ SR-Constant $\succ$ LA-Cluster $\approx$ SR-ParamILS $\succ$ SR-Contour $\succ$ ParamILS.

Similar to QAP, we also generate five parameter configurations using ParamILS and compare against five points generated from the contour plot, as shown in Table 5 (SR-ParamILS and LA-ParamILS columns). We again conclude that our proposed approach using the contour plot outperforms the portfolio with configurations generated by ParamILS.

Table 4. The parameter space of the TSP

Parameters	Initial range	DOE range	Step size	Contour plot	Clustering
Maximum # iterations ($Iter_{max}$)	[100, 900]	[400, 600]	50	5 values	2 values
Perturbation strength (P_s)	[1, 10]	[1, 3]	1	5 values	2 values
Non-improving moves tolerance (Tl_{nip})	[1, 10]	[4, 6]	1	5 values	2 values
Perturbation choice (P_c)	[3, 4]	3	-	-	-

Table 5. The performance of the tested approaches on the TSP instances with respect to the best known solutions (P: ParamILS, Cs: Constant, Ct: Contour, Cl: Cluster)

Metric	Methods								
	P	SR-P	LA-P	SR-Cs	LA-Cs	SR-Ct	LA-Ct	SR-Cl	LA-Cl
% Dev Avg (Test)	1.742	1.331	1.321	1.325	1.295	1.377	**1.291**	1.295	1.332
% Dev Best (Test)	0.852	0.752	0.787	0.792	0.704	0.768	**0.664**	0.749	0.880
% Dev Avg (All)	1.671	1.259	1.211	1.262	1.272	1.304	**1.207**	1.252	1.277
% Dev Best (All)	0.838	0.736	0.702	0.815	0.717	0.770	**0.684**	0.800	0.800

6   Conclusion

This paper shows that Design of Experiments (DOE) coupled with random sampling can automatically generate good portfolios of parameter configurations that can be used by an online algorithm selection process. The computational results on two classical combinatorial optimisation problems showed the strength of our proposed method compared to state-of-the-art configurators such as ParamILS. We show that the proposed approach lead to improvements to two combinatorial optimization problems, QAP and TSP, compared against single configurations.

Many interesting problems arise from this research. For example, how to set the learning rates in the learning automaton? How to improve our proposed schemes at generating portfolios? Will our generic approach perform well in other problems? How to speed up the process through parallelization?

Acknowledgements. This research is supported by the National Research Foundation, Prime Minister's Office, Singapore under its International Research Centres in Singapore Funding Initiative.

References

1. Rice, J.: The algorithm selection problem. Adv. Comput. **15**, 65–118 (1976)
2. Gomes, C., Selman, B.: Algorithm portfolios. AI **126**(1), 43–62 (2001)
3. Huberman, B., Lukose, R., Hogg, T.: An economics approach to hard computational problems. Science **275**(5296), 51 (1997)
4. Smith-Miles, K.: Cross-disciplinary perspectives on meta-learning for algorithm selection. ACM Comput. Surv. **41**(1), 1–25 (2008)
5. Hutter, F., Hoos, H.H., Leyton-Brown, K., Stützle, T.: ParamILS: an automatic algorithm configuration framework. JAIR **36**(1), 267–306 (2009)
6. Bergstra, J., Bengio, Y.: Random search for hyper-parameter optimization. JMLR **13**, 281–305 (2012)
7. Eiben, A., Michalewicz, Z., Schoenauer, M., Smith, J.: Parameter control in evolutionary algorithms. In: Lobo, F.G., Lima, C.F., Michalewicz, Z. (eds.) Parameter Setting in Evolutionary Algorithms. SCI, vol. 54, pp. 19–46. Springer, Heidelberg (2007). doi:10.1007/978-3-540-69432-8_2
8. Xu, L., Hutter, F., Hoos, H.H., Leyton-Brown, K.: SATzilla: portfolio-based algorithm selection for SAT. JAIR **32**(1), 565–606 (2008)

9. Xu, L., Hoos, H.H., Leyton-Brown, K.: Hydra: automatically configuring algorithms for portfolio-based selection. In: AAAI 2010, pp. 210–216 (2010)
10. Adenso-Diaz, B., Laguna, M.: Fine-tuning of algorithms using fractional experimental designs and local search. OR **54**(1), 99–114 (2006)
11. Birattari, M., Stützle, T., Paquete, L., Varrentrapp, K.: A racing algorithm for configuring metaheuristics. In: GECCO 2002, pp. 11–18 (2002)
12. Kadioglu, S., Malitsky, Y., Sellmann, M., Tierney, K.: ISAC–instance-specific algorithm configuration. In: ECAI 2010, pp. 751–756 (2010)
13. Lau, H.C., Xiao, F., Halim, S.: A framework for automated parameter tuning in heuristic design. In: MIC 2009, Hamburg, Germany, 13–16 June 2009
14. Seipp, J., Braun, M., Garimort, J., Helmert, M.: Learning portfolios of automatically tuned planners. In: ICAPS 2012, Atibaia/Sao Paulo, Brazil (2012)
15. Hutter, F., Hoos, H.H., Leyton-Brown, K.: Sequential model-based optimization for general algorithm configuration. In: Coello, C.A.C. (ed.) LION 2011. LNCS, vol. 6683, pp. 507–523. Springer, Heidelberg (2011). doi:10.1007/978-3-642-25566-3_40
16. Lindauer, M., Hoos, H.H., Hutter, F., Schaub, T.: AutoFolio: an automatically configured algorithm selector. J. Artif. Intell. Res. **53**, 745–778 (2015)
17. Mısır, M., Handoko, S.D., Lau, H.C.: ADVISER: a web-based algorithm portfolio deviser. In: Dhaenens, C., Jourdan, L., Marmion, M.-E. (eds.) LION 2015. LNCS, vol. 8994, pp. 23–28. Springer, Heidelberg (2015). doi:10.1007/978-3-319-19084-6_3
18. Kadioglu, S., Malitsky, Y., Sabharwal, A., Samulowitz, H., Sellmann, M.: Algorithm selection and scheduling. In: Lee, J. (ed.) CP 2011. LNCS, vol. 6876, pp. 454–469. Springer, Heidelberg (2011). doi:10.1007/978-3-642-23786-7_35
19. Hutter, F., Hoos, H.H., Stutzle, T.: Automatic algorithm configuration based on local search. In: AAAI 2007, vol. 22, p. 1152 (2007)
20. Ansótegui, C., Sellmann, M., Tierney, K.: A gender-based genetic algorithm for the automatic configuration of algorithms. In: Gent, I.P. (ed.) CP 2009. LNCS, vol. 5732, pp. 142–157. Springer, Heidelberg (2009). doi:10.1007/978-3-642-04244-7_14
21. KhudaBukhsh, A.R., Xu, L., Hoos, H.H., Leyton-Brown, K.: SATenstein: automatically building local search sat solvers from components. In: IJCAI 2009, pp. 517–524 (2009)
22. Montgomery, D.: Design and Analysis of Expeirments, 6th edn. Wiley, Hoboken (2005)
23. Gunawan, A., Lau, H.C., Lindawati, : Fine-tuning algorithm parameters using the design of experiments approach. In: Coello Coello, A. (ed.) LION 2011. LNCS, vol. 6683, pp. 278–292. Springer, Heidelberg (2011). doi:10.1007/978-3-642-25566-3
24. Mısır, M., Verbeeck, K., De Causmaecker, P., Vanden Berghe, G.: An investigation on the generality level of selection hyper-heuristics under different empirical conditions. Appl. Soft Comput. **13**(7), 3335–3353 (2013)
25. Rousseeuw, P.J.: Silhouettes: a graphical aid to the interpretation and validation of cluster analysis. J. Comput. Appl. Math. **20**, 53–65 (1987)
26. Mısır, M., Handoko, S.D., Lau, H.C.: OSCAR: online selection of algorithm portfolios with case study on memetic algorithms. In: Dhaenens, C., Jourdan, L., Marmion, M.-E. (eds.) LION 2015. LNCS, vol. 8994, pp. 59–73. Springer, Heidelberg (2015). doi:10.1007/978-3-319-19084-6_6
27. Pfahringer, B., Bensusan, H., Giraud-Carrier, C.: Tell me who can learn you and I can tell you who you are: landmarking various learning algorithms. In: Proceedings of the 17th International Conference on Machine Learning, pp. 743–750 (2000)
28. Wolpert, D., Macready, W.: No free lunch theorems for optimization. IEEE Trans. Evol. Comput. **1**, 67–82 (1997)

29. He, J., He, F., Dong, H.: Pure strategy or mixed strategy? In: Hao, J.-K., Middendorf, M. (eds.) EvoCOP 2012. LNCS, vol. 7245, pp. 218–229. Springer, Heidelberg (2012). doi:10.1007/978-3-642-29124-1_19
30. Lehre, P.K., Özcan, E.: A runtime analysis of simple hyper-heuristics: to mix or not to mix operators. In: FOGA 2013, Adelaide, Australia (2013)
31. Cowling, P., Kendall, G., Soubeiga, E.: A hyperheuristic approach to scheduling a sales summit. In: Burke, E., Erben, W. (eds.) PATAT 2000. LNCS, vol. 2079, pp. 176–190. Springer, Heidelberg (2001). doi:10.1007/3-540-44629-X_11
32. Mısır, M., Wauters, T., Verbeeck, K., Vanden Berghe, G.: A new learning hyper-heuristic for the traveling tournament problem. In: Caserta, M., Voß, S. (eds.) Metaheuristics: Intelligent Problem Solving - MIC 2009. Springer, Heidelberg (2010)
33. Thathachar, M., Sastry, P.: Networks of Learning Automata: Techniques for Online Stochastic Optimization. Kluwer Academic Publishers, Dordrecht (2004)
34. Ng, K.M., Gunawan, A., Poh, K.L.: A hybrid algorithm for the quadratic assignment problem. In: CSC 2008, Nevada, USA, 14–17 July 2008
35. Burkard, R., Karisch, S., Rendl, F.: Qaplib-a quadratic assignment problem library. J. Global Optim. **10**(4), 391–403 (1997)
36. Halim, S., Yap, R.H.C., Lau, H.C.: An integrated white+black box approach for designing and tuning stochastic local search. In: Bessière, C. (ed.) CP 2007. LNCS, vol. 4741, pp. 332–347. Springer, Heidelberg (2007). doi:10.1007/978-3-540-74970-7_25

Portfolios of Subgraph Isomorphism Algorithms

Lars Kotthoff[1], Ciaran McCreesh[2(✉)], and Christine Solnon[3]

[1] University of British Columbia, Vancouver, Canada
[2] University of Glasgow, Glasgow, Scotland
c.mccreesh.1@research.gla.ac.uk
[3] INSA-Lyon, LIRIS, UMR5205, 69621 Villeurbanne, France

Abstract. Subgraph isomorphism is a computationally challenging problem with important practical applications, for example in computer vision, biochemistry, and model checking. There are a number of state-of-the-art algorithms for solving the problem, each of which has its own performance characteristics. As with many other hard problems, the single best choice of algorithm overall is rarely the best algorithm on an instance-by-instance. We develop an algorithm selection approach which leverages novel features to characterise subgraph isomorphism problems and dynamically decides which algorithm to use on a per-instance basis. We demonstrate substantial performance improvements on a large set of hard benchmark problems. In addition, we show how algorithm selection models can be leveraged to gain new insights into what affects the performance of an algorithm.

1 Introduction

The subgraph isomorphism problem is to find an adjacency-preserving injective mapping from vertices of a small *pattern* graph to vertices of a large *target* graph. This NP-complete problem has many important practical applications, for example in computer vision [6,25], biochemistry [8], and model checking [24]. There exist various exact algorithms, which have been compared on a large suite of instances by McCreesh and Prosser [15]. These experiments indicated that the single best algorithm depends on the CPU time limit considered: for very small time limits, VF2 [5] is the best choice, whereas the GLASGOW algorithm [15] has better success rates for larger time limits. They also showed that on an instance by instance basis, other algorithms are often better.

The per-instance algorithm selection problem [21] is to select from an algorithm portfolio [9,10] the algorithm expected to perform best on a given problem instance. Algorithm selection systems usually build machine learning models of

---

L. Kotthoff – This work was supported by an NSERC E.W.R. Steacie Fellowship and under the NSERC Discovery Grant Program.

C. McCreesh– This work was supported by the Engineering and Physical Sciences Research Council (grant number EP/K503058/1).

C. Solnon – This work has been supported by the ANR project SoLStiCe (ANR-13-BS02-0002-01).

P. Festa et al. (Eds.): LION 2016, LNCS 10079, pp. 107–122, 2016.
DOI: 10.1007/978-3-319-50349-3_8

the algorithms or the portfolio which they are contained in to forecast which algorithm to use in a particular context. Using the predictions, one or more algorithms from the portfolio can be selected to be run sequentially or in parallel.

In our subgraph isomorphism context, algorithm performance is highly constrained by memory bandwidth (as pointed out by Sabharwal and Samulowitz [22] for SAT solvers). Therefore, we cannot simply run different algorithms in parallel, and we consider the case where exactly one algorithm is selected for solving the problem. One of the most prominent and successful systems that employs this approach is SATzilla [28], which defined the state of the art in SAT solving for a number of years. Other application areas include constraint solving [19], the travelling salesperson problem [13], and AI planning [23]. The interested reader is referred to a recent survey [12] for additional information on algorithm selection.

Overview of the Paper. We formally define the subgraph isomorphism problem in Sect. 2. In Sect. 3, we describe the main existing algorithms for solving this problem, and we also introduce two new algorithms which are derived from Solnon's LAD algorithm [26]. In Sect. 4, we experimentally compare eight state-of-the-art algorithms. We introduce a large benchmark set composed of 5725 instances grouped into twelve classes. Ten of these classes were considered in the experimental study reported by McCreesh and Prosser [15]; two are new. We evaluate the algorithms on this benchmark set, and show that they have very complementary performance. In particular, we show that depending on the CPU time limit, different algorithms achieve the best performance on the entire benchmark set. In Sect. 5, we discuss the features that are used to describe instances, and we describe our algorithm selection approach. It combines a presolving step, which allows us to easy instances very quickly, with an algorithm selection step that uses LLAMA [11]. In Sect. 6, we experimentally evaluate our selection approach and show that it is able to close more than 60% of the gap between the single best and the virtual best solver. We conclude and give directions for future work in Sect. 7.

2   Definitions and Notations

A *graph* $G = (N, E)$ consists of a *node set* N and an *edge set* $E \subseteq N \times N$, where an edge (u, u') is a pair of nodes. The number of neighbors of a node u is called the degree of u, denoted $d^\circ(u) = \#\{(u, u') \in E\}$. In this paper, we implicitly consider non-directed graphs, such that $(u, u') \in E \Leftrightarrow (u', u) \in E$. The extension to directed graphs is rather straightforward, and all algorithms compared in this paper can handle directed graphs as well.

Given a pattern graph $G_p = (N_p, E_p)$ and a target graph $G_t = (N_t, E_t)$, the *subgraph isomorphism problem* consists of deciding whether G_p is isomorphic to some subgraph of G_t. More formally, the goal is to find an injective matching $f : N_p \rightarrow N_t$, that associates a different target node to each pattern node, and preserves pattern edges, i.e. $\forall (u, u') \in E_p, (f(u), f(u')) \in E_t$. Note that the subgraph is not necessarily induced, so that two pattern nodes not linked by an edge may be mapped to two target nodes which are linked by an edge. We define $n_p = \#N_p$, $n_t = \#N_t$, $e_p = \#E_p$, $e_t = \#E_t$, and d_p and d_t to be the maximum degrees of the graphs G_p and G_t.

3 Subgraph Isomorphism Algorithms

Subgraph isomorphism problems may be solved by a systematic exploration of the search space consisting of all possible injective matchings from N_p to N_t: starting from an empty matching, one incrementally extends a partial matching by matching a non-matched pattern node to a non-matched target node until either some edges are not matched by the current matching (so the search must backtrack to a previous choice point and go on with another extension), or all pattern nodes have been matched (a solution has been found). To reduce the search space, this exhaustive exploration is combined with filtering techniques that aim at removing candidate pairs of non-matched pattern-target nodes $(u, v) \in N_p \times N_t$. Different filtering techniques may be considered; some are stronger than others (they remove more candidate pairs), but also have higher time complexities.

3.1 Filtering for Subgraph Isomorphism

The simplest form of filtering is to propagate difference constraints (which ensure that the matching is injective) and edge constraints (which ensure that the matching preserves pattern edges): each time a pattern node $u \in N_p$ is matched with a target node $v \in N_t$, one removes every candidate pair $(u', v') \in N_p \times N_t$ such that either $v' = v$ (difference constraint) or (u, u') is a pattern edge but (v, v') is not a target edge (edge constraint). This simple filtering (called *Forward-Checking*) is very fast to achieve: in $\mathcal{O}(n_p)$ for difference constraints, and in $\mathcal{O}(d_p \cdot n_t)$ for edge constraints. It is used, for example, in McGregor's algorithm [17] and in VF2 [5].

Régin [20] introduced a stronger filtering for difference constraints, which ensures that all pattern nodes can be matched with different target nodes, all together. This filtering (called *All-Different Generalized Arc Consistency*) removes more candidate pairs than when each difference constraint is propagated separately which Forward-Checking. However, it is also more time consuming as it requires $\mathcal{O}(n_p^2 \cdot n_t^2)$ time.

Various filtering techniques have been tried for edge constraints. Ullman [27] introduced a filtering which ensures that for each pattern edge $(u, u') \in E_p$ and each candidate pair $(u, v) \in N_p \times N_t$, there exists a candidate pair $(u', v') \in N_p \times N_t$ such that (v, v') is a target edge. Candidate pairs (u, v) that do not satisfy this property are iteratively removed until a fixed point is reached. This filtering (called *Arc Consistency*) removes more candidate pairs than Forward-Checking, but is also more time consuming as it runs in $\mathcal{O}(e_p \cdot n_t^2)$ when using AC4 [18].

Stronger filtering may be obtained by propagating edge constraints in a more global way, as proposed by Larrosa and Valiente [14]. The idea is to check for each candidate pair $(u, v) \in N_p \times N_t$ that the number of pattern nodes adjacent to u is smaller than or equal to the number of target nodes that are both adjacent to v and that may be matched with nodes adjacent to u. This is done in $\mathcal{O}(n_p^2 \cdot n_t^2)$. This idea was generalised by Solnon's LAD algorithm [26], where, for each candidate pair $(u, v) \in N_p \times N_t$, a redundant Local All-Different constraint ensures that each neighbour of u may be matched with a different neighbour of v. This is done in $\mathcal{O}(n_p \cdot n_t \cdot d_p^2 \cdot d_t^2)$.

3.2 Propagation of Invariant Properties

Some filtering techniques exploit invariant properties, i.e. properties associated with nodes such that nodes may be matched only if they have compatible properties. A classical property is the degree: a pattern node $u \in N_p$ may be matched with a target node $v \in N_t$ only if $d^{\circ}(u) \leq d^{\circ}(v)$. This property is usually used at the beginning of the search to reduce the set of candidate pairs to $\{(u, v) \in N_p \times N_t \mid d^{\circ}(u) \leq d^{\circ}(v)\}$. Other examples of invariant properties are the number of cycles of length k passing through the node, and the number of cliques of size k containing the node, which must be smaller for a pattern node than for its matched target node. Invariant properties may also be associated with pairs of nodes. For example, the number of paths of length k between two pattern nodes is smaller than or equal to the number of paths of length k between the target nodes with which they may be matched. These invariant properties are used, for example,

- by Battiti and Mascia [2], to remove candidate pairs $(u, v) \in N_p \times N_t$ such that the number of paths starting from pattern node u is greater than the number of paths starting from target node v;
- by Audemard et al. [1] to generalise the locally all-different constraint proposed by Solnon [26] so that it ensures that a subset of pattern nodes can be matched with all different compatible target nodes, where compatibility is defined with respect to invariant properties;
- by McCreesh and Prosser [15] to filter the set of candidate pairs before starting the search, and to generate additional implied adjacency-like constraints which are processed during search.

Audemard et al. [1] do not limit the length of paths considered, and iteratively increment the length until no more pairs are removed. Battiti and Mascia [2], and McCreesh and Prosser [15] parameterise their algorithms by the maximum path length considered when counting paths: larger values for this parameter remove more candidate pairs, but are also more time consuming. Battiti and Mascia's experiments show that the best setting depends on the instance considered, and that a portfolio running several randomised versions in time-sharing decreases the total CPU time needed to find a solution for feasible instances. McCreesh and Prosser simply set the parameter to 3, as this setting presented the best overall performance in their case.

4 Experimental Comparison of Individual Algorithms

We consider six algorithms from the literature and propose two novel ones.

4.1 Algorithms from the Literature

We selected the following algorithms from the literature, based on their performance.

- VF2 [5] performs weak filtering that is especially fast on trivially satisfiable instances;
- LAD [26] combines two strong but expensive filtering techniques (All-Different Generalized Arc Consistency and Locally All-Different);
- GLASGOW [15] does expensive preprocessing based on path length invariant properties to generate additional constraints, followed by weaker filtering (forward-checking, and a heuristic All-Different propagator which can miss deletions) and conflict-directed backjumping during search.

We have not considered the algorithm introduced in [29] because it is outperformed by LAD. Also, we have not considered MIP nor SAT solvers because they are not competitive with the selected algorithms [16].

The GLASGOW algorithm has a parameter, which controls the lengths of paths used when reasoning about non-adjacent vertices. In experiments reported by McCreesh and Prosser [15], the choice of paths of length 3 was used as a reasonable compromise—longer paths lead to prohibitively expensive preprocessing on larger, denser instances. This is often not the best choice on an instance by instance basis: sometimes path-based reasoning gives no benefit at all, sometimes considering only paths of length 2 suffices, occasionally paths of length 4 are helpful, and even looking at paths of length 3 is relatively expensive on some graphs. We thus consider all lengths up to 4, naming these variants GLASGOW1 through GLASGOW4.

4.2 New Algorithms

We introduce two new variants of LAD. The first, called INCOMPLETELAD, does weaker filtering which is applied once, without performing a backtracking search, and very quickly detects inconsistencies on many instances: for each pattern node u, we check if there exists at least one target node v such that for each neighbor u' of u there exists a different neighbor v' of v such that the degree of u' is smaller than or equal to the degree of v'. INCOMPLETELAD is an incomplete algorithm that checks a sufficient, but not necessary, condition for inconsistency: when it does not detect inconsistency, the instance may still be unsatisfiable. Its main benefit is that it runs very fast: its time complexity is $\mathcal{O}(n_p(n_t + e_t))$.

The second variant of LAD is called PATHLAD. It combines the locally all-different constraints introduced by Solnon [26] with the exploitation of path length properties proposed by Audemard et al. [1]. The idea is to label each edge (u, v) with the number of paths of length 2 between u and v, and each node u with the number of cycles of length 3 passing through u, and to add the constraint that the label of a pattern node (resp. edge) must be smaller than or equal to the label of its associated target node (resp. edge).

4.3   Problem Instances

We consider a large benchmark set of 5725 instances, which are available in a simple text format[1]. These instances are grouped into 12 classes.

- Class 1 contains randomly generated scale-free graphs [29].
- Classes 2 and 3 contain instances built from a database containing various kinds of graph gathered by Larrosa and Valiente [14]: class 2 contains small instances generated from the first 50 graphs of the database, and class 3 contains larger instances with pattern graphs from the first 50 graphs of the database and target graphs from the next 50 graphs.
- Classes 4 to 8 contain randomly generated graphs from a database of graphs commonly used for benchmarking subgraph isomorphism algorithms [7]: bounded-degree graphs for classes 4 and 5, regular meshes for classes 6 and 7, and random graphs with uniform edge probabilities for class 8. All of these instances are satisfiable.
- Classes 9 and 10 contain instances from segmented images [6,25].
- Class 11 contains instances from meshes modeling 3D objects [6].
- Class 12 contains random graph instances chosen to be close to the satisfiable-unsatisfiable phase transition—these instances are expected to be particularly challenging, despite their small size.

Note that Classes 3 and 12 were not considered in the previous experimental study by McCreesh and Prosser [15]. Our set of instances is much larger than that of Battiti and Mascia [2], who were the first to propose algorithm portfolios for subgraph isomorphism problems. Battiti and Mascia only considered a pure parallel portfolio consisting of two randomised solvers without a selection mechanism. Their problem set consisted entirely of satisfiable instances.

4.4   Experimental Setup

We measured runtimes on machines with Intel Xeon E5-2640 v2 CPUs and 64GBytes RAM, running Scientific Linux 6.5. We used the C++ implementation of the GLASGOW algorithm [15], the C implementation of LAD [26], and the VFLib C++ implementation of VF2 [5]. Software was compiled using GCC 4.9. Each problem instance was run with a timeout of 10^8 ms ($\approx$27.8 h).

4.5   Results

Figure 1 displays the evolution of the cumulative number of instances solved with respect to CPU time. It shows us that the best solver depends on the time limit

---

[1] http://liris.cnrs.fr/csolnon/SIP.html.

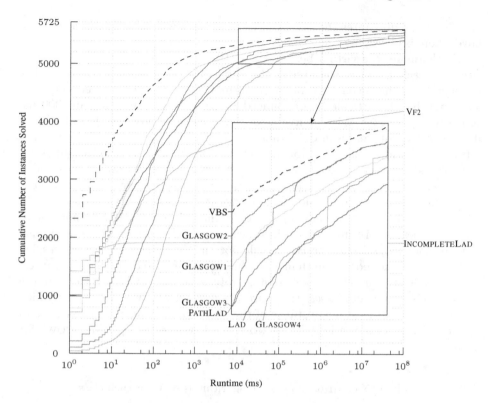

Fig. 1. Cumulative number of solved instances over CPU time for the eight algorithms we consider in this paper, and the virtual best solver (VBS) that shows the best solver on an instance by instance basis. (Color figure online)

considered. INCOMPLETELAD is able to solve easy unsatisfiable instances very quickly, in a few milliseconds. For time limits less than 5 ms, it is the best solver. However, it is not able to solve harder unsatisfiable instances, nor can it solve satisfiable instances.

PATHLAD and GLASGOW1 outperform INCOMPLETELAD for longer time limits: PATHLAD is the best solver for time limits greater than 5 ms and less than 40 ms, and GLASGOW1 is the best solver for time limits greater than 40 ms and less than 3000 ms.

GLASGOW2 becomes the best solver for time limits greater than 3000 ms. As we increase the CPU time limit, the performance of variants of GLASGOW with longer paths (GLASGOW3 and GLASGOW4) improves. This is what we expect, as more reasoning is expensive, but increases the potential reduction of the search space. Eventually, GLASGOW2 and GLASGOW3 become very closely matched, and with runtimes very close to the limit, GLASGOW4 nearly catches up. This behavior is class-dependent: for class 2, for example, the behavior is roughly monotone, with GLASGOW1 dominating for low runtimes, then GLASGOW2, then GLASGOW3, then GLASGOW4 each becoming best as the runtimes increase.

The figure illustrates the potential for portfolios and algorithm selection we have: there is clearly no single solver that dominates throughout.

Furthermore, the virtual best solver (VBS), which considers the best algorithm for each instance separately, obtains much better results, showing us that the algorithms have complementary performance. The difference between VBS and single best is particularly pronounced for CPU time limits less than 1000 ms. In many applications, it is important to have the fastest possible algorithm even if the absolute differences in CPU time are small. For example, in pattern recognition [6,25] and chemical [8] applications, we often have to solve the subgraph isomorphism problem repeatedly for a very large number of graphs (in order to find a pattern image or molecule in a large database of target images or compounds, for example), so having an algorithm that is able to solve an instance in 100 ms instead of 1000 ms makes a big difference. Therefore, it is important to select the best algorithm for each instance, even if the instance is an easy one. Furthermore, this selection process should not unduly penalise easy instances, i.e. it should not take more time than the solution process time for these instances.

Table 1 shows us that we cannot simply select algorithms based on the instance class. For all classes, there are always at least two algorithms which are the best for at least one instance of the class. In particular, for classes 2 and 3, each algorithm is the best for at least one instance (except GLASGOW4 for class 3).

Table 1. Number of times each algorithm is best, for each class.

Class	VF2	LAD			GLASGOW			
		INCOMPLETE	DEFAULT	PATH	1	2	3	4
1	0	20	0	0	80	0	0	0
2	201	92	189	270	520	180	53	15
3	112	1608	617	959	396	195	21	0
4	270	0	0	0	5	0	0	0
5	266	0	1	3	31	0	0	0
6	71	0	0	0	7	14	1	0
7	270	0	0	0	5	0	0	0
8	0	0	0	1	195	69	6	0
9	77	3	0	19	103	1	0	0
10	13	0	2	2	7	0	0	0
11	2	142	71	17	23	0	0	0
12	0	0	1	2	158	6	1	0
Total	1282	1865	881	1273	1530	465	82	15

5   Algorithm Selection Approach

Our approach is composed of three steps. First, we run two presolvers in a static way to quickly solve easy instances. This ensures that we achieve good performance on instances that can be solved in a small amount of time. Second, we extract features from instances which are not solved by the first step. Finally, we run algorithm selection to choose the algorithm to solve the instance with.

5.1   Presolving

Experimental results reported in Sect. 4.5 show that INCOMPLETELAD is very fast (7 ms on average) and able to solve 1919 instances from our benchmark set very quickly. Therefore, we first run INCOMPLETELAD: if unsatisfiability is detected, we do not need to process it further.

VF2 is also able to solve many easy instances very quickly: from the 3806 instances that are not solved by INCOMPLETELAD, 1470 are solved by VF2 in less than 50 ms. Therefore, after running INCOMPLETELAD, we run the VF2 solver for 50 ms. This solves easy instances without the overhead of running algorithm selection and avoids potentially making incorrect solver choices.

We also include VF2 in the portfolio, as it may solve an instance given more time, but not INCOMPLETELAD, as it is an incomplete solver that cannot solve satisfiable instances.

After the presolving step, we are left with 2336 hard instances that we consider for algorithm selection.

5.2   Feature Extraction

If presolving does not give us a solution, we extract features that characterize the instances. For both the pattern and the target graph, we consider some basic graph properties that can be computed very quickly:

- the number of vertices and edges;
- the density—we expect that some kinds of filtering (like those based upon locally all-different constraints) might be expensive and ineffective on dense graphs;
- how many loops (self-adjacent vertices) the graph contains—as loops must be mapped to loops, this could have a strong effect on how easy an instance is;
- the mean and maximum degrees, and whether or not every vertex has the same degree (the degree-based invariants used by LAD and GLASGOW do nothing at the top of search if every vertex has the same degree);
- whether or not the graph is connected;
- the mean and maximum distances between all pairs of vertices (if nearly all vertices are close together, path-based reasoning is likely to be ineffective) and the proportion of vertex pairs which are at least 2, 3 and 4 apart.

Alongside these basic features, we include information computed by INCOM-PLETELAD. To (try to) prove inconsistency, INCOMPLETELAD removes candidate pairs. The number of successfully removed pairs gives information on the distribution of edges (the fewer removed pairs, the more uniform the distribution). As well as the number of removed pairs, we also record the percentage with respect to all possible pairs, and the minimum and maximum percentages of removed values on a per-variable basis. Finally, we include the CPU time required to compute these features as features. However, those features were not more informative than the other ones.

5.3   Selection Model

We use LLAMA [11] to build our algorithm selection model. LLAMA supports the most common algorithm selection approaches used in the literature. We performed a set of preliminary experiments to determine the approach that works best here.

We use 10-fold cross-validation to determine the performance of the LLAMA models. The entire set of instances was randomly partitioned into 10 subsets of approximately equal size. Of the 10 subsets, 9 were combined to form the training set for the algorithm selection models, which were evaluated on the remaining subset. This process was repeated 10 times for all possible combinations of training and test sets. At the end of this process, each problem instance in the original set was used exactly once to evaluate the performance of the algorithm selection models.

LLAMA's pairwise regression approach with random forest regression gave the best performance. The idea is very similar to the pairwise classification models used by Xu et al. [28]. For each pair of algorithms in our portfolio, we train a model that predicts the performance difference between them. If the first algorithm is better than the second, the difference is positive, otherwise negative. The algorithm with the highest cumulative performance difference, i.e. the most positive difference over all other algorithms, is chosen to be run.

As this approach gives very good performance already, we did not tune the parameters of the random forest machine learning algorithm. It is possible that overall performance can be improved by doing so and we make no claims that the particular algorithm selection approach we use in this paper cannot be improved.

The data we use in this paper is available as ASlib [3] scenario GRAPHS-2015.

6   Experimental Evaluation of Algorithm Selection

Table 2 shows the performance of our algorithm selection approach, compared to two baselines, on the set of 2336 hard instances. The virtual best solver is the oracle predictor that, for each instance, chooses the best solver from our portfolio. This is the upper bound of what an algorithm selection approach can achieve. The single best solver is the one solver from the portfolio that has the

Table 2. Algorithm selection performance on the set of 2336 hard instances. MCP is the misclassification penalty; that is, the additional time required to solve an instance because of choosing solvers that perform worse than the best. Mean MCP and performance are over all 2336 instances; when an instance is not solved, its performance is set to the time limit (10^8 ms).

Model	Mean MCP	Solved instances	Mean performance
Virtual best	0	2219	5822809
LLAMA	705097	2203	6529563
GLASGOW2	1960683	2173	7783492

overall best performance across the entire set of instances, at the CPU time limit of 10^8 ms, i.e. GLASGOW2. We consider it a lower bound on the performance of the algorithm selection approach. We are able to solve 30 more instances than the single best solver within the timeout, with only an additional 16 to the virtual best. In terms of average performance, we are able to close 64% of the gap between the single best and the virtual best solver.

Figure 2 shows the cumulative number of solved instances over time for the individual solvers, the virtual best solver, and the LLAMA algorithm selection approach. The algorithm selection model does not perform well for instances that can be solved quickly because of the overhead incurred through feature computation. As the instances become more difficult to solve, its performance improves.

Table 2 shows the performance of the selection model on its own. The performance of the entire algorithm selection system, including the preprocessing, is shown in Table 3. Our system is able to close more than 60% of the gap between single and virtual best, similar to the results on the set of hard instances.

Figure 3 shows the cumulative number of solved instances over time for the algorithm selection system including INCOMPLETELAD and VF2 presolving on the full set of instances. The performance on small instances is much better than the LLAMA selector alone (cf. Fig. 2) and the region where LLAMA performs worse than the individual solvers is now limited to approximately 10^2 to 10^5 ms.

We train the algorithm selection model specifically for the timeout of 10^8 ms. In particular, we are interested in minimising the performance difference to the virtual best. Problem instances that take longer to solve contribute more to this difference than easy instances and therefore carry more weight for the algorithm selection model. That is, choosing the wrong solver for a hard instances is much worse than choosing the wrong solver for an easy instance.

Figures 2 and 3 show that for the easy instances from the set of hard instances, the performance improvement through algorithm selection is negated by the cost of computing the features. The presolving steps improve performance dramatically over the full set of instances (cf. Tables 2 and 3).

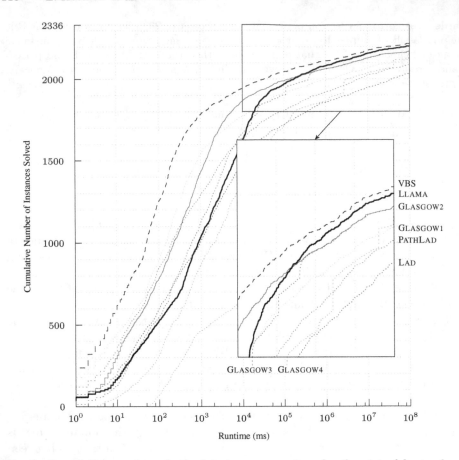

Fig. 2. Cumulative number of solved instances over time for the virtual best solver, LLAMA, and the single best solver GLASGOW2 on the set of 2336 hard instances. Other individual solvers are shown as dotted lines, in the same colors as Fig. 1. (Color figure online)

Table 3. Algorithm selection system performance on the full set of 5725 instances.

Model	Mean MCP	Solved instances	Mean performance
Virtual best	0	5608	2375913
LLAMA	287704	5592	2664293
GLASGOW2	798660	5562	3174573

6.1   Analysis of Features Used by the Model

Analysing the final model, we saw that the most important features were, in order:

- the maximum degree of the pattern graph;
- the mean degree of the target graph;

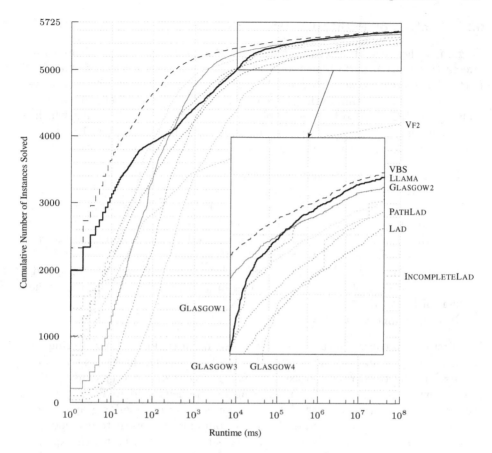

Fig. 3. Cumulative number of solved instances over time for the virtual best solver, LLAMA, and the single best solver GLASGOW2 on the full set of 5725 instances. Other individual solvers are shown as dotted lines, in the same colors as Fig. 1. (Color figure online)

– the proportion of target vertices that are at least distance 3 apart;
– the number of values removed during INCOMPLETELAD presolving.

We introduced the proportion of target vertices that are at least 3 apart as a feature expecting it to be helpful in distinguishing between GLASGOW variants— if few vertices are far apart, longer paths are unlikely to be useful. However, in practice this feature also gives a rough indication of how sparse the graph is—locally all-different filtering is weak and expensive on dense graphs, and the feature turned out to be helpful for selecting between GLASGOW and LAD variants too.

As expected, both the pattern graph and the target graph provide important features. We conclude that even basic graph properties are predictive of sophisticated algorithms' performance.

6.2   Analysis of PathLad Versus Glasgow2

To gain further insight into the behavior of the algorithms, we investigated what affects the relative performance of PATHLAD and GLASGOW2. This pair is of particular interest because they are the best "medium-case" algorithms that use strong and weak filtering during search, respectively. We used machine learning techniques (JRip [4]) to train a simple, human-understandable model which is able to distinguish these solvers for the 2336 hard instances and gives performance better than always choosing one of them. The model uses four rules:

1. If INCOMPLETELAD presolving removes at least 28.01% of the pairs, and at least 94.12% of the values from at least one domain, then pick PATHLAD.
2. If the target has at least 610 vertices, and if the maximum distance between any two pattern vertices is at most 8, and if the pattern is not regular, and if the time taken to compute the distance-based features on the target graph is no more than 1277 ms, then pick PATHLAD.
3. If INCOMPLETELAD filtering removed at least 5.90% of possible pairs, and if less than 84.66% of the pattern vertices are within distance 2 of each other, then pick PATHLAD.
4. Otherwise, pick GLASGOW2.

The first and third rules intuitively make sense: if INCOMPLETELAD filtering does well, it is likely that continuing with this kind of filtering during search will be successful. The third rule also excludes using PATHLAD on very dense pattern graphs, where locally all-different filtering is expensive and weak. The second rule is less obvious: while PATHLAD filtering is weak on regular graphs and it makes sense to exclude this case, the other components appear to exclude large and dense target graphs. The model suggests that it would be worth exploring *dynamically* enabling or disabling locally all-different filtering during search, based upon very simple features which could be recomputed as search progresses and conditions change.

This provides an interesting insight into the behavior of our algorithms, as well as giving indications for future work.

7   Conclusion and Future Work

The problem of identifying subgraph isomorphisms is a hard computational problem that has many applications in diverse areas. In this paper, we presented a portfolio of six algorithms from the literature and two new variants of the LAD algorithm. We introduced a set of novel features to characterise subgraph isomorphism problems and leveraged them to select the most appropriate algorithm from the portfolio for each instance.

We demonstrated that our algorithm selection approach achieves substantial performance improvements over the single algorithm that has the best performance on our benchmark set. We showed that combining an algorithm selection approach with a new incomplete variant of LAD that is able to detect inconsistencies and a presolver boosts performance even further. Finally, we showed how insights from machine learning can guide algorithm development.

Directions for future work include scheduling multiple solvers to run instead of a single one; in particular the GLASGOW algorithms provide a multi-core parallel implementation, which can use a configurable number of threads. It would also be interesting to investigate other variants of the subgraph isomorphism problem.

References

1. Audemard, G., Lecoutre, C., Samy-Modeliar, M., Goncalves, G., Porumbel, D.: Scoring-based neighborhood dominance for the subgraph isomorphism problem. In: O'Sullivan, B. (ed.) CP 2014. LNCS, vol. 8656, pp. 125–141. Springer, Heidelberg (2014). doi:10.1007/978-3-319-10428-7_12. http://dx.doi.org/10.1007/978-3-319-10428-7_12
2. Battiti, R., Mascia, F.: An algorithm portfolio for the sub-graph isomorphism problem. In: Stützle, T., Birattari, M., Hoos, H.H. (eds.) SLS 2007. LNCS, vol. 4638, pp. 106–120. Springer, Heidelberg (2007). doi:10.1007/978-3-540-74446-7_8
3. Bischl, B., Kerschke, P., Kotthoff, L., Lindauer, M.T., Malitsky, Y., Fréchette, A., Hoos, H.H., Hutter, F., Leyton-Brown, K., Tierney, K., Vanschoren, J.: ASlib: a benchmark library for algorithm selection. Artif. Intell. J. (2016, in press)
4. Cohen, W.W.: Fast effective rule induction. In: Twelfth International Conference on Machine Learning, pp. 115–123. Morgan Kaufmann (1995)
5. Cordella, L.P., Foggia, P., Sansone, C., Vento, M.: A (sub)graph isomorphism algorithm for matching large graphs. IEEE Trans. Pattern Anal. Mach. Intell. 26(10), 1367–1372 (2004). http://doi.ieeecomputersociety.org/10.1109/TPAMI.2004.75
6. Damiand, G., Solnon, C., de la Higuera, C., Janodet, J.C., Samuel, E.: Polynomial algorithms for subisomorphism of nD open combinatorial maps. Comput. Vis. Image Underst. (CVIU) 115(7), 996–1010 (2011)
7. De Santo, M., Foggia, P., Sansone, C., Vento, M.: A large database of graphs and its use for benchmarking graph isomorphism algorithms. Pattern Recogn. Lett. 24(8), 1067–1079 (2003). http://dx.doi.org/10.1016/S0167-8655(02)00253-2
8. Giugno, R., Bonnici, V., Bombieri, N., Pulvirenti, A., Ferro, A., Shasha, D.: Grapes: a software for parallel searching on biological graphs targeting multi-core architectures. PLoS ONE 8(10), e76911 (2013). http://dx.doi.org/10.1371%2Fjournal.pone.0076911
9. Gomes, C.P., Selman, B.: Algorithm portfolios. Artif. Intell. 126(1–2), 43–62 (2001)
10. Huberman, B.A., Lukose, R.M., Hogg, T.: An economics approach to hard computational problems. Science 275(5296), 51–54 (1997)
11. Kotthoff, L.: LLAMA: Leveraging learning to automatically manage algorithms. Technical report arXiv:1306.1031 June 2003
12. Kotthoff, L.: Algorithm selection for combinatorial search problems: a survey. AI Mag. 35(3), 48–60 (2014)
13. Kotthoff, L., Kerschke, P., Hoos, H., Trautmann, H.: Improving the state of the art in inexact TSP solving using per-instance algorithm selection. In: Dhaenens, C., Jourdan, L., Marmion, M.-E. (eds.) LION 2015. LNCS, vol. 8994, pp. 202–217. Springer, Heidelberg (2015). doi:10.1007/978-3-319-19084-6_18
14. Larrosa, J., Valiente, G.: Constraint satisfaction algorithms for graph pattern matching. Math. Struct. Compt. Sci. 12(4), 403–422 (2002)

15. McCreesh, C., Prosser, P.: A parallel, backjumping subgraph isomorphism algorithm using supplemental graphs. In: Pesant, G. (ed.) CP 2015. LNCS, vol. 9255, pp. 295–312. Springer, Heidelberg (2015). doi:10.1007/978-3-319-23219-5_21. http://dx.doi.org/10.1007/978-3-319-23219-5_21

16. McCreesh, C., Prosser, P., Trimble, J.: Heuristics and really hard instances for subgraph isomorphism problems. In: IJCAI (2016, to appear)

17. McGregor, J.J.: Relational consistency algorithms and their application in finding subgraph and graph isomorphisms. Inf. Sci. **19**(3), 229–250 (1979)

18. Mohr, R., Henderson, T.: Arc and path consistency revisited. Artif. Intell. **28**, 225–233 (1986)

19. O'Mahony, E., Hebrard, E., Holland, A., Nugent, C., O'Sullivan, B.: Using case-based reasoning in an algorithm portfolio for constraint solving. In: Proceedings of the 19th Irish Conference on Artificial Intelligence and Cognitive Science, January 2008

20. Régin, J.C.: A filtering algorithm for constraints of difference in CSPs. In: Proceeding of the 12th Conference American Association Artificial Intelligence, vol. 1, pp. 362–367. American Association Artificial Intelligence (1994)

21. Rice, J.R.: The algorithm selection problem. Adv. Comput. **15**, 65–118 (1976)

22. Sabharwal, A., Samulowitz, H.: Insights into parallelism with intensive knowledge sharing. In: O'Sullivan, B. (ed.) CP 2014. LNCS, vol. 8656, pp. 655–671. Springer, Heidelberg (2014). doi:10.1007/978-3-319-10428-7_48. http://dx.doi.org/10.1007/978-3-319-10428-7_12

23. Seipp, J., Braun, M., Garimort, J., Helmert, M.: Learning portfolios of automatically tuned planners. In: ICAPS (2012)

24. Sevegnani, M., Calder, M.: Bigraphs with sharing. Theor. Comput. Sci. **577**, 43–73 (2015). http://www.sciencedirect.com/science/article/pii/S0304397515001085

25. Solnon, C., Damiand, G., de la Higuera, C., Janodet, J.: On the complexity of submap isomorphism and maximum common submap problems. Pattern Recogn. **48**(2), 302–316 (2015)

26. Solnon, C.: Alldifferent-based filtering for subgraph isomorphism. Artif. Intell. **174**(12–13), 850–864 (2010). http://dx.doi.org/10.1016/j.artint.2010.05.002

27. Ullmann, J.R.: An algorithm for subgraph isomorphism. J. ACM **23**(1), 31–42 (1976)

28. Xu, L., Hutter, F., Hoos, H.H., Leyton-Brown, K.: SATzilla: Portfolio-based algorithm selection for SAT. J. Artif. Intell. Res. (JAIR) **32**, 565–606 (2008)

29. Zampelli, S., Deville, Y., Solnon, C.: Solving subgraph isomorphism problems with constraint programming. Constraints **15**(3), 327–353 (2010)

Structure-Preserving Instance Generation

Yuri Malitsky[1], Marius Merschformann[2], Barry O'Sullivan[3],
and Kevin Tierney[2(✉)]

[1] IBM T.J. Watson Research Center, New York, USA
yuri.malitsky@gmail.com
[2] Decision Support and Operations Research Lab, University of Paderborn,
Paderborn, Germany
{merschformann,tierney}@dsor.de
[3] Insight Centre for Data Analytics, University College Cork, Cork, Ireland
b.osullivan@insight-centre.org

Abstract. Real-world instances are critical for the development of state-of-the-art algorithms, algorithm configuration techniques, and selection approaches. However, very few true industrial instances exist for most problems, which poses a problem both to algorithm designers and methods for algorithm selection. The lack of enough real data leads to an inability for algorithm designers to show the effectiveness of their techniques, and for algorithm selection it is difficult or even impossible to train a portfolio with so few training examples. This paper introduces a novel instance generator that creates instances that have the same structural properties as industrial instances. We generate instances through a large neighborhood search-like method that combines components of instances together to form new ones. We test our approach on the MaxSAT and SAT problems, and then demonstrate that portfolios trained on these generated instances perform just as well or even better than those trained on the real instances.

1 Introduction

One of the largest problems facing algorithm developers is a distinct lack of industrial instances with which to evaluate their approaches. Yet, it is the use of such instances that helps ensure the applicability of new methods and procedures to the real-world. Algorithm configuration and selection techniques are particularly sensitive to the lack of industrial instances and are prone to overfitting, as it is difficult to build valid learning models when little data is present. Although a plethora of random instance generators exist, the structure of industrial instances tends to be different than that of randomly generated instances, as has been shown for the satisfiability (SAT) problem [3,4,20].

In this work, we therefore present a novel framework for instance generation that creates new instances out of existing ones through a large neighborhood search-like iterative process of destruction and reconstruction [28] of structures present in the instances. Specifically, given an instance to modify, m, and a pool of similar instances, P, we destroy elements of m that fit certain properties

© Springer International Publishing AG 2016
P. Festa et al. (Eds.): LION 2016, LNCS 10079, pp. 123–140, 2016.
DOI: 10.1007/978-3-319-50349-3_9

(such as variable connectivity) and merge portions of the instances in P into m, to create a set of new instances. We compute the features of each new generated instance and accept the instance if it falls into the cluster of instances defined by P. To the best of our knowledge, our framework is the first approach able to generate industrial-like instances directly from real data.

Aside from the immediate benefits of providing a good training set for portfolio techniques, this type of instance generation has the potential of opening new avenues for future research. In particular, the underlying assumption of most portfolio techniques is that a representative feature vector can be used to identify the best solver to be employed on that instance. Techniques like ISAC [19] take this idea further by claiming that instances with similar features are likely to have the same underlying structure, and can therefore be solved using the same solver. Structure-preserving instance generation uses and furthers this notion, that in order to predict the best solver for a given instance, one should create a plethora of instances with very similar features, train a model on them, and then make a prediction for the original instance. Additionally, given recent results regarding the benefits of having multiple, correlated instances for CSPs [14], our instance generator may be able to help solvers more quickly find solutions, as it can provide such correlated instances.

In theory, structure-preserving instance generation can even be used to generate instances significantly different from those observed before. This could in turn allow the targeted creation of portfolios that can anticipate any novel instances not part of the original training set. It would also allow for a systematic way of studying the problem space to identify regions of hard and easy problems, as in [31] but with a stronger basis in real instances. This would allow algorithm designers to create new approaches to specifically target challenging problems.

In this work we primarily focus on the well-known SAT problem, as well as its optimization version, maximum SAT (MaxSAT). These two problems pose ideal test beds for our approach, as although the number of industrial instances is low, it is still larger than what is available in most other domains. For example, the MaxSAT competition in 2013 [7] had only 55 industrial instances in the unweighted category, as opposed to 167 crafted and 378 random instances. We show that the instances generated by our method have similar runtime profiles as the industrial instances they are based on, that they have similar features, and that they can be used in algorithm selection techniques with no loss of performance and, in some cases, even provide small gains. We then show that our technique will even help for problems with larger available datasets, as is the case with SAT and the 300 available industrial instances from the 2013 SAT Competition [8]. We also evaluate our generator using the Q-score from [11], which is specifically designed for evaluating instance generators, and receive near perfect scores. Finally, our source code is available in a public repository at: https://bitbucket.org/eusorpb/spig.

2   Related Work

Numerous random instance generators exist for SAT/CSP problems, such as [9,16,32], to name a few[1]. Some generators try to hide solutions within the problem or generate a specific number of solutions ([21,27], respectively), whereas others convert problems from other fields to SAT/CSP problems (e.g., [1]). The approach of Slater (2002) [29] creates instances by connecting "modules" of 3SAT instances with a shared component, a structure that is often present in industrial instances. For MaxSAT, several generators exist, e.g. [12], which generates bin packing-like problems. The generator from Motoki [24] can create MaxSAT problems with a specific number of unsatisfied clauses. However, in all of these generators the structures inherent in industrial problems are not present.

The most industry-like SAT/MaxSAT instances are generated by Ansótegui et al. [2] through the modification of a random instance generator to use a power-law distribution. In contrast, our framework is able to specifically target certain types of industrial instances. Our approach is similar to instance morphing [15], the primary difference being our focus on instance features and a destroy-repair paradigm. Furthermore, morphing is meant to "connect" the structures of instances, while our goal is to also find new combinations of structures leading to new areas of the instances' feature space.

Burg et al. propose a way of generating SAT instances by clustering the variables based on their degree in a weighted variable graph in Burg et al. [13]. Several approaches are tested to try to "re-wire" the instance by adding and removing connections between the variables. The authors compute the features from Nudelman et al. [26] for the original instances and the generated ones, noting that several features of the generated instances no longer resemble the original instances. In contrast, the instances we generate have similar features to their original instances and stay in the same cluster as the original instance pool.

An evolutionary algorithm approach is used by Smith-Miles and van Hemert [31] to generate traveling salesman problem instances that are uniquely hard or easy for a set of algorithms. This approach starts from a random instance under some assumptions about the size and structure of the resulting instance. In Lopes and Smith-Miles [22], real-world-like instances for a timetabling problem are generated using an existing instance generator. While similar to our approach, their work focuses mainly on generating instances that are able to discriminate between solvers in terms of performance. Furthermore, our approach does not require an existing instance generator. TSP instances are also evolved in Nallaperuma et al. [25] for a parameter prediction model for an ant colony optimization model. However, all these works focus on creating hard instances for particular solvers, rather than instances that resemble industrial instances.

---

[1] An extended version of this work provides a more extensive literature review; see: https://bitbucket.org/eusorpb/spig/.

The most similar work to ours is from Smith-Miles and Bowly [30], in which instances are generated for the graph coloring problem targeting specific instance features. The authors project their features into a two dimensional space with a principal component analysis, and then check if the instance features to be generated are in a feasible region of the space. A genetic algorithm is then used to try to find an instance matching the input features. This approach is more general than ours, but it is not known how well it works with industrial instances, or other problem types.

3   Structure-Preserving Instance Generation

Our instance generation algorithm is motivated by the differing structures found in SAT instances, especially between the industrial, crafted and random categories of instances. Figure 1 provides some visualizations of SAT instances based on their clause graphs. In these graphs, each node represents a clause in the formula, with an edge specifying that the two clauses share at least one variable. The nodes are also color coded from red to blue, where nodes with only a few edges are colored red and those with the most edges are colored blue. A force-based algorithm is used to spread nodes apart. In this way, nodes that share many edges between each other are pulled together into clumps, while the others are pushed away.

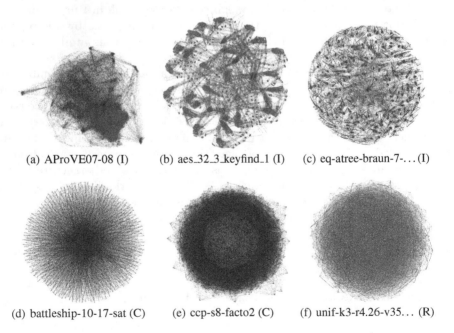

(a) AProVE07-08 (I)   (b) aes_32_3_keyfind_1 (I)   (c) eq-atree-braun-7-...(I)

(d) battleship-10-17-sat (C)   (e) ccp-s8-facto2 (C)   (f) unif-k3-r4.26-v35... (R)

Fig. 1. Visualizations of industrial (I), crafted (C) and random (R) instances made with Gephi [10]. Nodes are clauses, and an edge exists if the corresponding clauses share a variable. (Color figure online)

Algorithm 1. Structure-preserving instance generation algorithm.

```
 1: function SPIG(m, P, α, β)
 2:      gen ← ∅, m' ← m
 3:      repeat
 4:          do
 5:              m' ← DESTROY(m', SELECT-STRUCT(m'))
 6:          while SIZE(m') > β · SIZE(m)
 7:          do
 8:              i ← random instance in P
 9:              d ← max(0, VARS(m) − VARS(m'))
10:              m ← REPAIR(m', SELECT-STRUCT(i), d)
11:          while SIZE(m') < α · SIZE(m)
12:          if ACCEPT(m', P ∪ {m}) then
13:              gen ← gen ∪ {m'}
14:      until TERMINATE
15: return gen
```

Note that the industrial instances, which in Fig. 1 are (a), (b) and (c), tend to contain a core set of clauses that share at least one variable with many other clauses. In addition, a large number of small subsets of clauses are built on the same set of variables. A few variables link the subsets of clauses to the common core. In contrast, instances (d) and (e), which are from the crafted category from the SAT competition, and (f), which is from the random category, show significantly more connectivity between clauses and less modules or groupings of nodes within the graph.

Given a pool of instances, P, and an instance to modify, m, structure-preserving instance generation works as shown in Algorithm 1 to create a set of generated instances gen. The instance pool should be a set of homogeneous instances, such as the instances in a particular cluster from the ISAC method [19]. While using a heterogeneous pool would still result in instances, recall that in this work we aim to create instances with similar properties. In cases where an industrial instance has no similar instances, our method can still be used with a pool consisting of only the instance to modify. The parameters $\alpha \in [0, 1]$ and $\beta \in [1, \infty]$ define the minimum and maximum of the size of the generated instances in proportion to m, respectively. We use these values as general guidelines rather than hard constraints in order to prevent the instance from growing too large or too small.

Our proposed algorithm can be thought of as a modified large neighborhood search [28], in which the incumbent solution, in this case the instance to modify, is iteratively destroyed and repaired. The DESTROY function identifies and removes a particular structure or component of m. The destroy process is run at least once, and is continued until the instance drops below its maximum size. The REPAIR function then identifies and extracts structures from one or more randomly chosen instances from P and inserts them into m. This is repeated until m is larger than the minimum instance size. The d parameter taken by the repair method makes sure that the total number of variables in the problem also stays constant. An acceptance criterion determines whether or not the modified instance should be added to the dataset of instances being built. We base this acceptance on the features of the instance and check whether each individual feature is close to the features of the cluster formed by $P \cup \{m\}$. For problems like SAT or CSP, where an unsatisfiable component or tautology could be introduced, a check can be performed to ensure that the instance did not become trivial to solve. The algorithm terminates when enough instances are generated.

Here it may be argued that an alternate search strategy may also work as well or even better than the one outlined in this section. While alternatives are certainly possible, they are beyond the scope of this work (although some have been tried). For example, one can imagine a combination of local search strategies where each method adjusts an instance to match some combination of features. This modifies the internals of an instance while keeping the instance features relatively unchanged, or moves them back if they change too much. The issue with this method is that the features of interest for problems like SAT are highly interdependent, making any fine grained control over them an arduous task at best. Furthermore, it is frequently very easy to make an instance trivial to solve by introducing an infeasibility, a position that is very difficult to remedy.

Alternatively, one can argue that as long as the provided acceptance criteria is utilized as is, it is possible to employ a local search to just try a number of instantiations. While possible in theory, this approach can take a considerable amount of time before stumbling over even a single seemingly useful instance. The problem space of instance generation is simply too vast. Therefore, while we do not claim that the approach presented here is the only way of generating instances or even the best way, it is a systematic procedure that allows rich datasets to be quickly generated that we can empirically demonstrate works well in practice.

4   Application to SAT and MaxSAT

We present an instantiation of the structure-preserving instance generation framework on the NP-complete SAT problem and NP-hard MaxSAT problem. A SAT problem consists of a propositional logic formula F given in conjunctive normal form. The goal of the SAT problem is to find an assignment to the variables of F such that F evaluates to true. MaxSAT is the optimization version of SAT, in which the goal is to find the largest set of clauses of F that have some satisfying assignment. A version of MaxSAT can also have a weight associated with each clause, with the objective then being to maximize the sum of satisfied clauses. In this work, however, we concentrate only on the unweighted variant of MaxSAT and describe our structure identification routines (SELECT-STRUCT), destroy, repair and acceptance operators. Due to the similarity of the SAT and MaxSAT problems, our instance generation procedure is the same with the exception of the acceptance criteria, which we modify to avoid trivial SAT instances.

Algorithm 2. Variable based structure selection heuristic.

```
1: function SELECT-STRUCT-VAR(i)
2:     E ← ∅, f ← 0
3:     a ← MEAN-VAR-IN-CLAUSES(i)
4:     while E = ∅ do
5:         E ← {v ∈ VARS(i) | |CLAUSES(v) − a| ≤ f}
6:         f ← f + 1
7: return IN-CLAUSES(random variable in E)
```

Algorithm 3. Clause based structure selection heuristic.

```
1: function SELECT-STRUCT-CLAUSE(i, σ)
2:     C = ∅
3:     mcod ← MEAN-CLAUSE-OUT-DEGREE(i)
4:     scod ← STD-CLAUSE-OUT-DEGREE(i)
5:     while C = ∅ do
6:         c ← random clause in i
7:         cod ← CLAUSE-OUT-DEGREE(c)
8:         if |cod − mcod| < σ · scod then
9:             C ← {c' ∈ CLAUSES(i) | c and c' share at least one variable.}
10: return C
```

Structure Identification. Many industrial SAT/MaxSAT instances consist of a number of connected components that are bound together through a core of common variables (see Fig. 1). Our goal is to identify one of these components in an instance at random and remove it. To this end, we propose two heuristics for identifying such structures that we use in both the destroy and repair functions with 50% probability in each iteration. The first heuristic identifies a set of clauses shared by a particular variable, whereas the second identifies a clause and selects all of the clauses it shares a variable with.

Our goal in the *variable-based selection* heuristic is to identify components of instances with a shared variable, as shown in Algorithm 2. We first calculate the mean number of clauses that a variable is in. Variables in many clauses are likely to be a part of the "core" of an instance that connects various sub-components, whereas variables in the average number of clauses are more likely to be part of the sub-components themselves. The algorithm selects a set of variables, E, in the average number of clauses, if there are any. If E is empty, the algorithm relaxes its strictness of how many clauses a variable should be in until some clauses are found. Finally, the algorithm selects a variable from E at random and returns all of the clauses that variable is present in.

In contrast to our previous heuristic, the *clause-based selection* heuristic focuses on clauses with an average out degree. The out degree of a clause is defined as the number of clauses sharing at least one variable in common with the clause. This corresponds to the out degree of the clause's node in the clause-variable graph. Algorithm 3 accepts an instance i and a parameter $σ$, described below. The heuristic selects a random clause and compares its out degree to the average out degree of all the clauses. If the clause's out degree is within σ standard deviations of the average clause out degree, we accept the clause and return all of the clauses it is connected to in the clause-variable graph.

Destroy. Our destroy function accepts an instance i and a set of clauses C selected by SELECT-STRUCT-VAR or SELECT-STRUCT-CLAUSE that are to be removed from the instance. First, all clauses in C are removed, i.e., CLAUSES$(i) \leftarrow$ CLAUSES$(i) \setminus C$, and then all variables that no longer belong to any clause are removed from VARS(i).

Repair. Our repair procedure maps the variables contained within a previously selected set of clauses to the variables present in the instance to modify, and adds new variables with some probability. To avoid confusion, we refer to the instance being modified as the *receiver*, and the instance providing clauses as the *giver*. Algorithm 4 shows the repair process, which is initialized with the receiving instance r, the set of clauses to add, C, and some number of variables to add to the instance, d. The parameter d is used to increase the size of the receiver if too many variables are deleted during the destruction phase. Additionally, we note that C contains clauses from the giver, meaning the variables in those clauses do not match those in the receiving instance. Thus, the main action of the repair method is to find a mapping, M, that allows us to convert the variables in C into similar variables in the receiver.

We map the variables in C into the variables of r by computing the following three features for each variable in the VAR-FEATURES function. We use these features because our goal is to map variables with similar connectivity to other parts of the instance with each other and they are easy to compute.

Algorithm 4. SAT/MaxSAT instance repair procedure.

1: **function** REPAIR(r, C, d)
2:    $V_g \leftarrow \bigcup_{c \in C}$ VARS(c)
3:    $F_r \leftarrow$ VAR-FEATURES(r)
4:    $F_g \leftarrow$ VAR-FEATURES(V_g)
5:    $M \leftarrow \emptyset$
6:    **for** $v_g \in V_g$ **do**
7:        **if** $\neg$TRIVIAL(v_g) and RND$(0, 1) < d/|V_g|$ **then**
8:            $v' \leftarrow$ new variable
9:            VARS$(r) \leftarrow$ VARS$(r) \cup \{v'\}$
10:           $M \leftarrow M \cup \{v_g \mapsto v'\}$
11:       **else**
12:           $dists \leftarrow \{\|F_r(v) - F_g(v_g)\|^2, \forall v \in$ VARS$(r)\}$
13:           $v' \leftarrow argmin_{v \in \text{VARS}(r)}\{dists(v) \mid v \notin M\}$
14:           $M \leftarrow M \cup \{v_g \mapsto v'\}$
15:    CLAUSES$(r) \leftarrow$ CLAUSES$(r) \cup$ MAP-VARS(C, M)
16: **return** r

---

1. Number of clauses the variable is in divided by the total number of instance clauses.
2. Percent of clauses the variable is in, in which the variable is positive.
3. Average of the number of variables of each clause the variable v is in.

On Line 7 of Algorithm 4 we decide whether to map v_g to an existing variable in r or to a new variable. The function TRIVIAL(v_g) returns true if v_g is not (i) positive in at least one clause in C, and (ii) negated in at least one clause C. This ensures that if we map v_g to a new variable, a valid assignment of v_g is not entirely obvious. We assign v_g to a new variable with probability $d/|V_g|$, as with this probability we add roughly d new variables in the absence of trivial variables.

We compute the $L2$ norm between the giver variables and the receiver's variables on line 12, and should we decide not to add a new variable to r, we map v_g to the variable in r that most closely resembles its features that is not yet assigned to a different variable. This is a greedy procedure, that finds the best match for each variable individually. Finally, the algorithm performs the variable mapping and merges the clauses of C into r. We omit the details of the merging process as it is straight forward.

Acceptance Criteria. We compute a set of well known features for SAT and MaxSAT[2] problems from [26] in order to determine whether to accept a modified instance. We compute the average and standard deviation for each feature across the entire pool of instances (including the instance to modify). An instance is accepted if all of its features are within three standard deviations of the mean. That is, we compare a feature to the cluster center on a feature by feature basis. However, some features do not vary at all in a cluster, meaning they have a standard deviation of 0. In such cases, even small changes to an instance can result in a rejection of all generated instances, although the instance is for the most part within the cluster. Thus, when absolutely no instance could be generated we relax the conditions for features that do not vary within the cluster, and allow them to vary by some epsilon value. We note that in our experiments such instances were still well situated within clusters when measured with the Euclidean distance to the cluster center.

For SAT problems, we extend this acceptance criteria with an execution of the instance with a SAT solver. If the instance is solvable in under 30 s it is discarded. We do this because our generation procedure sometimes introduces unsatisfiable components to satisfiable problems that are easily found and exploited by solvers. This clearly breaks the structure of the instance that we are striving to preserve, thus the instance must be discarded. Note that this does not guarantee that the instance will be satisfiable, all we are checking is that the generated instance is not trivially solvable. This issue generally only happens to a couple of instances per generation procedure.

---

[2] We do note use local search probing features in this work.

5   Computational Evaluation

We perform an evaluation of structure-preserving instance generation on instances from the MaxSAT and SAT competitions. We show that the instances generated using our method preserve structure well enough such that they are effectively solved using the same algorithms as the original instances. To evaluate this, we train an algorithm selection approach on the generated data and evaluate it on the subset of original test instances that were neither part of the training nor the generation. It is assumed that if our generated instances can allow us to train a portfolio to identify the most appropriate solver for the instance at hand, then they successfully embody the same key structures as the original industrial data. For SAT, we also use the Q-score method of [11] to show the quality of our instance generator. All experiments were performed on a cluster of Intel Xeon E5-2670 processors with 4 GB of RAM for random instances and 12 GB for industrial instances for MaxSAT (as industrial MaxSAT instances are very large), and 4 GB of RAM for all SAT instances.

5.1   MaxSAT

For MaxSAT, we evaluate our technique on both the random as well as industrial instances from the MaxSAT 2013 competition [7]. Using the random dataset in addition to the industrial dataset shows that our method can be used for any group of similar instances, even though our main target is industrial instances. Here we generate our datasets according to a manually established similarity measure based on the filenames of the instances. For each pool of instances, we perform 10 instance generations for each instance of the pool with a different random seed. For the experiments presented in this section, we limit each generation attempt to 25 destroy/repair iterations. This process generated 5,306 instances based off of 378 random instances, and 2,606 instances based off of 42 industrial instances[3].

One measure to ensure the quality of the generated instances is to compare the runtime of a solver on both the original and the new dataset. Should the runtime performance of an algorithm be similar on both the original and the new dataset, we can conclude that the new dataset is similar to the old one. This is a desirable property for our instance generator, and is based on a fundamental argument on which algorithm portfolios are built and what makes them so successful in practice: that a solver/algorithm performs analogously well or poorly on instances that are similar.

Figure 2 shows the average solution time of the original instances and their generated counterparts for the akmaxsat and msuncore2013 solvers for several clusterings of instances on the random and industrial datasets, respectively. The clusters were generated based on the categories of the instances. The solutions times are comparable for both random and industrial instances, with the exception of clusters 2 and 5 on the industrial dataset in which the generated instances

---

[3] The MaxSAT 2013 dataset contains 55 instances, but we remove instances over 110 MB after performing unit propagation, as SpIG cannot fit them in RAM.

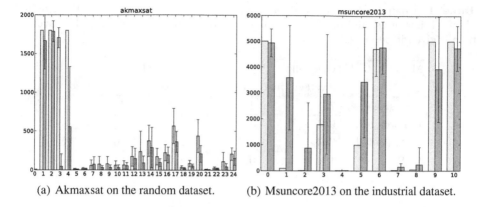

(a) Akmaxsat on the random dataset.   (b) Msuncore2013 on the industrial dataset.

Fig. 2. The average solution time in CPU seconds and standard deviation for each cluster in terms of original (left bar, light gray) and generated (right bar, dark gray) instances.

are too hard. This runtime performance similarity strongly indicates that our generator preserves the structure of instances during generation.

Note that by comparable runtimes, we do not mean identical, which would be an undesirable quality since generated instances should be slightly different from the originals. Furthermore, even when running a solver on the same instance runtimes can vary. The results displayed for the industrial dataset are somewhat noisy, due to the fact that very few original instances exist as a basis for comparison. For example, if we had more instances for cluster 2 (in reality we only have a single instance), it could very well be the case that they are hard to solve as well, but the one instance we have turned out to be solved through a smart (or lucky) branching decision by msuncore2013.

Another important result of our CPU runtime experiments is that "industrial" solvers perform well on our generated industrial instances, whereas "random" solvers tend to timeout. The opposite is also true; when we run an industrial solver on our generated random instances the industrial solvers tend to timeout, but the random solvers perform well. This means that our instance generation framework is able to preserve instance structure nearly regardless of what type of instance it is used on. We note, however, that we do not intend for our instance generator to be used on random or crafted instances, as perfectly good generators already exist for these categories of instances. We show results from these categories only to serve as an evaluation of the overall approach.

One might not even expect our generator to work at all on random instances, as they tend to have little structure. We believe the effectiveness of our approach for such instances is simply due to random changes to a random instance not having a huge effect. Industrial instances (or any instance with some kind of global/local structure), however, require an approach like the one we provide so that generated instances do not get malformed through completely random changes.

Table 1. Comparison of a portfolio trained (leave-one-out) on only the original MaxSAT instances, and one that is trained on the generated instances. The average time is given in seconds.

Model	Original		Generated	
	Average time	Unsolved	Average time	Unsolved
Best single	735	2	735	2
Random forest	988	5	599	2
SVM (radial)	734	2	591	1
VBS	184	0	184	0

Our final experimental comparison on the MaxSAT dataset observes the effect of training a simple portfolio on only the generated instances as opposed to the original ones. Table 1 shows the performance of a portfolio trained and evaluated using leave-one-out cross validation. Note here that for evaluating the generated dataset, none of the instances generated from the test instances were included in the training set. We compare the performance of only using the overall best solver to a portfolio that uses either a random forest or a support vector machine (SVM) to predict the runtime of each solver, selecting the solver with the best expected performance. There are of course a plethora of other popular and more powerful portfolio techniques that could be used and compared, but random forests and SVMs are readily available to anyone and have been previously shown to be effective for runtime prediction, and here we show that even they are able to perform well in our case. The virtual best solver (VBS) gives the performance of an oracle that always picks the fastest solver. Due to the limited training set, the best the portfolio can do is match the performance of a single solver. However, if we train the same solvers on the generated instances and evaluate on the original instances, we are able to see improvements over the best solver, meaning the extra generated instances provide value to the portfolio approach. Our generator is able to help fill in gaps between training instances in the feature space, allowing learning algorithms to avoid having to make a guess as to which algorithm will work best within such gaps. Instead, learning approaches have data on the instances in these gaps and can make informed decisions for their portfolio.

5.2   SAT

We next use the instances from the 2013 SAT competition [8] to conduct further experiments. For the competition, these instances are split into three categories each consisting of 300 instances: application (industrial), crafted, and random. And although this competition has taken place annually for the last decade, it is important to note that the majority of the industrial instances repeat each year, which means our experiments use most of the instances available. We first evaluate our generated dataset using the Q-score technique from [11]. We then use an algorithm selection approach to confirm the usefulness of our generator.

Algorithm Selection with CSHC. Due to the increased number of available instances over the Max-SAT scenario, we apply a more automated technique for grouping SAT instances for generation. In MaxSAT, the instances were grouped based on a manually defined similarity metric associated with the filenames. For SAT, however, the industrial instances are clustered based on their features using the g-means algorithm from [17], an approach that automatically determines the best number of clusters based on how Gaussian distributed each cluster is. Limiting the minimum size of a cluster to 50 resulted in a total of 7 clusters. We typically use 50 instances for a cluster to ensure we have a reasonable statistical evidence that a particular solver works better than another. We sample 15% of the instances from each cluster to compose our training set, and to form the subsets of similar instances for generation.

We perform a more standard portfolio evaluation of the generator for SAT since so many instances are available. For this evaluation we use the top solvers from the 2013 SAT Competition: glucose, glue bit, lingeling 587f, lingeling aqw, MIPSat, riss3g, strangenight, zenn, CSHCapplLC, and CSHCapplLG.

Our test set for all the subsequent experiments is the collection of 254 industrial instances remaining after the subset of training instances is removed. This list is then further reduced by removing those instances for which no solver finds a solution within 1,800 s. This leaves a total of 195 instances. We compare the portfolios based on three metrics: average time without timeouts (Average), PAR10, and number of instances solved (Solved). PAR10 is a penalized average where a timeout counts as having taken 10 times the timeout time.

For our underlying portfolio technique, we utilized CSHC [23], the technique that won the "Open Track" at the 2013 SAT Competition and was behind the portfolio of ISAC++ [6] that won the MaxSAT Evaluation in 2013 and 2014. The core premise of this portfolio technique is a branching criteria for a tree that ensures that after each partition, the instances assigned to sub-node maximally prefer a particular solver other than the one used in the parent node. Training a forest of such trees then ensures the robustness of the algorithm.

The results of the experiments are presented in Table 2. The best stand-alone solver is Lingeling aqw, which solves a total of 157 instances. We trained

Table 2. Comparison of a portfolio evaluated on 195 Industrial SAT instances when trained on either 300 randomly generated instances, 300 crafted instances, a subset of 46 industrial instances, 300 generated instances, or 1500 generated instances.

	Average	PAR10	Solved
Best Single	453	3,872	157
Random (300)	541	5,107	144
Crafted (300)	386	4,090	154
Industrial (46)	348	3,426	161
Generated (300)	502	3,463	162
Generated (1,500)	437	3,049	166
VBS	364	364	195

the portfolio on a variety of training sets: 300 random instances, 300 crafted instances, 46 industrial instances, 300 generated (industrial) instances and 1500 generated (industrial) instances. Not surprisingly, training on random or crafted instances does not perform well. In both cases, less instances can be solved than just using the best single solver, and the PAR10 scores are significantly higher. This further confirms a well-known result in the algorithm selection literature that training and test sets need to be similar in order for the learning algorithm to be successful. We include these results to emphasize the fact that if our generated instances were significantly different from the datasets they were generated from, we would expect similarly bad performance on the test set. Indeed, using even just 46 industrial instances already results in better performance than the best single solver in terms of average time, PAR10 and the number of instances solved.

Using 300 generated industrial instances shows similar performance to the original industrial instances in terms of the number of the PAR10 score and number of instances solved, although the average runtime is higher. This is already enough evidence to further confirm that our instance generation routine is successful at preserving instance structures in SAT, as in Max-SAT. However, because we are not limited by the number of instances we generate, we can create much larger training samples. We therefore evaluate our portfolio trained on 1,500 generated instances and observe that 166 instances can be solved, 5 more than with the original training set. In a competition setting, this improvement is often the difference between first place and finishing outside the top three solvers. This provides further support that our approach fills in gaps in the instance feature space, and that this provides critical information to selection algorithms that improves their performance.

Q-score. The Q-score, introduced by Bayless et al. in [11], provides a mechanism for assessing whether or not a dataset of instances can act as a proxy for some other set of instances. In other words, using the Q-score we can check whether the instances we generate share similar properties with the original dataset of industrial instances. The score is based on the performance of parameters found through algorithm configuration using a method like [5] or [18]. We use SMAC [18] as it was previously used for calculating the Q-score in [11]. We configure the Lingeling and Spear solvers each three times for five days on the same 1500 generated instances used in our algorithm selection experiments and all 300 original industrial instances, which we label S and T, respectively. Adopting the notation of [11] (which we refer to for full details), the Q-score is computed by $c(A(\theta'_T), T)/c(A(\theta'_S), T)$, where c is the PAR10 score of a parameterization on the specified dataset, A specified an algorithm configuration, and θ'_T and θ'_S are the best performing configurations (on the test set) of all tuned configurations and on the generated set, respectively.

We found the Q-score 0.9177 for lingeling and 0.9978 for Spear on our generated instances. We note that 1.0 is the best possible score. This indicates that the datasets we generate lead to high quality parameter configurations that generalize to the original instances. Interestingly the best parameter configuration for Lingeling on the test set was one of the parameterizations trained on the generated instances. However, its training set evaluation was beaten by another parameterization, thus we do not use it in the calculation of the Q-score for the set S. This is especially noteworthy given that our generated instance set is not even based on all of the industrial instances, but is nonetheless being compared to parameters specifically tuned on all 300 industrial instances.

5.3   Structure Comparison

As a final evaluation of our instance generation methodology we present a comparison of the original and generated instances when their features are projected into a two dimensional space. We do this using a standard principal component analysis (PCA). Figure 3 presents the results for both the MaxSAT and SAT datasets. The figure shows that there is not a perfect matching between the generated and original instances. While future work can focus on reducing the spread between these instances, we note that a perfect matching is not desirable as we do not want exact replicas of our instance pool. Instead, we want to cover a range of scenarios of similar instances, which can be seen in many parts of the projection. This subsequently leads to a better trained portfolio. Furthermore, note that the generated instances tend to be close to their original counterparts in this projected space. This means that although they are not completely identical, the generated instances are still fairly representative of their originals.

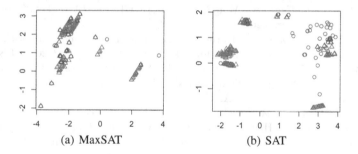

(a) MaxSAT   (b) SAT

Fig. 3. Projection of the instances into 2D using PCA on their features. Original training instances are represented as blue circles, the generated instances are represented by red triangles. (color figure online)

6   Conclusion and Future Work

One of the current key problems in solver development is the limited number of instances on which algorithms can be compared. This is especially the case for industrial instances, where datasets are extremely limited and difficult to expand. To remedy this, this paper presented a novel methodology for generating new instances with structures similar to a given dataset. We then demonstrated the quality of the generated datasets by training portfolios on them and evaluating them on the original instances. This showed that not only do the instances have similar structures as the originals, but that those structures also allow a portfolio (and algorithm configuration) to correctly learn the best solver for provided instances. For future work, will evaluate our instance generation framework on other types of problems, such as CSPs and MIPs, as well as explore how to improve the generated instances' coverage of the feature space.

Acknowledgements. We thank the Paderborn Center for Parallel Computing for the use of the OCuLUS cluster for the experiments in this paper. Barry O'Sullivan was supported in part by Science Foundation Ireland (SFI) under Grant Number SFI/12/RC/2289.

References

1. Ansótegui, C., Béjar, R., Fernàndez, C., Mateu, C.: Edge matching puzzles as hard SAT/CSP benchmarks. In: Stuckey, P.J. (ed.) CP 2008. LNCS, vol. 5202, pp. 560–565. Springer, Heidelberg (2008). doi:10.1007/978-3-540-85958-1_42
2. Ansótegui, C., Bonet, M.L., Levy, J., Li, C.M.: Analysis and generation of pseudo-industrial maxsat instances. In: CCIA. FAIA, vol. 248, pp. 173–184. IOS Press (2012)
3. Ansótegui, C., Giráldez-Cru, J., Levy, J.: The community structure of SAT formulas. In: Cimatti, A., Sebastiani, R. (eds.) SAT 2012. LNCS, vol. 7317, pp. 410–423. Springer, Heidelberg (2012). doi:10.1007/978-3-642-31612-8_31
4. Ansótegui, C., Levy, J.: On the modularity of industrial sat instances. In: Fernández, C., Geffner, H., Manyà, F. (eds.) CCIA. FAIA, vol. 232, pp. 11–20. IOS Press, Amsterdam (2011)
5. Ansótegui, C., Sellmann, M., Tierney, K.: A gender-based genetic algorithm for the automatic configuration of algorithms. In: Gent, I.P. (ed.) CP 2009. LNCS, vol. 5732, pp. 142–157. Springer, Heidelberg (2009). doi:10.1007/978-3-642-04244-7_14
6. Ansotegui, C., Malitsky, Y., Sellmann, M.: MaxSAT by improved instance-specific algorithm configuration. In: AAAI (2014)
7. Argelich, J., Li, C.M., Manyà, F., Planes, J.: Eighth Max-SAT evaluation (2013)
8. Balint, A., Belov, A., Heule, M., Järvisalo, M.: Proceedings of SAT competition 2013; solver and benchmark descriptions. Technical report, University of Helsinki (2013)
9. Barták, R.: On generators of random quasigroup problems. In: Hnich, B., Carlsson, M., Fages, F., Rossi, F. (eds.) CSCLP 2005. LNCS (LNAI), vol. 3978, pp. 164–178. Springer, Heidelberg (2006). doi:10.1007/11754602_12
10. Bastian, M., Heymann, S., Jacomy, M.: Gephi: an open source software for exploring and manipulating networks. In: AAAI Conference on Weblogs and Social Media (2009)
11. Bayless, S., Tompkins, D.A.D., Hoos, H.H.: Evaluating instance generators by configuration. In: Pardalos, P.M., Resende, M.G.C., Vogiatzis, C., Walteros, J.L. (eds.) LION 2014. LNCS, vol. 8426, pp. 47–61. Springer, Heidelberg (2014). doi:10.1007/978-3-319-09584-4_6
12. Bejar, R., Cabiscol, A., Manya, F., Planes, J.: Generating hard instances for MaxSAT. In: International Symposium on Multiple-Valued Logic (ISMVL 2009), pp. 191–195, May 2009

13. Burg, S., Kottler, S., Kaufmann, M.: Creating industrial-like SAT instances by clustering and reconstruction. In: Cimatti, A., Sebastiani, R. (eds.) SAT 2012. LNCS, vol. 7317, pp. 471–472. Springer, Heidelberg (2012). doi:10.1007/978-3-642-31612-8_40

14. Dinur, I., Goldwasser, S., Lin, H.: The computational benefit of correlated instances. In: Proceedings of the 2015 Conference on Innovations in Theoretical Computer Science. pp. 219–228. ACM (2015)

15. Gent, I.P., Hoos, H.H., Prosser, P., Walsh, T.: Morphing: combining structure and randomness. In: Hendler, J., Subramanian, D. (eds.) AAAI, pp. 654–660 (1999)

16. Gomes, C.P., Selman, B.: Problem structure in the presence of perturbations. In: Kuipers, B., Webber, B.L. (eds.) AAAI, pp. 221–226 (1997)

17. Hamerly, G., Elkan, C.: Learning the k in k-means. In: Neural Information Processing Systems (NIPS) (2003)

18. Hutter, F., Hoos, H.H., Leyton-Brown, K.: Sequential model-based optimization for general algorithm configuration. In: Coello, C.A.C. (ed.) LION 2011. LNCS, vol. 6683, pp. 507–523. Springer, Heidelberg (2011). doi:10.1007/978-3-642-25566-3_40

19. Kadioglu, S., Malitsky, Y., Sellmann, M., Tierney, K.: ISAC - instance-specific algorithm configuration. In: ECAI. FAIA, vol. 215, pp. 751–756. IOS Press (2010)

20. Katsirelos, G., Simon, L.: Eigenvector centrality in industrial sat instances. In: Milano, M. (ed.) CP 2012. LNCS, vol. 7514, pp. 348–356. Springer, Heidelberg (2012). doi:10.1007/978-3-642-33558-7_27

21. Krzakala, F., Zdeborová, L.: Hiding quiet solutions in random constraint satisfaction problems. Phys. Rev. Lett. **102**(23), 238701 (2009)

22. Lopes, L., Smith-Miles, K.: Generating applicable synthetic instances for branch problems. Oper. Res. **61**(3), 563–577 (2013)

23. Malitsky, Y., Sabharwal, A., Samulowitz, H., Sellmann, M.: Algorithm portfolios based on cost-sensitive hierarchical clustering. In: IJCAI (2013)

24. Motoki, M.: Test instance generation for MAX 2SAT. In: Beek, P. (ed.) CP 2005. LNCS, vol. 3709, pp. 787–791. Springer, Heidelberg (2005). doi:10.1007/11564751_65

25. Nallaperuma, S., Wagner, M., Neumann, F.: Parameter prediction based on features of evolved instances for ant colony optimization and the traveling salesperson problem. In: Bartz-Beielstein, T., Branke, J., Filipič, B., Smith, J. (eds.) PPSN XIII 2014. LNCS, vol. 8672, pp. 100–109. Springer, Heidelberg (2014). doi:10.1007/978-3-319-10762-2_10

26. Nudelman, E., Leyton-Brown, K., Hoos, H.H., Devkar, A., Shoham, Y.: Understanding random SAT: beyond the clauses-to-variables ratio. In: Wallace, M. (ed.) CP 2004. LNCS, vol. 3258, pp. 438–452. Springer, Heidelberg (2004). doi:10.1007/978-3-540-30201-8_33

27. Pari, P.R., Lin, J., Yuan, L., Qu, G.: Generating 'random' 3-SAT instances with specific solution space structure. In: McGuinness, D.L., Ferguson, G. (eds.) AAAI, pp. 960–961 (2004)

28. Shaw, P.: Using constraint programming and local search methods to solve vehicle routing problems. In: Maher, M., Puget, J.-F. (eds.) CP 1998. LNCS, vol. 1520, pp. 417–431. Springer, Heidelberg (1998). doi:10.1007/3-540-49481-2_30

29. Slater, A.: Modelling more realistic SAT problems. In: McKay, B., Slaney, J. (eds.) AI 2002. LNCS (LNAI), vol. 2557, pp. 591–602. Springer, Heidelberg (2002). doi:10.1007/3-540-36187-1_52

30. Smith-Miles, K., Bowly, S.: Generating new test instances by evolving in instance space. Comput. Oper. Res. **63**, 102–113 (2015)

31. Smith-Miles, K., van Hemert, J.: Discovering the suitability of optimisation algorithms by learning from evolved instances. Ann. Math. Artif. Intell. **61**(2), 87–104 (2011)
32. Van Gelder, A., Spence, I.: Zero-one designs produce small hard SAT instances. In: Strichman, O., Szeider, S. (eds.) SAT 2010. LNCS, vol. 6175, pp. 388–397. Springer, Heidelberg (2010). doi:10.1007/978-3-642-14186-7_37

Feature Selection Using Tabu Search with Learning Memory: Learning Tabu Search

Lucien Mousin$^{(\boxtimes)}$, Laetitia Jourdan, Marie-Eléonore Kessaci Marmion, and Clarisse Dhaenens

Univ. Lille, CNRS, Centrale Lille, UMR 9189 - CRIStAL - Centre de Recherche en Informatique Signal et Automatique de Lille, 59655 Lille, France
lucien.mousin@ed.univ-lille1.fr, {laetitia.jourdan,me.kessaci, clarisse.dhaenens}@univ-lille1.fr

Abstract. Feature selection in classification can be modeled as a combinatorial optimization problem. One of the main particularities of this problem is the large amount of time that may be needed to evaluate the quality of a subset of features. In this paper, we propose to solve this problem with a tabu search algorithm integrating a learning mechanism. To do so, we adapt to the feature selection problem, a learning tabu search algorithm originally designed for a railway network problem in which the evaluation of a solution is time-consuming. Experiments are conducted and show the benefit of using a learning mechanism to solve hard instances of the literature.

1 Introduction

A lot of computational challenges are linked to the big-data context and knowledge discovery represents a very active research domain. Classification is one of the critical tasks of knowledge discovery. In a classification context, a dataset is composed by a set of observations. Each observation is defined by a set of features and a class. The goal is to learn a model on those data in order to predict classes of new observations. However, the high number of features complicates the learning of the model, and, as a result, makes difficult the correct prediction of new observations. Consequently, a preliminary phase is applied to help the construction of the model, the feature selection phase.

The feature selection problem consists in choosing a subset of features, among a larger set. It may be used (i) to simplify the understanding of a model in order to facilitate its comprehension by users, (ii) to reduce the computational time of algorithms that exploit those data, (iii) to reduce overfitting, in other words, to reduce the specialization of the model to known observations.

The feature selection problem in classification can be modeled as a combinatorial optimization problem [4], first because it consists in choosing a subset of features among N (2^N possible subsets exist), and secondly because the quality of a subset may be evaluated (by the quality of the classification model constructed with this subset, for example). However, the use of a classifier to

© Springer International Publishing AG 2016
P. Festa et al. (Eds.): LION 2016, LNCS 10079, pp. 141–156, 2016.
DOI: 10.1007/978-3-319-50349-3_10

construct the model may be time expensive if an elaborate one is used. This may be a difficulty for optimization approaches to deal with large datasets.

In this paper we investigate an optimization approach able to jointly deal with large datasets and time-consuming classifiers. This approach based on Tabu Search integrates a learning mechanism in order to evaluate only promising subsets of features.

The remainder of this paper is organized as follows. Section 2 introduces the feature selection problem. Section 3 presents the Learning Tabu Search approach proposed. Section 4 drives experiments and compares results with the classical Tabu Search approach in order to appreciate the contribution of the learning mechanism. Finally, Sect. 5 gives some conclusions and perspectives for future works.

2   The Feature Selection Problem in Classification

2.1   Problem Description

In a classification problem, a set of observations with known classes is used to learn a classification model to predict the class of any new observations. A feature selection process may be used to select information that may help the classification. In this context, a dataset (in the following called *instance*) is represented by a set of d observations. Each observation i is characterized by n features and one class. Hence an instance is represented by a matrix A of d rows and n columns which represents the value of each feature for each observation, and a vector C of size d which represents the class of each observation, as follows:

$$A = \begin{bmatrix} a_{11} & \cdots & a_{1n} \\ \vdots & \ddots & \vdots \\ a_{d1} & \cdots & a_{dn} \end{bmatrix} , \quad C = \begin{bmatrix} c_1 \\ \vdots \\ c_d \end{bmatrix} \tag{1}$$

where $c_i \in \{1, ..., k\}$ with k the number of classes.

An instance is composed by two sets. The first one, called *training set*, allows resolution approaches to learn a model and, the second one, called *validation set*, is used to evaluate that model on new observations.

2.2   Resolution Approaches

For this problem, resolution approaches may be classified in three major types according to the way the search procedure and the classifier are combined:

- **Filter Approaches**: Select features independently of the classification method used.
- **Wrapper Approaches**: Exploit the classifier performance to select features. This type of approaches is used in this paper, and detailed hereafter.
- **Embedded Approaches**: Combine filter and wrapper approaches. They are used to reduce overfitting.

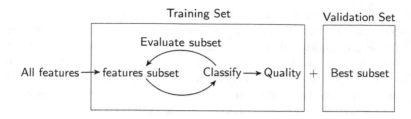

Fig. 1. Wrapper approach

The wrapper model, initiated by Kohavi [13], applies a search procedure to find different subsets that are evaluated with a classifier on the training set. The best subset found during the search procedure is then evaluated on the validation set (see Fig. 1). An advantage of this approach is to be able to deal with correlations between features and to find relevant associations of them. However, this kind of approaches may generate overfitting, *i.e.,* the specialization of the model to observations used to build the model. Moreover, the computing time may become large with regard to the classifier used, when the dataset contains a large number of observations and/or features.

2.3 State of the Art

Finding the best subset of features can be viewed as a combinatorial optimization problem. Hence, a lot of methods, such as metaheuristics have been proposed to solve it. Table 1 presents some metaheuristics from the literature, to tackle this problem together with the type of approach used for resolution.

Table 1. Metaheuristics for the feature selection problem. The bibliographic reference, the date and the resolution approach are also given.

Ref	Date	Algorithm	Approach
[20]	1998	Genetic algorithm with DistAl	Wrapper
[7]	2000	Niched pareto genetic algorithm	Wrapper
[15]	2006	Genetic algorithm	Wrapper
[14]	2007	HillClimbing	Filter + Wrapper
[12]	2007	NSGA II	Wrapper
[6]	2009	Genetic algorithm + iterated local search	Embedded
[1]	2010	Multi-cluster feature selection	Wrapper
[8]	2010	Simulated annealing and genetic algorithm	Wrapper
[3]	2012	Particle swarm optimization	Wrapper
[19]	2013	Particle swarm optimization	Wrapper
[18]	2014	Modified micro genetic algorithm	Wrapper

This table shows that very recent methods have been proposed. Most of them are wrapper approaches. The population-based algorithms are mainly applied and, in particular, genetic algorithms which seem to be the favored metaheuristics for this problem. On the contrary, very few local search algorithms exist.

While using an efficient classifier, such as *SVM* (Support Vector Machine) [17], on large datasets, the evaluation of a subset may be time consuming. In this context, population based metaheuristics that need to make many evaluations at each generation, are not any more good candidates and local search approaches may be privileged. Indeed, local search approaches benefit from neighborhood relationships, exploit them to guide the search and to spare some evaluations.

Following these remarks, this paper proposes a local search that integrates a mechanism to learn about these neighborhood relationships to guide the search efficiently.

3   The Feature Selection Problem with Learning Tabu Search

Learning Tabu Search is an efficient local search integrating a learning mechanism. This section presents the steps needed to adapt this method to the Feature Selection (FS) problem. First, the modeling of this problem is described. Then, the integration of the learning mechanism into a tabu search is explained. Finally, each component of the method is detailed to understand the adaptation.

3.1   Feature Selection Problem Modeling

Representation of Solutions. A solution s is a subset of features. It is represented by a bit string of size n, the total number of features: $s = [a_1, ..., a_n]$ with $a_i \in \{0, 1\}, \forall i \in \{1, ..., n\}$. The ith bit a_i indicates if the feature i is chosen ($a_i = 1$) or, on the contrary, if it is not ($a_i = 0$).

Evaluation of Solutions. For the FS problem in classification, several criteria are commonly used to measure the quality of a solution. First, it may be measured by the quality of the classification realized using the selected features. Most of classifiers propose to compute the *accuracy*, which is defined as the ratio between the well-classified observations and the total number of observations tested. The *accuracy* is computed as follows:

$$accuracy = \frac{\text{number of well-classified observations}}{\text{total number of observations}}$$

Secondly, the number of selected features is an important criterion for FS problem. Indeed, in order to obtain more interpretable models, the number of selected features should be minimized. This criterion is defined as the ratio between the number of selected features (# S_Features) and the total number of features (# Features). In order to obtain a maximization criterion, the criterion, noted *features*, is defined as follows:

$$features = 1 - \frac{\# \, S\_Features}{\# \, Features}$$

This paper considers these two maximization criteria, *accuracy* and *features*. Note that, in the literature, other criteria are also used such as sensitivity or specificity. In this work, the FS problem is presented as a single-objective combinatorial optimization problem.

Consequently, the fitness function f is defined as a weighted sum between *accuracy* and *features*:

$$f = \alpha * accuracy + (1 - \alpha) * features$$

where $\alpha \in [0, 1]$ is a weighting coefficient (set to 0.75 in the experiments). The goal is to find the subset of features that maximizes f.

Neighborhood. For the FS problem, we consider the well-known *one-flip neighborhood* defined, for all s in the search space, as follows:

$$\mathcal{N}_1^0(s) = \{s' \mid \exists i \in \{1, ..., n\} \text{ s.t. } a_i' \neq a_i \text{ and } \forall j \neq i, \, a_j' = a_j\}$$

As the number of selected features has to be minimized, a *good* solution is represented with most of bits equal to 0. Hence the probability of flipping a bit from 0 to 1 is higher than flipping a bit from 1 to 0. Consequently, in order to give the same chance to both flips *0 to 1* and *1 to 0*, we divided the neighborhood into two sub-neighborhoods. The *add neighborhood* ($\mathcal{N}_A$) is the set of neighboring solutions where one bit has been flipped from 0 to 1. The *drop neighborhood* ($\mathcal{N}_D$) is the set of neighboring solutions where one bit has been flipped from 1 to 0. Then, $\mathcal{N}_1^0(s) = \mathcal{N}_A(s) \cup \mathcal{N}_D(s)$ and $\mathcal{N}_A(s) \cap \mathcal{N}_D(s) = \emptyset$. The neighborhoods $\mathcal{N}_A$ and $\mathcal{N}_D$ are mathematically defined as:

$$\mathcal{N}_A(s) = \{s' \mid \exists i \in \{1, ..., n\} \text{ with } a_i' = 1 \text{ and } a_i = 0 \text{ and } \forall j \neq i, \, a_j' = a_j\}$$

$$\mathcal{N}_D(s) = \{s' \mid \exists i \in \{1, ..., n\} \text{ with } a_i' = 0 \text{ and } a_i = 1 \text{ and } \forall j \neq i, \, a_j' = a_j\}$$

3.2   From Tabu Search to Learning Tabu Search

In local search algorithms and in particular in Tabu Search, the exploration of the neighborhood of a solution can be time-consuming. Indeed, in the original Tabu Search method, all the non-tabu neighbors of a solution are evaluated at each iteration. In the FS problem, the evaluation of a solution is computed by applying a classification procedure (*KNN, SVM, ...*). This one can be computationally expensive when the number of observations and/or features becomes large. Hence, the evaluation of the whole neighborhood at each iteration can not be considered.

Schindl, and Zufferey designed the Learning Tabu Search (LTS) [16] in order to avoid this. Hence, the exploration of the neighborhood is divided into two steps: (i) the quality of all neighbors is estimated and, (ii) the Q most promising ones are *fully* evaluated. LTS is based on an *estimation function* used to estimate the potential quality of each neighboring solution. The computation of this estimation is based on this idea: "if, some combinations of characteristics often belong to good solutions during the search process, such combinations of characteristics should be favored when generating new solutions". The estimation of the quality of one combination is computed from the quality of solutions where this combination appears. Therefore, LTS needs a memory to save the quality of each features combination.

The performance of LTS rests on the definition of this memory that represents the learning mechanism. This mechanism is related to the pheromones concept of ant colony optimization (ACO) algorithms [5]. The quality of one combination is then called its *trail* value. The higher the trail value of a combination, the better is its quality. Like in ACO, the memory has to be updated to increase the trail of promising combinations and to decrease those that are associated to bad ones. An *evaporation* procedure is used to forget them.

In LTS, the *update* procedure is applied at regular intervals, called *cycles*. The quality of the best solution found during each cycle is used to update the trail values. The size of the cycle is a sensible parameter of LTS., as it impacts the performance of the learning mechanism.

The update procedure aims to concentrate the search in regions containing high quality solutions. In order to visit new regions of the solutions space,

---

Algorithm 1. Learning Tabu Search (LTS)

---

begin
    $s \leftarrow$ initial solution;
    $s* \leftarrow s$;
    repeat
        Estimate the quality of non-tabu neighbors of $\mathcal{N}(s)$;
        $N_Q \leftarrow Q$ most promising neighbors of $\mathcal{N}(s)$ according to the diversification policy;
        $s \leftarrow \max\limits_{s' \in N_Q} f(s')$;
        if $s > s^*$ **then**
            $s^* \leftarrow s$;
        if $s > \hat{s}$ **then**
            $\hat{s} \leftarrow s$;
        Update the tabu list;
        if *End of cycle* **then**
            Update trails of each combination with $\hat{s}$;
    until *Stopping condition is met*;
    return s^*

a *diversification* procedure has been introduced. This procedure modifies the policy of choosing the Q most promising neighbors to be evaluated during the neighborhood exploration. Usually, a learning mechanism favors the neighbors with the highest estimation values but, it may lead to a premature convergence of LTS. To avoid this issue, when diversification is triggered, the combinations with the lowest estimation values are favored.

Algorithm 1 gives an insight of LTS. From an initial solution, different steps are applied until the stopping criterion is met. Every non-tabu neighbors are estimated and then, the Q most promising neighbors are evaluated. *Most promising neighbors* stands for neighbors with the highest estimation when *diversification* is disabled but, ones with the lowest estimation when *diversification* is triggered. At the end of the neighborhood exploration, the best solutions s^* of the search, $\hat{s}$ of the current cycle and the tabu list are updated. At the end of each cycle, trail values are updated from the fitness of $\hat{s}$ according to the *diversification* policy.

3.3   Learning Tabu Search for Feature Selection

In the following, we explain the adaptation of LTS to the FS problem.

Definition of Trail. This paper proposes to consider the combination of two features. A combination of two features is interesting if these features are both selected in good solutions *i.e.,* the combination of these two features brings information for the classification task. The trail value $tr(a_i, a_j)$ associated to features a_i and a_j, indicates if the combination of a_i and a_j is promising, thanks to the observations of the search history.

Estimation of Neighbors. A solution s and each neighbor s' differ from one bit a_i. The estimation of a neighbor ($Estim(s, a_i)$) (*i.e.,* its potential quality), is computed from the relevance of selecting the feature a_i in relation to other features in s: $Estim(s, a_i) = \sum_{a_j \in s} tr(a_j, a_i)$.

Neighborhoods Exploration. $\mathcal{N}_A$ and $\mathcal{N}_D$ are the neighborhoods composed with *add* flips and *drop* flips respectively. A *promising* add flip is to add a feature a_i to a solution s, if $Estim(s, a_i)$ is high in order to select a feature which brings the most information to the solution. A *promising* drop flip is to delete a feature a_i from a solution s, if $Estim(s, a_i)$ is low. During the exploration of the neighborhood in LTS, only the Q best promising neighbors are evaluated. Then, A_q (resp. D_q) is the subset of non-tabu neighbors of $\mathcal{N}_A$ (resp. $\mathcal{N}_D$) composed of the q neighbors with the highest (resp. lowest) estimations. Finally, all neighbors of $A_q \cup D_q$ are evaluated and the best one is chosen.

Update Procedure. As mentioned before, the trail values $tr(a_i, a_j)$ are updated at the end of each cycle from the best solution $\hat{s}$ found during this cycle: $tr(a_i, a_j) = \rho * tr(a_i, a_j) + \Delta tr(a_i, a_j)$, where $\rho \in [0, 1]$ is the *evaporation* *rate* and $\Delta tr(a_i, a_j)$ is proportional to the fitness of $\hat{s}$, if a_i and a_j both belong to $\hat{s}$, and is equal to 0 otherwise.

Diversification Procedure. It is used to escape from a region of the search space. Therefore, when the mechanism is triggered, the construction of the sets A_q and D_q during the exploration of the neighborhood is inverted *i.e.*, A_q (resp. D_q) is the subset of non-tabu neighbors of $\mathcal{N}_A(s)$ (resp. $\mathcal{N}_D$) composed of the q neighbors with the lowest (resp. highest) trail values. This mechanism depends on two parameters t_1 and t_2: the mechanism is triggered after t_1 iterations without improving s^* (the best solution found during the search), and is disabled as soon as s^* has improved, or after t_2 iterations with diversification.

4   Experiments

4.1   Experimental Protocol

We choose to compare LTS to other local search algorithms: a *Hill Climbing* (HC) and a *Tabu Search* (TS). Hill Climbing is a classic local search algorithm, that has the major inconvenient, to stop the search when it falls in a local optimum. In order to give the same chance for all algorithms, when HC falls in a local optimum, it restarts the search with a random solution until the stopping time is reached.

The Tabu Search is a local search that uses a memory to escape from local optima. The memory is used to store recently visited solutions that are qualified as *tabu*. At each step, the tabu search moves to the best non-tabu solution of the neighborhood. Hence, the tabu search is able to escape from a local optimum by moving to the least deteriorating neighbor. In the literature, Tabu Search applies the *best improvement* strategy for the neighborhood exploration. Nevertheless, this strategy may be time-consuming when the evaluation is costly, therefore in this paper, we choose a *first improvement* strategy.

Instances used for experiments are divided into two parts. The first one is the *training set*, used by the algorithm to look for the best subset of features. The second one is the *validation set*, used to evaluate the ability of the subset of features previously found, to well classify new data.

For each instance with their training and validation sets, we performed for each algorithm the following different steps: (*i*) the search algorithm is performed on the training set, (*ii*) the best solution found is selected and its accuracy on the validation set is computed, (*iii*) these two steps are executed 30 times per instance per algorithm, (*iv*) the statistical Wilcoxon test is performed on fitness obtained on the training set to compare algorithms, and (*v*) the statistical Wilcoxon test is performed on accuracy obtained on the validation set.

4.2   Description of Instances

Experiments are computed using six instances from the literature. Each line of these instances represents an observation. Table 2 details information about the instances used for experiments (well-balanced binary classes).

Table 2. Instances description. The total number of features (# Features), the size of the training $|T|$ and validation $|V|$ sets (*i.e.,* number of observations), the run-time (in seconds) needed by SVM to build and evaluate a model on each training set (without feature selection), and the runtime (in seconds) allocated to each optimization algorithm are given. Instances are divided into two groups according to the SVM runtime.

Name	# Features	$|T|$	$|V|$	SVM runtime	Allocated runtime
Schizophrenia [2]	410	56	30	0.01	500
Colon [21]	2000	62	32	0.052	120
Breast [21]	24481	78	26	0.734	500
Arcene [10]	10000	100	100	1.123	3000
DNA [9]	180	1400	600	1.172	500
Madelon [11]	500	2000	600	38.089	5000

An important point is the classifier used to compute the accuracy. In this paper, we used *SVM* [17] (Support Vector Machine) that constructs hyperplanes to separate data into two classes. This procedure becomes time consuming when the number of observations increases and when data are difficult to separate into two classes. Hence, for such instances, the runtime needed by SVM to construct and then evaluate a model is expensive.

In consequence, we choose to distinguish two groups of instances (low evaluation cost vs. high evaluation cost) according to the SVM runtime when applied on the training sets. The first one groups *Schizophrenia*, *Colon* and *Breast* instances (SVM runtime lower than 1 s) and the second one groups *Arcene*, *DNA* and *Madelon* instances. Note that, SVM requires more than 38 s on *Madelon* instance to compute the accuracy on the whole training set.

Preliminary experiments helped to set the allocated runtime given to HC, TS and LTS. This allocated runtime is the same for the three methods, and is partially dependent on SVM runtime, since it is used within the evaluation to compute the accuracy of a solution. Let us remark, that even if *Arcene* instance requires less than 2 s to compute the accuracy, preliminary experiments showed that the convergence is quite low but happened for each algorithm before 3000 s.

4.3 Parameters

Different parameter settings were studied before deciding which one to use for the final experiments. Table 3 shows parameters involved in this study.

Two parameters deserve special attention. The first one is q that tunes the number of promising estimated neighbors from each set, A_q and D_q, that will be evaluated. Indeed, if q is small, LTS converges quickly because the first best solutions are often the same. Otherwise, if q is large, LTS becomes time-consuming because many solutions are evaluated. Note that q could be adapted to the instance size, but preliminary experiments show that $q = 10$ appears to be

Table 3. Learning tabu search parameters. Gives each parameter together with its setting value.

Parameter	Value
Size of Tabu List	7
Size of A_q and D_q (q)	10
Cycle (I)	10
Evaporation rate (ρ)	0.9
Number of iterations with diversification ($t1$)	10
Number of iterations without diversification ($t2$)	10
Aggregation factor (α)	0.75

a good trade-off for these instances. The second one is the size of a cycle (I). If I is small, the learning mechanism will make overfitting because the search has not enough time to find a new best solution. Otherwise, if I is large, the learning mechanism will take much time to discover good combinations and to forgot bad ones. Preliminary experiments show that $I = 10$ appears to be also a good trade-off.

4.4   Performance Analysis

This analysis is organized in two parts. The first part deals with the optimization perspective (capacity of the method to find a good subset of selected features *i.e.*, with a high fitness value) and evaluates its performance on the training set with the single-objective function defined in Sect. 3.1. The second part concerns the datamining perspective (capacity of the model to predict class of unknown observations) and evaluates results obtained on the validation set.

Analysis of the Optimization Approach: Table 4 shows a comparative study between the proposed approach LTS and the other approaches. For each instance, the accuracy computed with SVM from the whole features is pointed out in order to exhibit the benefit of the feature selection.

This table shows that concerning results about the fitness, LTS gives in most of the cases the best results with a standard deviation close to zero. In details, we can see that LTS often gives the best accuracy and selects always the least number of features.

This may be explained by the neighborhood exploration strategy. Indeed, LTS selects for evaluation the q best add flips as well as the q best drop ones. Consequently, drop flips have as much chances to be chosen as add flips. On the contrary, other approaches have a random neighborhood. As the number of selected features is small, the probability to find a drop flip is low and may required the evaluation of many neighbors. As a result, LTS can find a solution with a good accuracy with the least number of features faster than other algorithms.

Table 4. Average and standard deviation (in brackets) of Fitness, Accuracy and # S_Features values obtained on training sets for HC, TS and LTS. For each algorithm, the fitness values have been computed from the Accuracy and # S_Features, the number of selected features (see Sect. 3.1). Fitness values in bold stand for algorithms outperforming the other one(s) according to the Wilcoxon test. For each instance, the value of the accuracy obtained by SVM without any feature selection is pointed out in brackets. The statistical comparison between algorithms is given under the instance name.

Instance	Algorithm	Fitness	Accuracy (%)	# S_Features
Schizophrenia (69.64%) $LTS > HC > TS$	HC	$0.992_{(0)}$	$99.946_{(0.097)}$	$11.788_{(3.735)}$
	TS	$0.968_{(0)}$	$97.132_{(4.173)}$	$16.939_{(26.246)}$
	LTS	$\mathbf{0.995}_{(0)}$	$100_{(0)}$	$8.758_{(0.627)}$
Colon (87.09%) $(LTS = HC) > TS$	**HC**	$\mathbf{0.998}_{(0)}$	$99.951_{(0.079)}$	$10.909_{(11.835)}$
	TS	$0.982_{(0)}$	$97.752_{(3.405)}$	$10.97_{(6.905)}$
	LTS	$\mathbf{0.996}_{(0)}$	$99.609_{(0.655)}$	$6.788_{(3.047)}$
Breast (67.3%) $HC > (LTS = TS)$	**HC**	$\mathbf{0.98}_{(0)}$	$97.319_{(7.382)}$	$18.394_{(44.684)}$
	TS	$0.94_{(0)}$	$92.308_{(16.18)}$	$13.121_{(7.86)}$
	LTS	$0.94_{(0)}$	$92.075_{(16.817)}$	$11.879_{(8.172)}$
Arcene (83%) $LTS > HC > TS$	HC	$0.999_{(0)}$	$99.879_{(0.11)}$	$21.97_{(40.405)}$
	TS	$0.971_{(0)}$	$96.273_{(10.392)}$	$22.273_{(18.08)}$
	LTS	$\mathbf{0.999}_{(0)}$	$100_{(0)}$	$14.97_{(9.905)}$
DNA (89.57%) $LTS > (HC = TS)$	HC	$0.941_{(0)}$	$95.71_{(0.364)}$	$19.152_{(27.82)}$
	TS	$0.941_{(0)}$	$95.762_{(0.343)}$	$19.485_{(21.633)}$
	LTS	$\mathbf{0.945}_{(0)}$	$95.234_{(0.46)}$	$13.606_{(8.434)}$
Madelon (56,45%) $LTS > (HC = TS)$	HC	$0.712_{(0)}$	$64.135_{(0.331)}$	$37.273_{(92.017)}$
	TS	$0.714_{(0)}$	$63.885_{(0.453)}$	$31.03_{(46.905)}$
	LTS	$\mathbf{0.731}_{(0)}$	$65.152_{(0.107)}$	$15.606_{(10.246)}$

Table 4 also shows that, for each instance, LTS improves results obtained by the original Tabu Search. These results show the improvement obtained by the introduction of the learning mechanism.

In order to analyze the behavior of the different algorithms, we computed their evolution over time. Figure 2 shows the evolution of the average fitness of each approach over the time and gives the box-and-whisker plots (after one third, two thirds, and at the end of the allocated runtime) on *Madelon* instance, which is the most difficult instance to solve. For this one, LTS has a quick progression compared to the two other methods. Indeed, *Madelon* is a high-cost instance, so thanks to the estimation function, LTS avoids a large number of evaluations.

Consequently, LTS finds the potential good solutions more quickly than other approaches. These results show the interest of the estimation function.

To understand the behavior of the learning mechanism, we also investigate the dynamic of add and drop flips over time. Thus, Fig. 3 shows, for one execu-

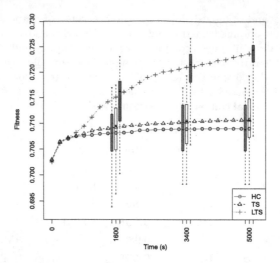

Fig. 2. Evolution of algorithms on *Madelon* instance

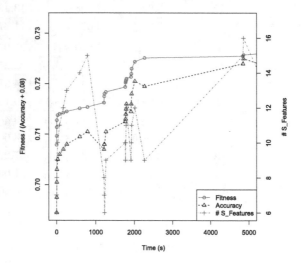

Fig. 3. Evolution of Fitness, Accuracy and # S_Features values for LTS on *Madelon* instance. For more readability, the accuracy curve has been translated by +0.08.

tion, the evolution of the different metrics (Fitness, Accuracy and # S_Features) for LTS on *Madelon* instance. We can observe several phases on this figure. The first one (from the beginning to time 1000 s approximately) adds features to increase the accuracy, and in consequence also the fitness. The second one starts when fitness is high. In this phase, the learning mechanism chooses the worst features to remove thanks to the trail values. As LTS removes features, which bring the least information, the accuracy decreases slightly while the second *part* of the fitness that favors small subsets of features increases. Consequently

LTS makes a good trade-off between the accuracy and the number of selected features.

Analysis of the Datamining Approach: Table 5 shows the results about the accuracy values on both training and validation sets for each instance. The objective is to analyze the ability to make a good classification on the validation set, using features selected on the training set.

A first observation is that performance decreases between the training set and the validation one. This difference reveals overfitting, that is to say, the solution

Table 5. Average and standard deviation (in brackets) of Accuracy values obtained on both training and validation sets for HC, TS and LTS. Accuracy values in bold stand for algorithms outperforming the other one(s) according to the Wilcoxon test. The statistical comparison between algorithms is given. The double line shows the separation between low and high evaluation time cost.

Instance	Algorithm	Accuracy (%) Training	Accuracy (%) Validation
Schizophrenia	HC	$\mathbf{99.946}_{(0.097)}$	$60.208_{(87.454)}$
	TS	$97.132_{(4.173)}$	$55.457_{(119.31)}$
	LTS	$\mathbf{100}_{(0)}$	$\mathbf{61.319}_{(66.62)}$
		$(LTS = HC) > TS$	$LTS > HC > TS$
Colon	HC	$\mathbf{99.951}_{(0.079)}$	$\mathbf{94.318}_{(26.523)}$
	TS	$97.752_{(3.405)}$	$91.004_{(39.524)}$
	LTS	$99.609_{(0.655)}$	$\mathbf{94.127}_{(23.655)}$
		$HC > LTS > TS$	$(LTS = HC) > TS$
Breast	HC	$\mathbf{97.319}_{(7.382)}$	$54.079_{(105.34)}$
	TS	$92.308_{(16.18)}$	$50.116_{(106.77)}$
	LTS	$92.075_{(16.817)}$	$52.098_{(98.087)}$
		$HC > (LTS = TS)$	$LTS = HC = TS$
Arcene	HC	$\mathbf{99.879}_{(0.11)}$	$72.18_{(16.028)}$
	TS	$96.273_{(10.392)}$	$71.69_{(22.15)}$
	LTS	$\mathbf{100}_{(0)}$	$\mathbf{74.60}_{(21.43)}$
		$(LTS = HC) > TS$	$LTS > HC > TS$
DNA	HC	$\mathbf{95.71}_{(0.364)}$	$93.63_{(2.50)}$
	TS	$\mathbf{95.762}_{(0.343)}$	$93.25_{(2.57)}$
	LTS	$95.234_{(0.46)}$	$94.09_{(2.37)}$
		$(HC = TS) > LTS$	$LTS = HC = TS$
Madelon	HC	$64.135_{(0.331)}$	$57.07_{(4.84)}$
	TS	$63.885_{(0.453)}$	$56.56_{(4.23)}$
	LTS	$\mathbf{65.152}_{(0.107)}$	$\mathbf{59.69}_{(1.485)}$
		$LTS > (HC = TS)$	$LTS > (HC = TS)$

built on the training set is specific for these data. Consequently, the solution looses in prediction quality for new data. This is especially true for instances with few observations and confirms the difficulty to find a good classification model. The standard deviations obtained with the validation set on these instances are high and show a bad stability of the results produced. Conversely, in instances with a large numbers of observations, the standard deviations obtained with the validation sets are reasonable. Solutions are less sensitive to the data used for the validation.

As for the training set, LTS manages to obtain better or equivalent results than other approaches on the validation set. In particular, we can observe an improvement about the results obtained by LTS compared to TS.

In conclusion, these experiments show the good performance of the Learning Tabu Search regarding both the optimization and the datamining perspectives. In particular, these experiments show the contribution of the learning mechanism, as the Learning Tabu Search is able to find better subset of features than the classical Tabu Search although they are based on the same components.

5   Conclusion and Perspectives

This work proposes to consider the Feature Selection problem for classification as a combinatorial optimization one and presents an adaptation of the Learning Tabu search to solve it. The objective was to be able to jointly deal with large datasets and efficient classifiers. Indeed, these two elements may lead to an expensive objective function, which is a difficult aspect for optimization methods that require a large amount of evaluations.

Therefore, to solve the Feature Selection problem with a local search, we first propose of modelization, including the definition of neighborhood operators. Then we propose some adaptations of the Learning Tabu Search, previously proposed to solve a railway network problem, to the FS problem. Some specificities for the FS problem are presented and explained.

Experiments are conducted in order to analyze the benefit of the integration of a learning mechanism. Then, the Learning Tabu Search is mainly compared to the classical Tabu Search, using exactly the same components except the learning mechanism. Datasets from the literature are used. Some of them have a low cost evaluation time, whereas others are more costly. The main conclusions are that according to the optimization perspective (the ability to obtain good fitness solutions), the Learning Tabu Search obtained better results than the classical Tabu Search, especially for high evaluation cost instances. This is due to the use of the estimation function that avoids many evaluations and allows the Learning Tabu Search to progress faster. Regarding the datamining perspective (the ability to find solutions that can lead to good classifications on other datasets), a same observation is done: the Learning Tabu Search obtained better results than the classical Tabu Search. This may be explained by the small number of features selected by the Learning tabu Search compared to other methods. Hence, these experiments show the contribution of the learning mechanism.

This work is very encouraging and perspectives of future works are interesting. As far as the Feature Selection problem is concerned, such perspectives may deal either with the extension of the learning mechanism (other definition of the trail, for example) for LTS, or with the integration of the proposed learning mechanism in other metaheuristics that may benefit from the estimation function. Other perspectives deal with the definition of such a learning mechanism for other optimization problems with a high cost evaluation function.

References

1. Cai, D., Zhang, C., He, X.: Unsupervised feature selection for multi-cluster data. In: Proceedings of the 16th ACM SIGKDD international conference on Knowledge discovery and data mining, pp. 333–342. ACM (2010)
2. Calhoun, V.: MLSP 2014 schizophrenia classification challenge (2014). https://www.kaggle.com/c/mlsp-2014-mri
3. Cervante, L., Xue, B., Zhang, M., Shang, L.: Binary particle swarm optimisation for feature selection: a filter based approach. In: 2012 IEEE Congress on Evolutionary Computation (CEC), pp. 1–8. IEEE (2012)
4. Corne, D., Dhaenens, C., Jourdan, L.: Synergies between operations research and data mining: the emerging use of multi-objective approaches. Eur. J. Oper. Res. **221**(3), 469–479 (2012)
5. Dorigo, M., Birattari, M.: Ant colony optimization. In: Sammut, C., Webb, G.I. (eds.) Encyclopedia of Machine Learning, pp. 36–39. Springer, Heidelberg (2010)
6. Duval, B., Hao, J.-K., Hernandez Hernandez, J.C.: A memetic algorithm for gene selection and molecular classification of cancer. In: Genetic and Evolutionary Computation Conference, GECCO 2009, Proceedings, Montreal, Québec, Canada, 8–12 July 2009, pp. 201–208 (2009)
7. Emmanouilidis, C., Hunter, A., MacIntyre, J.: A multiobjective evolutionary setting for feature selection and a commonality-based crossover operator. In: Evolutionary Computation, 2000, vol. 1, pp. 309–316. IEEE (2000)
8. Gheyas, I.A., Smith, L.S.: Feature subset selection in large dimensionality domains. Pattern Recogn. **43**(1), 5–13 (2010)
9. Guerra-Salcedo, C., Whitley, L.D.: Genetic approach to feature selection for ensemble creation. In: GECCO 1999, Orlando, Florida, USA, 13–17 July 1999, pp. 236–243 (1999)
10. Guyon, I., Gunn, S., Nikravesh, M., Zadeh, L.A.: Feature Extraction: Foundations and Applications, vol. 207. Springer, Heidelberg (2008)
11. Guyon, I., Gunn, S.R., Ben-Hur, A., Dror, G.: Result analysis of the NIPS 2003 feature selection challenge. In: Advances in Neural Information Processing Systems 17, NIPS 2004, Vancouver, British Columbia, Canada, 13–18 December 2004, pp. 545–552 (2004)
12. Hamdani, T.M., Won, J.-M., Alimi, A.M., Karray, F.: Multi-objective feature selection with NSGA II. In: Beliczynski, B., Dzielinski, A., Iwanowski, M., Ribeiro, B. (eds.) ICANNGA 2007. LNCS, vol. 4431, pp. 240–247. Springer, Heidelberg (2007). doi:10.1007/978-3-540-71618-1_27
13. Kohavi, R., John, G.H.: Wrappers for feature subset selection. Artif. Intell. **97**(1–2), 273–324 (1997)

14. Long, N., Gianola, D., Rosa, G.J.M.: Machine learning classification procedure for selecting SNPs in genomic selection: application to early mortality in broilers. J. Anim. Breed. Genet. **124**(6), 377–389 (2007)
15. Oliveira, L.S., Morita, M., Sabourin, R.: Feature selection for ensembles using the multi-objective optimization approach. In: Oliveira, L.S., Morita, M., Sabourin, R. (eds.) Multi-objective Machine Learning. SCI, vol. 16, pp. 49–74. Springer, Heidelberg (2006)
16. Schindl, D., Zufferey, N.: Solution methods for fuel supply of trains. INFOR **51**(1), 23–30 (2013)
17. Schölkopf, B., Smola, A.: Support vector machines. Encycl. Biostat. (1998). http://onlinelibrary.wiley.com/doi/10.1002/0470011815.b2a14038/abstract
18. Tan, C.J., Lim, C.P., Cheah, Y.-N.: A multi-objective evolutionary algorithm-based ensemble optimizer for feature selection and classification with neural network models. Neurocomputing **125**, 217–228 (2014)
19. Xue, B., Zhang, M., Browne, W.N.: Particle swarm optimization for feature selection in classification: a multi-objective approach. IEEE Trans. Cybern. **43**(6), 1656–1671 (2013)
20. Yang, J., Honavar, V.: Feature subset selection using a genetic algorithm. In: Liu, H., Motoda, H. (eds.) Feature Extraction, Construction and Selection, vol. 453, pp. 117–136. Springer, Heidelberg (1998)
21. Zhu, Z., Ong, Y.-S., Dash, M.: Markov blanket-embedded genetic algorithm for gene selection. Pattern Recogn. **40**(11), 3236–3248 (2007)

The Impact of Automated Algorithm Configuration on the Scaling Behaviour of State-of-the-Art Inexact TSP Solvers

Zongxu Mu[1]([✉]), Holger H. Hoos[1]([✉]), and Thomas Stützle[2]

[1] Department of Computer Science, University of British Columbia,
Vancouver, BC, Canada
{zongxumu,hoos}@cs.ubc.ca
[2] IRIDIA, CoDE, Université Libre de Bruxelles, Brussels, Belgium
stuetzle@ulb.ac.be

Abstract. Automated algorithm configuration is a powerful and increasingly widely used approach for improving the performance of algorithms for computationally hard problems. In this work, we investigate the impact of automated algorithm configuration on the scaling of the performance of two prominent inexact solvers for the travelling salesman problem (TSP), EAX and LKH. Using a recent approach for analysing the empirical scaling of running time as a function of problem instance size, we demonstrate that automated configuration impacts significantly the scaling behaviour of EAX. Specifically, by automatically configuring the adaptation of a key parameter of EAX with instance size, we reduce the scaling of median running time from root-exponential (of the form $a \cdot b^{\sqrt{n}}$) to polynomial (of the form $a \cdot n^b$), and thus, achieve an improvement in the state of the art in inexact TSP solving. In our experiments with LKH, we noted overfitting on the sets of training instances used for configuration, which demonstrates the need for more sophisticated configuration protocols for scaling behaviour.

1 Introduction

The travelling salesperson problem (TSP) is a well known and widely studied $\mathcal{NP}$-hard problem. Given a set of cities and pair-wise distances between them, the objective of the TSP is to find the shortest round trip that visits each city exactly once. TSP algorithms are usually categorised into two kinds: exact algorithms, which are guaranteed to find an optimal solution to any TSP instance and can prove the optimality of the solution, and inexact algorithms, which may find optimal solutions but cannot prove optimality. Presently, Concorde [2] represents the long-standing state of the art among exact TSP algorithms. In terms of inexact TSP algorithms, LKH [5,6] had been the best available solver until the recent introduction of EAX [20], an evolutionary algorithm that makes uses of an improved edge assembly crossover operator [19] for recombining short tours. Empirical results show that EAX tends to perform better than LKH on a broad range of TSP instances [20]; however, it has been shown recently that

© Springer International Publishing AG 2016
P. Festa et al. (Eds.): LION 2016, LNCS 10079, pp. 157–172, 2016.
DOI: 10.1007/978-3-319-50349-3_11

LKH is not dominated by EAX in that there are many instances for which LKH finds optimal solutions more efficiently than EAX [14].

In theoretical computer science, time complexity is arguably the most important concept for analysing and understanding the difficulty of problems and the performance of algorithms. The time complexity of an algorithm is characterised by the scaling of the time required for solving a problem instance as a function of instance size. In spite of the significant role that theoretical methods play in understanding the complexity of problems and algorithms, many high-performance algorithms are beyond the reach of such methods, and therefore have to be studied using principled empirical approaches.

In this work, we investigate the question whether and to which extent the empirical scaling of the running time of state-of-the-art inexact TSP solvers EAX and LKH changes as the parameter settings of these solvers are optimised. This question is particularly relevant as automated procedures for optimising parameter settings (so-called algorithm configurators) are now readily available and used increasingly frequently in the development of state-of-the-art solvers for computationally challenging problems as well as for customisation of such solvers for particular application contexts (see, e.g., [1,3,10,11]). To study the scaling behaviour of EAX and LKH, we use an advanced empirical scaling analysis approach that challenges automated fitted scaling models by extrapolation and uses bootstrap re-sampling to statistically assess scaling models [7,18]. Our main findings are as follows:

- automated algorithm configuration can significantly improve the scaling behaviour of EAX, and by adapting the population size with instance size, the empirical time complexity of EAX is reduced from root-exponential (of the form $a \cdot b^{\sqrt{n}}$) to polynomial (of the form $a \cdot n^b$);
- the state of the art in inexact TSP solving can thus be improved: for instance, we reduce the median running time of EAX for solving instances of size $n = 4500$ (three times larger than those used for training the configurator) by about a factor of 1.13;
- automated algorithm configuration can significantly impact the scaling of LKH, but configuring LKH suffers from overfitting that leads to improved running times for small instances but worse performance for larger ones.

In the remainder of this work, we first describe the benchmark instances, algorithms and methods that we use in our experiments (Sect. 2); next, we present in detail our results (Sect. 3); and finally, we draw some general conclusions and briefly outline avenues for future work (Sect. 4).

2   Instances, Algorithms and Methods

2.1   Benchmark Instances

2D Euclidean TSP instances, i.e., instances where the locations to be visited correspond to points in the Euclidean plane, often occur in practical applications.

A particularly widely studied type of 2D Euclidean TSP instances are obtained by placing cities uniformly at random in a square. The so-called RUE instances thus obtained are known to have properties similar to a broad range of other 2D Euclidean instances and represent a challenging, widely used benchmark for TSP solvers [12,13]. In the following, we use the benchmark sets of RUE instances generated and studied earlier by Hoos and Stützle [8,9]. These instances were generated using the portgen generator from the 8th DIMACS implementation challenge for TSP, which places n points in a $100\,000 \times 100\,000$ square uniformly at random and computes the Euclidean distances between pairs of points. There are 1000 instances for each instance size $n = 500, 600, \ldots, 1500, 2000$ and 100 instances for each $n = 2500, 3000, \ldots, 4500$.

2.2 Inexact Algorithms for TSP

The inexact TSP solvers we selected for our study are the latest versions of LKH and EAX. LKH [5,6], Helsgaun's variant of the Lin-Kernighan TSP heuristic, is a variable-depth search algorithm that performs sophisticated heuristically guided local search moves based on sequences of five or more edge exchanges. It restarts the local search based on perturbations of previously found solutions using various strategies. LKH represents a milestone in the development of inexact TSP solvers and is arguably the most prominent method for finding finding high-quality solutions to challenging TSP instances. In this work, we used LKH version 2.0.7, keeping all parameters at their default values, except for PATCHING_A and PATCHING_C, which we set to 2 and 3, respectively, to include patching of cycles in searching for improving moves. These values were also adopted in the example parameter file for solving TSPLIB instance pr2392 and used in earlier work studying LKH [4].

EAX [20] is a recent evolutionary algorithm that makes use of improved variants of the edge assembly cross-over recombination operator. It also exploits diversity preservation techniques and initialises the initial population by local optimisation. For our experiments, we used EAX with default parameter settings, namely with population size set to 100 and the number of offsprings generated per recombination attempt set to 30.

For our analyses, we used the same modified implementations of LKH and EAX as Dubois-Lacoste et al. [4], who enhanced the original solvers with a restart mechanism to achieve improved performance. This type of modification is a simple, yet effective means for overcoming stagnation behaviour often encountered in stochastic local search algorithms [21]; in the case of LKH and EAX, the added restart mechanism helps considerably in finding optimal solutions more efficiently.

2.3 Algorithm Configurator

To automatically configure the parameters of EAX and LKH, we used SMAC, a prominent, state-of-the-art algorithm configurator [10]. SMAC is based on a sequential model-based optimisation procedure that builds and iteratively refines a statistical model mapping parameter configurations of a given algorithm to performance predictions. This empirical performance model is used to select promising configurations in each iteration of SMAC; these configurations are then run, and the performance values thus observed are used to update the model. The standard version of SMAC, as used in our experiments for configuring EAX and LKH, uses random regression forests to model performance.

2.4 Scaling Analysis

We use a recent boostrap approach for studying the empirical scaling of algorithm performance with input size [7]. It automatically fits scaling models to a set of support data and then challenges these models by testing the performance predictions obtained from them for larger input sizes. Most importantly, it uses a re-sampling approach to assess the models and their predictions in a statistically meaningful way. This approach has been used to characterise the scaling behaviour of state-of-the-art exact and inexact TSP algorithms [4,8] and to study the empirical time complexity of state-of-the-art solvers for the propositional satisfiability problem (SAT) [18]. In this latter work, Mu and Hoos extended the approach to compare scaling models based on bootstrap confidence intervals for predicted and observed running times and to assess differences in the scaling models for two given algorithms. To perform this type of scaling analysis for different configurations of EAX and LKH, we used the ESA system [17].

2.5 Computing Environment and Experimental Setup

For the automatic configuration of EAX and LKH, we used a standard protocol, according to which we performed 25 independent runs of SMAC for each scenario and selected the best parameter configuration according to the performance on the given set of training instances for scaling analysis. The cut-off time for each run of the algorithm being configured and the overall time budget for each run of SMAC differ between our experiments for EAX and LKH, and we report them as we discuss each experiment.

For our scaling analysis, we considered three parametric models:

- $Exp\,[a,b]\,(n) = a \cdot b^n$ (2-parameter exponential);
- $RootExp\,[a,b]\,(n) = a \cdot b^{\sqrt{n}}$ (2-parameter root-exponential);
- $Poly\,[a,b]\,(n) = a \cdot n^b$ (2-parameter polynomial).

Models were fitted to performance observations in the form of medians of the distributions of running times over sets of instances for given n. Compared to the mean, the median has two advantages: it is statistically more stable and immune to the presence of a certain number of timed-out runs. We performed 10 independent runs per instance and used the median over those 10 running times as the running time for the respective instance. Our approach could be easily extended to other scaling models, but, as we will show in the following, these models jointly characterise the scaling observed in all our experiments, and, thus, we saw no need to consider different or more complex models. For fitting parametric scaling models to observed data, the ESA system we used for scaling analysis uses the non-linear least-squares Levenberg-Marquardt algorithm.

Following previous work [4,7,8], we computed 95% bootstrap confidence intervals for the performance predictions obtained from our scaling models, based on 1000 bootstrap samples per instance set and 1000 automatically fitted variants of each scaling model. For collecting running time data for our TSP solvers, we used the Compute Canada/Westgrid cluster orcinus (DDR), each node of which is equipped with two 3.0 GHz Intel Xeon E5450 quad-core CPUs and 16 GB of RAM, running 64-bit Red Hat Enterprise Linux Server 5.3.

3 Experimental Results

3.1 Treatment of Running Time Data

EAX and LKH cannot prove the optimality of the solutions they find; in our experiments, they therefore need access to the optimal solution qualities of the instances we consider, in order to measure the running time required to reach optimal solutions and to terminate runs once an optimal solution has been found. The optimal solutions to the RUE instance we used have been determined in an earlier study of the scaling of Concorde [8]. However, Concorde did not solve all instances within the allotted time in that study. To make more instances available for EAX and LKH to solve, we ran Concorde on the previously unsolved instances with different seeds and/or on faster machines. In addition, we performed multiple runs of EAX and LKH on those instances that were then still not solved by Concorde. For some of these instances, EAX and LKH found the same best solution in every run; we conjecture these solutions to be optimal and we refer to them as *pseudo-optimal*. For our analysis, we use data for both optimal and pseudo-optimal instances, but we note that qualitatively similar conclusions are obtained when excluding instances with pseudo-optimal solutions.

We took special care in dealing with the eight instances for which we did not succeed in establishing pseudo-optimal solutions (two of size 4000 and six of size 4500). As reported by [4], the pairwise performance correlations between EAX, LKH and Concorde are very low, and the instances for which we were unable to determine even pseudo-optimal solutions may still be easy for one of the inexact solvers. Thus, they are treated using an optimistic/pessimistic estimation, as done by [4]. More precisely, we treat these instances as easy with smaller-than-median running times in the optimistic estimation, and as timed-out instances in the pessimistic treatment. This gives us intervals for the median running times on those instance sizes for which some instances are lacking even pseudo-optimal solutions ($n = 4000$ and $n = 4500$). We note that these intervals are not confidence intervals, but bounds on the median running times, as they must contain the true median running times.

3.2   Scaling of EAX and LKH with Default Parameters

We first repeat the scaling analyses for EAX and LKH of [4, 16] with new sets of running time data collected using our machines. This ensures that the comparisons described below are not affected by differences between the machines used. The results we thus obtained are qualitatively similar to those reported in [4, 16]. More precisely, we found that the scaling of EAX is reasonably well described by a root-exponential model, while that of LKH is bounded from below and above by a polynomial and a root-exponential model, respectively. We report detailed results for EAX, labelled EAX (default), including the best fitted models and the bootstrap confidence intervals for the model parameters, in Tables 1 and 2, respectively. Analogous results for LKH, labelled LKH (default), can be found in Tables 6 and 7. Comparing our models to those reported in [16], our new results lead to a larger value of b in the root-exponential model for EAX, while large overlaps are seen in the bootstrap confidence intervals for b in the root-exponential and polynomial models of LKH, respectively.

3.3   Impact of Automated Configuration on Scaling of EAX

Next, we automatically configured the two parameters exposed by EAX: N_{pop} (fNumOfPop in the source code), the population size, and N_{ch} (fNumOfKids in the source), the number of offsprings generated per recombination attempt, which also affects the way EAX switches between different search strategies. The default values for these parameters are $N_{pop} = 100$ and $N_{ch} = 30$. In our experiments, we enforced a cut-off of one CPU day for each SMAC run. To ensure that SMAC could perform at least 1000 runs of EAX, we further enforced a cut-off time of 86 s for each EAX run. We note that even though EAX has only two parameters to configure, which makes the use of SMAC seemingly excessive, we still chose to use SMAC, because we saw no harm in doing so and because this allowed us to use the same configuration protocol as for LKH, whose configuration space is much larger. After configuration on a set of RUE instances with $n = 1500$, SMAC determined a parameter setting with N_{pop} (167 vs 100) and smaller N_{ch} (20 vs 30).

Table 1. Most accurate scaling models (according to RMSE on challenge data) for the median running times required by different variants of EAX for finding optimal solutions to RUE instances and corresponding RMSE values (in CPU sec). EAX (configured) uses a parameter configuration determined by SMAC, and EAX (configured + var pop) additionally determines N_{pop} as a linear function of instance size. All models were fitted using performance data obtained on RUE instances of size 500...1500 and challenged by performance data for instances of size 2000...4500.

Solver	Model		RMSE (support)	RMSE (challenge)
EAX (default)	RootExp.	$0.086254 \times 1.1439^{\sqrt{n}}$	0.12518	$[73.961, 116.47]$
EAX (configured)	RootExp.	$0.19910 \times 1.1193^{\sqrt{n}}$	0.22625	$[22.278, 32.899]$
EAX (configured + var pop)	Poly.	$1.6194 \times 10^{-8} \times n^{2.8364}$	0.027288	$[37.049, 44.847]$

Table 2. 95% bootstrap confidence intervals for the parameters of the scaling models from Table 1.

Solver	Model	Confidence interval for a	Confidence interval for b
EAX (default)	RootExp.	$[0.08287, 0.08991]$	$[1.1424, 1.1452]$
EAX (configured)	RootExp.	$[0.19024, 0.20586]$	$[1.1182, 1.1208]$
EAX (configured + var pop)	Poly.	$[1.3424 \times 10^{-8}, 1.9355 \times 10^{-8}]$	$[2.8110, 2.8623]$

We then investigated the scaling behaviour of EAX with these optimised parameters. The most accurate scaling model, according to the root mean squared error (RMSE) on challenge data (*i.e.*, instance sets not used for model fitting), is shown in Table 1, where this version of EAX is labelled EAX (configured). Similar to results for the default version of EAX, a root-exponential model characterises the scaling most accurately, while the best exponential and polynomial models can be rejected with 95% confidence. We cross-checked the predictions of the root-exponential model for the default version of EAX against the observed running times for the optimised version of EAX. Our results, illustrated in Fig. 1, clearly show that even though the optimised version loses some performance on small instances, the performance on larger instances is significantly improved. Thus, the original model over-estimates the running times for the optimised version of EAX. In addition, from the confidence intervals of the model parameters, as shown in Table 2, there is evidence that algorithm configuration significantly improves the scaling of EAX, since it reduces the value of b in the 2-parameter root-exponential models from 1.144 to 1.119, with non-overlapping confidence intervals ($[1.1424, 1.1451]$ vs. $[1.1182, 1.1208]$).

In preliminary experiments, we noted that the performance for larger instances could be improved by using larger population sizes, N_{pop}. Furthermore, the README file distributed with the source code of EAX recommends to use $N_{pop} = 300$ for instances with $n > 10\,000$ cities. We therefore considered the possibility of increasing N_{pop} with n and performed an experiment to quantify the impact of varying the population size as a simple linear function of n, *i.e.*,

$$N_{pop}(n) := \alpha \cdot n.$$

To obtain an estimate for α, we divided the optimised value of N_{pop} by the instance size $n = 1500$ at which this setting was obtained, resulting in $\alpha = 0.111$. We then analysed the empirical scaling behaviour of EAX with $N_{pop}(n)$ detemined in this manner and the optimised value of $N_{ch} = 20$ from our previous experiment and refer to this setting as EAX (configured + var pop). The most accurate scaling model is presented in Table 1, with bootstrap confidence intervals for the model parameters shown in Table 2. To our surprise, the best accuracy is now achieved by a polynomial model, while the best exponential and root-exponential models are rejected with 95% confidence, as illustrated in Fig. 2. Furthermore, from Fig. 1, it is obvious that this also stands in contrast with the scaling behaviour of EAX (default).

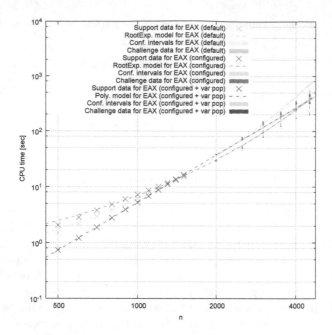

Fig. 1. Most accurate scaling models for the median running times required by three variants of EAX for finding optimal solutions to RUE instances and confidence intervals for model predictions, along with observed performance data; see Tables 1 and 2 for details.

As a result of its improved scaling behaviour, the optimised version of EAX significantly improves the previous state of the art in inexact TSP solving, as represented EAX (default); this can also be seen from the respective observed and predicted running times in Table 3.

We also compared this optimised version of EAX with EAX (default) on solving a set of TSPLIB instances with n between 500 and 4500, based on median running times determined from 10 independent runs per instance. Our results indicate that EAX (configured + var pop) performs better in some cases, but worse in others, and overall does not achieve significantly improved performance. We believe that this is primarily as most TSPLIB instances are actually easier to solve than RUE instances of a similar size, and smaller population size values are sufficient to solve these instances. We note that for the two TSPLIB instances that are harder than similarly-sized RUE instances, EAX (configured + var pop) does perform significantly better than EAX (default). Two other TSPLIB instances were not solved by EAX with any of the two parameter settings.

In addition, we compared LKH with EAX (configured + var pop). As illustrated in Fig. 3, the median running times of EAX (configured + var pop) for $n = 4500$ show substantially less variability than those of LKH (analogous observations hold for other n); furthermore, although EAX performs better than LKH on aggregate, there are many instances on which the converse is true.

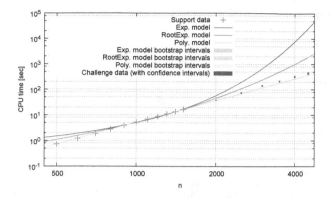

Fig. 2. Scaling models for the median running times required by EAX (configured + var pop) for finding optimal solutions to RUE instances and confidence intervals for predictions obtained from these models, along with observed median running times. All models were fitted using performance data obtained on RUE instances of size 500...1500 and challenged by performance data for instances of size 2000...4500.

Table 3. Improvement of the state of the art in inexact TSP solving, as demonstrated by the bootstrap confidence intervals for observed (for $n = 4500$) and predicted (for $n = 6000, 10000$) median running times for the default and optimised (configured + var pop) versions of EAX.

n	Median running time for EAX			75th percentile of running time for EAX		
	Default	Configured + var pop	Speedup	Default	Configured + var pop	Speedup
4500	$[364.1, 685.6]$	$[404.8, 478.5]$	$\approx 1.2\times$	$[685.6, 2350.3]$	$[482.8, 573.8]$	$\approx 2.9\times$
6000	$[2711, 3026]$	$[809, 880]$	$\approx 3.4\times$	$[5052, 9908]$	$[873, 1090]$	$\approx 7.6\times$
10000	$[54511, 64275]$	$[3401, 3800]$	$\approx 16.5\times$	$[132120, 364836]$	$[3613, 4826]$	$\approx 58.9\times$

We performed an analogous performance comparison for the previously mentioned set of TSPLIB instances. Both EAX and LKH fail to solve two of the 28 instances, and LKH fails to solve one instance solved by EAX. As seen in Fig. 3, once again, EAX (configured + var pop) performs better on aggregate, but LKH is considerably faster in solving several of these instances and therefore still contributes substantially to the state of the art in inexact TSP solving.

Analysing the performance results for EAX (configured + var pop) in more detail, we found that by increasing the population size for large instances, the success probability of each restart segment (*i.e.*, the part of a run between two restarts, between initialisation and the first restart and between the last restart and termination) is significantly increased. Figure 4, showing the distributions of the success probabilities of restart segments from all runs of EAX on RUE instances with $n = 1500$ and 4500, clearly illustrates this finding. Increasing population size, however, also increases the running time of the restart segments. The two effects can be clearly seen from medians of success probabilities and

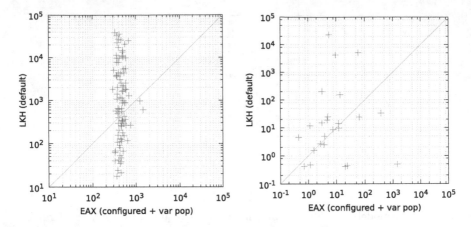

Fig. 3. Running time required by EAX (configured + var pop) *vs* the default version of LKH for solving RUE instances of size $n = 4500$ (left) and a set of TSPLIB instances (right); all running times are reported in CPU seconds. For RUE instances, the median running time for EAX and LKH lies within $[430.8, 454.8]$ and $[870.3, 1131.1]$, respectively. For TSPLIB instances, the median running time for EAX and LKH is 6.6 and 13.0, respectively.

Table 4. Median success probability and running time per restart segment of EAX (default) and EAX (configured + var pop) on sets of RUE instances.

n	Success probability per restart segment		Running time per restart segment	
	Default	Configured + var pop	Default	Configured + var pop
500	1.00	0.77	1.524	0.930
1500	0.77	0.91	16.265	17.991
2500	0.53	0.91	37.986	77.950
3500	0.30	0.91	70.406	218.735
4500	0.16	1.0	119.231	450.940

times to restart shown in Table 4. Hence, there is a tradeoff between the two effects. To achieve better scaling for EAX, it is critical to set the population size at the right point balancing the two effects.

To summarise, our experiments indicate that adapting the population size, N_{pop}, with instance size can significantly improve the scaling behaviour of EAX on RUE instances. After optimising the adaptation mechanism using automated configuration of α (together with N_{ch}), the scaling of EAX is best captured by a polynomial model, and the resulting version of EAX represents a significant improvement in the state of the art for inexact TSP solving. In particular, we observed an $\approx 1.13\times$ improvement in the median running time for EAX and even more substantial improvements for higher percentiles when solving instances

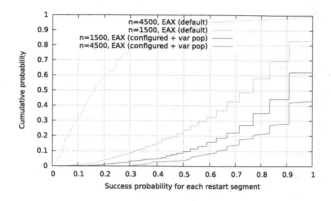

Fig. 4. Distributions of the success probability of each restart segment within EAX solving RUE instances with $n = 1500, 4500$. (For details, see text.)

of size $n = 4500$. Based on our scaling analysis, we expect the performance advantage of EAX (configured + var pop) over EAX (default) to grow further with instance size.

3.4 Impact of Automated Configuration on Scaling of LKH

We also attempted to configure LKH for improved scaling of running time. With over 40 parameters, LKH is arguably much more configurable than EAX. Out of these parameters, some require additional information, such as initial tours or sub-division of a given TSP instance, which are not available in our case. Thus, we selected the 21 parameters (12 numerical and 9 categorical) listed in Table 5, which we could configure without additional information or code modification. We used the same default values as specified in the user guide, except for PATCHING_A and PATCHING_C, as mentioned in Sect. 2.2. The ranges or sets of settings for all parameters were determined based on the user guide; when in doubt, we used large ranges to create a large configuration space for SMAC to explore. We enforced a cut-off of 2 CPU days for each SMAC run. To ensure that SMAC could perform at least 1000 runs of LKH, we further enforced a cut-off time of 172 s for each LKH run.

We first configured LKH analogously to EAX, that is, we performed 25 SMAC runs using a set of 100 RUE instances of size $n = 1500$ randomly selected from the full set, and selected from the 25 configurations the one with the best performance across the 100 training instances. The resulting configuration achieved improved performance only for instances up to size 2000, but did not scale to larger instance sizes, suggesting overfitting on smaller instance sizes.

Next, we attempted to improve the scaling performance by following a protocol proposed by Styles et al. [23]. More precisely, we performed 25 SMAC runs to configure LKH using a set of 100 instances with instance size $n = 1000$ and selected the best configuration based on performance on a set of 50 validation instances selected uniformly at random from the the set of RUE instances of

Table 5. List of numerical (N) and categorical (C) parameters for LKH considered in our configuration experiments.

Parameter name	Type	Domain
ASCENT_CANDIDATES	N	[10, 500]
BACKBONE_TRIALS	N	[0, 5]
BACKTRACKING	C	{YES, NO}
CANDIDATE_SET_TYPE	C	{ALPHA, DELAUNAY, NEAREST-NEIGHBOR, QUADRANT}
EXTRA_CANDIDATES	N	[0, 20]
EXTRA_CANDIDATE_SET_TYPE	C	{NEAREST-NEIGHBOR, QUADRANT}
GAIN23	C	{YES, NO}
GAIN_CRITERION	C	{YES, NO}
INITIAL_STEP_SIZE	N	[1, 5]
INITIAL_TOUR_ALGORITHM	C	{BORUVKA, GREEDY, MOORE, NEAREST-NEIGHBOR, QUICK-BORUVKA, SIERPINSKI, WALK}
KICK_TYPE	N	{0} ∪ [4, 20]
KICKS	N	[0, 5]
MAX_CANDIDATES	N	[3, 20]
MOVE_TYPE	N	[2, 20]
PATCHING_A	N	[1, 5]
PATCHING_C	N	[1, 5]
POPULATION_SIZE	N	[0, 1000]
RESTRICTED_SEARCH	C	{YES, NO}
SUBGRADIENT	C	{YES, NO}
SUBSEQUENT_MOVE_TYPE	N	{0} ∪ [2, 20]
SUBSEQUENT_PATCHING	C	{YES, NO}

size $n = 1500$. We then analysed the scaling of LKH with the configuration such obtained and found a root-exponential model to fit best, with a value of b very similar to that obtained for the default configuration (1.2077 vs 1.2067). Based on these scaling models and observed running times, we determined that this configuration performs better than the default configuration of LKH up to $n = 3500$, but not beyond. This indicates that using the modified configuration protocol reduces, but does not completely eliminate overfitting on smaller instance sizes.

After inspecting the parameter settings obtained from the two configuration experiments described so far in more detail, we concluded that they help LKH to more carefully check possible local search steps in a way that is effective only for smaller TSP instances. Based on this observation, we fixed KICKS, MOVE_TYPE, POPULATION_SIZE and RESTRICTED_SEARCH to their default values and performed another round of configuration following the protocol by Styles et $al..$ Results from the analysis of the scaling behaviour of

Table 6. Most accurate (bounding) scaling models for the median running times required by variants of LKH for finding optimal solutions to RUE instances and corresponding RMSE values (in CPU sec). LKH (scaling configuration) uses parameter settings obtained using SMAC and the protocol by Styles *et al.*; LKH (scaling configuration w/fewer para) uses a configuration obtained analogously using a reduced parameter space (for details, see text). All models were fitted using performance data obtained on RUE instances of size 500...1500 and challenged by performance data for instances of size 2000...4500.

Solver	Model		RMSE (support)	RMSE (challenge)
LKH (default)	RootExp.	$0.010186 \times 1.2067^{\sqrt{n}}$	0.34977	[859.21, 968.83]
	Poly.	$8.2328 \times 10^{-10} \times n^{3.2255}$	0.30801	[190.79, 293.36]
LKH (scaling configuration)	RootExp.	$0.0066414 \times 1.2077^{\sqrt{n}}$	0.13341	[533.13, 1165.9]
LKH (scaling configuration w/fewer para)	RootExp.	$0.0048097 \times 1.2010^{\sqrt{n}}$	0.15081	[1394.8, 2090.6]

Table 7. 95 % bootstrap confidence intervals for the parameters of the scaling models from Table 6.

Solver	Model	Confidence interval for a	Confidence interval for b
LKH (default)	RootExp.	$[0.0047391, 0.01955]$	$[1.1836, 1.2333]$
	Poly.	$\left[4.3312 \times 10^{-11}, 1.0215 \times 10^{-8}\right]$	$[2.8684, 3.6341]$
LKH (scaling configuration)	RootExp.	$[0.0040961, 0.010936]$	$[1.1908, 1.2245]$
LKH (scaling configuration w/fewer para)	RootExp.	$[0.0027107, 0.008814]$	$[1.1802, 1.2210]$

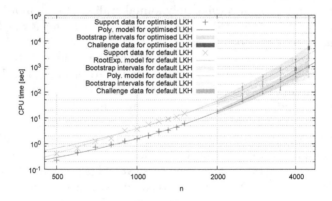

Fig. 5. Scaling models for default LKH and optimised LKH (scaling configuration w/fewer para) from Table 6, along with observed performance data and bootstrap confidence intervals for predicted performance obtained from the scaling models.

the configuration of LKH thus obtained are shown in Tables 6 and 7. We note that the confidence interval for b in the root-exponential model largely overlaps with that for LKH with default parameters ($[1.1802, 1.2210]$ *vs* $[1.1836, 1.2333]$). Again, the root-exponential model gives the best fit (according to RMSE on challenge data), and fits the running times well up to $n = 4000$, as seen in Fig. 5. The value of $b = 1.2010$ in this model is very similar to that for the default version of LKH ($b = 1.2077$), with a large overlap in the respective confidence intervals ($[1.1802, 1.2210]$ *vs* $[1.1908, 1.2245]$). Comparing the running times and scaling models between these two configurations of LKH, as illustrated in Figure 5, we noticed that the new configuration decreases the running times for $n \leq 4000$, but performs worse for $n = 4500$. In other words, the Styles *et al.* protocol applied to our reduced parameter space seems to overfit less, but still suffers from some overfitting to the instance sizes used for configuration.

4   Discussion

In this work, we investigated the impact of parameter settings and automated configuration on the scaling of inexact TSP algorithms. For EAX, algorithm configuration helps improve the scaling, which can be further improved by adapting the population size with instance size. In particular, we achieved an improvement in the median running time for EAX on RUE instances of size $n = 4500$ of a factor of ≈ 1.13 and of a factor of ≈ 1.87 for the 75th percentile of the distribution of running times over sets of RUE instances; based on our scaling models, we expect the improvement to be even more significant for larger instances.

Surprisingly, when adapting the population size with instance size, we obtain polynomial scaling of the median running time with instance size, compared to root-exponential scaling for the default configuration of EAX. To the best of our knowledge, this is the first time, polynomial scaling of the empirical median running time for an inexact TSP solver has been reported for a widely used set of challenging 2D Euclidean TSP instances.

Overall, our work complements earlier work on the scaling of state-of-the-art TSP solvers [4, 8, 9] and indicates potential for improvements in scaling behaviour through automated algorithm configuration and through setting certain parameters in dependence of features of the problem instance to be solved.

We see significant potential in developing automated configuration procedures for better scaling behaviour. Such procedures can make algorithm configuration more applicable to real-world situations, as problem instances from the target distribution may take a long time to solve. This poses a substantial challenge to existing algorithm configuration techniques, which require many runs of the target algorithms with different parameter settings. There is some previous work on configuration protocols addressing this challenge [15, 22, 23], but based on the findings for LKH reported in our study here, we believe that there is the need (and, indeed much room) for further improvements. In particular, we see the tight integration of empirical scaling analysis into the configuration process as a promising avenue for future research in this area.

Acknowledgements. HH and ZM acknowledge support through an NSERC Discovery Grant. TS acknowledges support from the Belgian F.R.S.-FNRS, of which he is a senior research associate. This work received support from Compute Canada/Westgrid and from the COMEX project within the Interuniversity Attraction Poles Programme of the Belgian Science Policy Office.

References

1. Ansótegui, C., Sellmann, M., Tierney, K.: A gender-based genetic algorithm for the automatic configuration of algorithms. In: Gent, I.P. (ed.) CP 2009. LNCS, vol. 5732, pp. 142–157. Springer, Heidelberg (2009). doi:10.1007/978-3-642-04244-7_14

2. Applegate, D.L., Bixby, R.E., Chvátal, V., Cook, W.J.: The traveling salesman problem, concorde TSP solver (2012). http://www.tsp.gatech.edu/concorde

3. Birattari, M., Yuan, Z., Balaprakash, P., Stützle, T.: F-race and iterated F-race: an overview. In: Bartz-Beielstein, T., Chiarandini, M., Paquete, L., Preuss, M. (eds.) Experimental Methods for the Analysis of Optimization Algorithms, pp. 311–336. Springer, Heidelberg (2010)

4. Dubois-Lacoste, J., Hoos, H.H., Stützle, T.: On the empirical scaling behaviour of state-of-the-art local search algorithms for the Euclidean TSP. In: Proceedings of GECCO 2015, pp. 377–384. ACM Press (2015)

5. Helsgaun, K.: An effective implementation of the Lin-Kernighan traveling salesman heuristic. Eur. J. Oper. Res. **126**(1), 106–130 (2000)

6. Helsgaun, K.: General k-opt submoves for the Lin-Kernighan TSP heuristic. Math. Program. Comput. **1**(2–3), 119–163 (2009)

7. Hoos, H.H.: A bootstrap approach to analysing the scaling of empirical run-time data with problem size. Technical report, TR-2009-16, Department of Computer Science, University of British Columbia (2009)

8. Hoos, H.H., Stützle, T.: On the empirical scaling of run-time for finding optimal solutions to the travelling salesman problem. Eur. J. Oper. Res. **238**(1), 87–94 (2014)

9. Hoos, H.H., Stützle, T.: On the empirical time complexity of finding optimal solutions vs proving optimality for Euclidean TSP instances. Optim. Lett. **9**, 1247–1257 (2015)

10. Hutter, F., Hoos, H.H., Leyton-Brown, K.: Sequential model-based optimization for general algorithm configuration. In: Coello, C.A.C. (ed.) LION 2011. LNCS, vol. 6683, pp. 507–523. Springer, Heidelberg (2011). doi:10.1007/978-3-642-25566-3_40

11. Hutter, F., Hoos, H.H., Leyton-Brown, K., Stützle, T.: ParamILS: an automatic algorithm configuration framework. J. Artif. Intell. Res. **36**(1), 267–306 (2009)

12. Johnson, D.S., McGeoch, L.A.: The traveling salesman problem: a case study in local optimization. In: Aarts, E.H.L., Lenstra, J.K. (eds.) Local Search in Combinatorial Optimization, pp. 215–310. Wiley, Chichester (1997)

13. Johnson, D.S., McGeoch, L.A.: Experimental analysis of heuristics for the STSP. In: Gutin, G., Punnen, A. (eds.) The Traveling Salesman Problem and Its Variations, pp. 369–443. Kluwer Academic Publishers, New York (2002)

14. Kotthoff, L., Kerschke, P., Hoos, H., Trautmann, H.: Improving the state of the art in inexact TSP solving using per-instance algorithm selection. In: Dhaenens, C., Jourdan, L., Marmion, M.-E. (eds.) LION 2015. LNCS, vol. 8994, pp. 202–217. Springer, Heidelberg (2015). doi:10.1007/978-3-319-19084-6_18

15. Mascia, F., Birattari, M., Stützle, T.: Tuning algorithms for tackling large instances: an experimental protocol. In: Nicosia, G., Pardalos, P. (eds.) LION 2013. LNCS, vol. 7997, pp. 410–422. Springer, Heidelberg (2013). doi:10.1007/978-3-642-44973-4_44

16. Mu, Z.: Analysing the empirical time complexity of high-performance algorithms for SAT and TSP. Master's thesis, University of British Columbia, Vancouver, Canada (2015)

17. Mu, Z., Hoos, H.H.: Empirical scaling analyser: an automated system for empirical analysis of performance scaling. In: GECCO 2015, Companion, pp. 771–772 (2015)
18. Mu, Z., Hoos, H.H.: On the empirical time complexity of random 3-SAT at the phase transition. In: Proceedings of IJCAI 2015, pp. 367–373 (2015)
19. Nagata, Y.: Edge assembly crossover: a high-power genetic algorithm for the traveling salesman problem. In: Proceedings of ICGA 1997, pp. 450–457 (1997)
20. Nagata, Y., Kobayashi, S.: A powerful genetic algorithm using edge assembly crossover for the traveling salesman problem. INFORMS J. Comput. **25**(2), 346–363 (2013)
21. Stützle, T., Hoos, H.H.: Analysing the run-time behaviour of iterated local search for the travelling salesman problem. In: Hansen, P., Ribeiro, C. (eds.) Essays and Surveys on Metaheuristics, pp. 589–611. Kluwer Academic Publishers, New York (2001)
22. Styles, J., Hoos, H.: Ordered racing protocols for automatically configuring algorithms for scaling performance. In: Proceedings of GECCO 2013, pp. 551–558. ACM (2013)
23. Styles, J., Hoos, H.H., Müller, M.: Automatically configuring algorithms for scaling performance. In: Hamadi, Y., Schoenaur, M. (eds.) LION 6. LNCS, vol. 7219, pp. 205–219. Springer, Heidelberg (2012). doi:10.1007/978-3-642-34413-8_15

Requests Management for Smartphone-Based Matching Applications Using a Multi-agent Approach

Gilles Simonin$^{(\boxtimes)}$ and Barry O'Sullivan

Insight Centre for Data Analytics, Department of Computer Science,
University College Cork, Cork, Ireland
{gilles.simonin,barry.osullivan}@insight-centre.org

Abstract. We present a new multi-agent approach to managing how requests are sent between users of smartphone-based applications for reaching bi-lateral agreements. Each agent is modelled as having a selfish behaviour based on his preferences and an altruist behaviour with respect to the links between the agent and his neighbours. The objective is to maximise the likelihood of an acceptable match while minimising the burden on the users due to unnecessary messaging. We provide a dynamic algorithm using this architecture and we present an empirical evaluation with various mathematical models of user behaviour and altruism. The evaluation shows that our approach can reduce the risks of rejections and the number of requests while increasing the likelihood of acceptable matches.

1 Introduction

Recently the use of smartphone-based applications has received attention due to the emergence of a number of new online optimisation problems that involve reaching bi-lateral agreements, such as in ride-sharing applications. The majority of such apps require sending a large number of requests or notifications to users. A typical goal of providers of these apps is to reduce the inconvenience to users by minimising the number of requests sent. The interaction with users represents the single greatest challenge in these applications. A social model/approach cannot be used for this problem because of the lack of information from the users. The system knows only the positive/negative answers from each user who received a request, and when a response was received. The system can only decide dynamically when it must send a new request. We can only simulate realistic behaviour of users to help the system in his decision.

Several robotics systems that propose an easy dialogue with, and a limited number of interactions for, each user have emerged in the literature. Many models and self-satisfaction architectures have also been developed and proposed for reactive robotic multi-agents systems [2,4,7]. These authors focus their models on communication and cooperation between learning situated agents. Robotic applications require distributed solutions and adaptive cooperation techniques.

© Springer International Publishing AG 2016
P. Festa et al. (Eds.): LION 2016, LNCS 10079, pp. 173–186, 2016.
DOI: 10.1007/978-3-319-50349-3_12

The agents learn to select behaviours that are well adapted to their neighbours' activities. Multi-agent systems for solving complex combinatorial problems are often formulated as Distributed Constraint Optimisation Problems (DCOPs) [3,5,6]. These authors propose search algorithms for DCOPs that are analogous to a reactive multi-agent architecture. These methods and algorithms are not applicable in our case since we are concerned with predicting an appropriate time to send a request between users. For our problem, we only need the principle of distributed communication but not the cooperation aspect. Our agents will not act to help a global mission or achieve a goal. We are using a self-satisfaction architecture and a model of altruistic behaviour for the purpose of simulating the potential stress of each user and helping the system to decide when and whom to send a new request.

The remainder of the paper is organised as follows. Section 2 presents our requests problem and how it relates to the Satisfaction-Altruism-based architecture for the multi-agent problem. In Sect. 3 we present this architecture for a multi-agent system, along with our hybrid model with the different concepts and functions. Section 4 focuses our study on the specific application of ride-sharing, where we propose a more detailed model for this specific problem. Finally, in Sect. 5 we present experiments with this specific model and compare the effect of varying the critical parameters of the model with an greedy system used.

2   The General Requests Problem

One of the advantages of smartphone-based applications is their ability to reduce the demands placed on users and their ability to reach decisions in a timely manner. In many applications (ride-sharing, on-line service between users, games, sales, etc.) users must find a bi-lateral agreement, e.g. a driver is happy to offer a ride to someone who is happy to accept that offer. Each user sends a request to other users and waits for a positive/negative response. In many applications this leads to a proliferation of notifications and requests sent to and by users. For this reason many automated systems that take decisions on behalf of users have emerged. In these systems, users can only initiate the agreement process, but they cannot control how an agreement is reached.

This requests problem can be modelled as a graph: the set of vertices represents users and we define an edge between two users if there exists a potential bi-lateral agreement, a deal, between them. A weight is assigned to each edge in order to quantify the quality of the potential deal. The edge weight encodes importance, but not likelihood of a match. The objective is to minimise the risk of rejection and the number of notifications/requests sent between users. Therefore, the goal is to select a maximum number of edges (deals) that maximises the sum of the weights. It is obvious that this selection is constrained by the structure of the problem and by the nature of the smartphone app in question. For example, in the ride-sharing problem, we want to select the potential riders in the same car. Therefore, we want three edges from a driver, if we want to fill the car, and one edge for a rider. From a deterministic global solution (see [9]), requests are sent by the system to each user. If a user does not respond quickly, the system has to propose an alternative solution to another potential partner who is waiting to make a deal. However, this alternative partner could be a poor quality match in comparison to the first ones, thus it might be advantageous to allow additional time to the partners already contacted before sending a request to an alternative new partner. The question is when should the system send such a new request? The quality of the solution depends on the reactivity and the type of answers from users. This problem can be seen as an online one in which we need a good protocol to satisfy our objectives.

One can see the requests problem as a Satisfaction-Altruism-based architecture for multi-agent problems [1,8] where each user is represented by an agent. Each agent has a personal behaviour based on his preferences and altruistic behaviour with respect to the links between the agent and his neighbours. Such a behaviour-based approach provides a basis to design each agent's actions over time. This perspective allows us to define agents that are able to evolve in dynamic and partially unknown environments.

3   A Multi-agent Approach

The main principle of our approach is to model, in real time, a realistic behaviour for smartphone users and to manage the automatic systems that control the sending of notifications and requests. This architecture uses three concepts of agent satisfaction: the **personal satisfaction** which measures the level of satisfaction (e.g. stress, wait); the **interactive satisfaction** which describes the benefit of the interaction values between an agent and its neighbours; and the **altruistic reactions** which involve monitoring personal/interactive satisfaction between agents/neighbours and in transforming them into specific actions. When an agent perceives a prior signal of mutual interaction from a neighbour and possess a negative personal satisfaction, it can have an altruistic reaction in order to satisfy the protagonists (agent/neighbour) involved. The general architecture is presented in Fig. 1.

In our setting, actions are the requests sent from a user to his potential partners during a given period of time. When an agent has a negative personal satisfaction and perceives from one of his partners a signal of potential interaction,

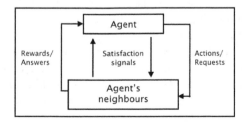

Fig. 1. Illustration of the requests managing architecture

the user can change its action or start a specific one. The interactive satisfactions of neighbours are represented by the weight of edges between potential partners.

3.1 Personal Satisfiability

Each user u (agent) is active between a starting time t_u^0 and an End of Time (EoT_u) and is looking for one/several deal(s), or matches, with other users. A deal is represented as a mutual agreement to a request between the user and one of his partners. We want to simulate the fact that the closer a user gets to his time limit, the greater his stress. We can define a threshold under which the system may send a new request. Under this threshold, we can assume that the user is under stress. Therefore, the personal satisfiability $Sat_u(t)$ measures, at each moment in time, a simulated behaviour/stress of the user u. This function is bounded by $[-1, 1]$ and is decreasing exponentially according to a variation function $\Delta_{Sat_u}(t)$ at each elapsed time point, e.g. every second (see Fig. 2). In delimiting the threshold by 0, every time the function has a negative value, the user's stress is enough to let the system send a new request if there exists a potential partner. In this case, we increase the value $Sat_u(t)$ by 1.

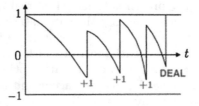

Fig. 2. Illustration of the personal satisfaction. Each spike represents the sending of a new request to a potential partner (we are adding $+1$ to $Sat_u(t)$).

This variation function can change during the horizon time. For example, one can consider the very natural behaviour which is that the closer the user gets to time EoT_u, if a deal has not been made, the higher his stress level is likely to be. Likewise if he has already a deal and waits for another possiblity, his stress decreases more slowly. It is important to limit the decrease in the variation function, nevertheless in worse case the user's satisfiability will be always be equal to -1. This limitation depends on the application and the context; therefore we can define a maximum variation.

We also define a minimum waiting time duration between two new potential actions (sending a new request, that means a negative personal satisfaction). Let T_{min} be the minimum time for the personal satisfaction function to go from 1 to 0. We propose a particular variation function for the personal satisfiability of a user limited by the upper bound $1/T_{min}$.

Definition 1 (Variation Function). *The variation function* $\Delta_{Sat_u}(t)$ *for the personal satisfiability of a user is defined by:*

$$\Delta_{Sat_u}(t) = \min\left(\frac{1}{T_{min}}, \frac{\beta_u}{EoT_u - t}\right). \tag{1}$$

Here β_u is a positive stress impact factor that is dependant on EoT_u. The impact factor is an important parameter in our study. The higher its value, the faster the variation function will decrease. Observe that with $\beta_u \geq 0$ and $t_u^0 < t < EoT_u$, we can bound the variation function, $\Delta_{Sat_u}(t) \in [0, 1/T_{min}]$. From Definition 1 we can give the definition of personal satisfaction as follows.

Definition 2 (Personal Satisfaction). *The personal satisfaction* $Sat_u(t)$ *of an active user u at time t is defined by:*

$$Sat_u(t) = Sat_u(t-1) - \Delta_{Sat_u}(t). \tag{2}$$

A second function, which is somewhat more accurate, could take into account the number of deals the user is already involved in. In this setting his stress/behaviour score will decrease more slowly if he is already partially satisfied. Thus we propose to reduce the variation with the number of the user's deals $NbDeals_u$ as follows.

Definition 3 (Variation Function with Deals). *The variation function with deals* $\Delta^D_{Sat_u}(t)$ *for the personal satisfaction of a user is defined by:*

$$\Delta^D_{Sat_u}(t) = \min\left(\frac{1}{T_{min}}, \frac{\beta_u}{EoT_u - t} \times \frac{1}{NbDeals_u + 1}\right). \tag{3}$$

From the fact that $\beta_u \geq 0$ and $t_u^0 < t < EoT_u$, we can still bound the variation function by $\Delta^D_{Sat_u}(t) \in [0, 1/T_{min}]$. In Sect. 4 we will define a specific personal satisfaction function with different values to describe the impact factor β_u and T_{min} for the ride-sharing version of our general problem.

3.2 Interactive Satisfaction for Partners

Each active user u has a list of potential partners $P = \{p_1, p_2, \ldots, p_k\}$. For each pair $\langle u, p_i \rangle$, we have a weight representing the preference $w_{\overrightarrow{up_i}}$ from u to p_i. Thus we can define a list of weights for the potential partners: $W_{\overrightarrow{uP}} = \{w_{\overrightarrow{up_1}}, w_{\overrightarrow{up_2}}, \ldots, w_{\overrightarrow{up_k}}\}$. Weights are defined on the range $[0, MaxW]$, where $MaxW$ is the highest value. The higher the value of a weight, the better the chances of a deal. In the following, we will call **best partner** the one with the highest weight from the list of the unrequested partners.

The interactive satisfaction $\boldsymbol{IntSat_{up_i}(t)}$ measures, at every moment in time, the potential interaction between user u and his potential partner p_i during the user's time window $]t_u^0, EoT_u[$. This function is bounded by $[0, MaxW]$ and is decreasing differently according to a variation function $\Delta_{IntSat_{\overrightarrow{up_i}}}(t)$ at each elapsed time period, say every second (see Fig. 3).

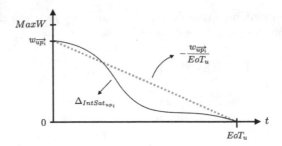

Fig. 3. Illustration of the interactive satisfaction function.

According to the satisfaction of u and his partners, $\Delta_{IntSat_{\overrightarrow{up_i}}}(t)$ will change over time. At every time step, this variation value decreases (resp. increases) if a request has been already sent to p_i (resp. not sent):

- In the decreasing case, the interactive value is decreasing from t_u^0 to EoT_u according to faster/slower variations around the constant decreasing ratio $\frac{w_{\overrightarrow{up_i}}}{EoT_u}$.
- In the increasing case, the interactive satisfaction remains at the same value or can increase slowly according to the partner's personal satisfaction. If his personal satisfaction is negative, the interaction value will increase. The increasing part of the variation function for the interactive satisfaction is upper bounded by a value in $[0, \frac{1}{EoT_u}]$.

Definition 4 (Variation Function for Interactive Satisfaction). *The variation function for the interactive satisfaction between an user u and his partner p_i is defined by:*

$$\Delta_{IntSat_{\overrightarrow{up_i}}}(t) = x_{\overrightarrow{up_i}} \times \left(\frac{w_{\overrightarrow{up_i}} \times \beta_{p_i}(t)}{EoT_u} \right)$$
$$+ [1 - x_{\overrightarrow{up_i}}] \times \left(\frac{\min\{0, Sat_{pi}(t)\}}{EoT_u} \right) \tag{4}$$

$$\text{where } x_{\overrightarrow{up_i}} = \begin{cases} 1, \text{ if a request has already been send to } p_i, \\ 0, \text{ if no request has not been send to } p_i. \end{cases}$$

and $\beta_{p_i}(t)$ is a positive impact factor of stress depending of the partner's personal satisfaction and the time. The higher its value, the faster the variation function will decrease. From (4), we can give the definition of the interactive satisfaction:

Definition 5 (Interactive Satisfaction). *The interactive satisfaction Int $Sat_{\overrightarrow{up_i}}(t)$ of an active user u at time t is define by:*

$$IntSat_{\overrightarrow{up_i}}(t) = IntSat_{\overrightarrow{up_i}}(t-1) - \Delta_{IntSat_{\overrightarrow{up_i}}}(t). \tag{5}$$

A second, more accurate, function could take into account the number of deals already owned by the user or his partners. His/their stress/behaviour will decrease/increase slower if he is already partially satisfied. Thus we propose to reduce the variation based on the number of the user's deals $NbDeals_u$ ($NbDeals_{p_i}$ for the partner p_i) already in place.

Definition 6 (Variation Function with Deals). *The variation function with deals for the interactive satisfaction of a user u and his partner p_i is:*

$$\Delta^D_{IntSat_{\overrightarrow{up_i}}}(t) = x_{\overrightarrow{up_i}} \times \left(\frac{w_{\overrightarrow{up_i}} \times \beta_{p_i}(t)}{EoT_u[NbDeals_u + 1]} \right)$$
$$+[1 - x_{\overrightarrow{up_i}}] \times \left(\frac{\min\{0, Sat_{pi}(t)\}}{EoT_u[NbDeals_{p_i} + 1]} \right). \tag{6}$$

In Sect. 4 we will propose a specific definition of the impact factor $\beta_{p_i}(t)$ for a ride-sharing version of our general problem.

3.3   Altruistic Reaction

According to the updated values for the personal/interactive satisfactions, we can take into account the level of user stress and define a protocol to allow agents to modify their actions. In our problem, the system is authorised to send a new request from a user to his partners when specific conditions are satisfied. We define two conditions required to allow at time t a new request to be sent from a user u:

- The value of the personal satisfaction of u is negative.
- One of his unrequested partners has an interactive satisfaction higher than the current highest one. Specifically, the weight $w_{\overrightarrow{up_i}}$ increased enough and is now higher than the current highest weight from requested partners.

If both conditions are respected, the system is authorised to send a new request. The first condition allows to minimise the delay between two messages. Indeed, when a notification is sent the system increases $sat_u(t)$ by 1. The second condition allows more time to the best partners who might answer positively to the user u.

We are interested in finding a global solution over all users in our graph. When the system receives an authorisation to send a new request for a user u (vertex) to p_i, that means the edge $\{u, p_i\}$ can be selected in the global solution. If the computation of a global solution contains this edge, the system can send a new request to p_i from u.

From the point of view of the model, the system gives a reward to $Sat_u(t)$. This bonus can depend of the number of deals required by the user $MaxNbDeals_u$, and thus $Sat_u(t)$ increases by $1/MaxNbDeals_u$.

3.4 The Algorithm

We present the algorithm that monitors a user and his partners using the three multi-agent concepts, presented above, over time in order to reduce the number of requests sent while also minimising the risk of rejection (requests sent to partners with too low a weight).

Data: A user u, his partners list $P = \{p_1, \ldots, p_k\}$,
His bounds t_u^0 and EoT_u, and a boolean $Change$

```
1  begin
2  |   Request sent to the best partner;
3  |   while t_u^0 < t < EoT_u do
4  |   |   Change ← False;
5  |   |   while Not Change do
6  |   |   |   Values updated;
7  |   |   |   if Negative answer received then
8  |   |   |   |   Remove the partner;
9  |   |   |   if Positive answer received then
10 |   |   |   |   if Deal accepted by u then
11 |   |   |   |   |   Sat_u(t) ← 1;
12 |   |   |   |   |   NbDeals_u ← NbDeals_u + 1;
13 |   |   |   |   |   if u fully satisfied then
14 |   |   |   |   |   |   End of process;
15 |   |   |   |   end
16 |   |   |   |   Remove the partner;
17 |   |   |   if Altruism reaction for one partner p_i then
18 |   |   |   |   Change ← True;
19 |   |   |   end
20 |   |   |   t ← t + 1;
21 |   |   end
22 |   |   New Global Solution with 1 more edge from u;
23 |   |   if the edge {u, p_i} is selected in the solution then
24 |   |   |   Send a new request to p_i;
25 |   |   |   Sat_u(t) ← Sat_u(t) + 1;
26 |   |   end
27 |   end
28 end
```

Algorithm 1: Monitoring of requests sending for a user u.

Algorithm 1 comprises a while loop inside another general loop. At Line 3, we start a general loop on the time windows of user u. This while loop represents the active time of the agent/user u, during this time interval the system monitors the personal/interactive satisfaction functions.

In Lines 5–21 we are in the Decision-Action loop. This while loop involves first computing and updating, after every time unit, the user's personal satisfaction Sat_u and the interactive satisfactions of his partners. Second we check every case

Table 1. Example of the system between a user u and three partners. The boxes represent the sending of a request.

Time	$w_{\overrightarrow{up_1}}$	$w_{\overrightarrow{up_2}}$	$w_{\overrightarrow{up_3}}$	Sat_u
$t = 0\,$s	$\boxed{10}$	5	3	0
$t = 600\,$s	7.458	6.458	4.458	-0.49
$t = 900\,$s	6.09	$\boxed{7.091}$	5.091	$-0.836 \rightarrow 0.164$
$t = 1200\,$s	4.53	5.53	5.53	-0.406
$t = 1320\,$s	$\underline{NO} \rightarrow 0$	4.855	$\boxed{5.655}$	$-0.721 \rightarrow 0.279$
$t = 1380\,$s	0	$\underline{YES}$	5.342	$0.062 \rightarrow 1$

when a partner answers negatively or positively. If we have a positive answer and a deal with the user, it is necessary to increase values and to check if the user is fully satisfied. If he need more deals, we continue the process, else the system stops the monitoring. Third, we test if the required conditions to send another request are satisfied, that means the personal satisfaction is negative and there exists a partner who has not yet received a request and has a weight higher than the best weight from partners already requested.

In Lines 22–26 we compute a new global solution with the new constraints and we send a new request if it's relevant. The new request leads to a lower level of stress for the user and to an increase in $Sat_u(t)$.

An example of the algorithm in operation is presented in Table 1 for a user u with three partners $\{p_1, p_2, p_3\}$ during the time window $[0, 1500]$. The ending of the process can advance if the user is fully satisfied. We monitor user behaviour every five minutes, with the first request sent to p_1 with the highest weight. After 900s, the required conditions are satisfied to send a new request to the new best partner. At 1320s, the request by the partner p_1 is declined and leads two actions: p_1 is removed from the list of potential partners and a new request can be sent if the conditions are satisfied. At $1380s$, p_2 answer positively to u. If u accepts the deal and is fully satisfied, the process stops and we remove him.

4   The Ride-Sharing Case

In this section, we focus on the specific case of our problem where users are drivers and riders, and where the smartphone-based application is for ride-sharing providing an automatic system to users to propose matchings through bi-lateral agreements. Each driver can have three riders in his car, and the system tries to find a match for each user u on the time windows $[t_u^0, EoT_u]$. Using data collected, in association with an industry partner, from the application users and their comments, we provide the best functions representing their potential stress and behaviour. The system automatically computes the matching and request messages. Users do not know what the system is doing (sending requests to their partners). Therefore, we cannot model and represent their true behaviour and stress. We use this partial model to create a good protocol to the system to know what to do and when.

4.1 Greedy Model

Previously the application offered by our industry partner was using a greedy strategy consisting in sending repeatedly more requests to other partners if the first ones did not receive an answer. This method provides solutions for each user but the quality can be poor, and the number of requests sent increases exponentially. For a large number of users, it can be critical in terms of communication and management. This system does not take into account the quality of the weights. If none of the first partners answers quickly, a new one can be selected from a new request and his fast positive answer can lead to a matching of bad quality (low weight). In the experimental section, we will use the minimal delay required to send a new request in the greedy system as a lower bound for the value T_{min}.

4.2 Multi-agent Model

We developed a multi-agent model for the ride-sharing requests problem as presented in the previous section. As one can see in the algorithm, the decision to send a new request depends on the values of the personal and interactive satisfaction functions. We define a specific function for each satisfaction for the ride-sharing version. Their update leads to an altruistic reaction from users to their partners. This section focuses on the specificities of these functions and on their implications for the objectives to control the risk of rejection while minimising the number of requests. In the following, we present for each function the different parameters which will vary in the simulations.

4.3 Updated Values: Personal Satisfaction

In Definition 1 we gave the definition of the personal satisfaction variation $\Delta_{Sat_u}(t)$ for a user u (and the deals version in Definition 3). This function depends on a stress impact factor β_u which depends on EoT_u. The higher its value, the faster the variation function will decrease.

A way to compute this impact factor involves choosing the number of times that we want the function to arrive at -1. For this, one can imagine that if every time the function arrives at -1 we add 2, we could compute the number of times the personal satisfaction function is reaching -1. Using the lower bound variation $-1/T_{min}$, we can compute this number from EoT_u. From the fact that we need $2 \times T_{min}$ to decrease the function by -2 (see Fig. 4), and that the personal satisfaction function is reaching -1 at least k times, we have the following definition.

Definition 7 (Impact Factor). *The impact factor β_u is define by:*

$$\beta_u = \max\left(1, \frac{EoT_u}{2 \times k \times T_{min}}\right). \tag{7}$$

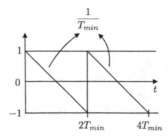

Fig. 4. Illustration of the lower bound.

From Definition 7 one can see that the only parameters that vary for the personal satisfaction function (with or without the deals) are k and T_{min}.

4.4 Updated Values: Interactive Satisfaction

In Definition 4 we defined the interactive satisfaction function $\Delta_{IntSat_{\overrightarrow{up_i}}}(t)$ for a user u and his partners (and the deals version in Definition 6). This function depends of a stress impact factor $\beta_{p_i}(t)$ which depends on the user u and his partner p_i. The higher its value, the faster the variation function will decrease. We tried several types of behaviour model and functions for the interactive satisfaction variation, and we kept the most relevant according to our problem.

Definition 8 (Ride-Sharing Requests Problem). *In the Ride-Sharing requests problem, the stress impact factor depending on the user u and his partner p_i is equal to $\beta_{pi} = \max\{0, 1 - Sat_u - Sat_{p_i}\} \in [0,3]$. The greater the level of stress, the faster other partners will receive a request. Therefore, the interactive satisfaction varies according to the personal satisfactions from protagonists at rate*

$$-\frac{w_{\overrightarrow{up_i}} \times (1 - Sat_u - Sat_{p_i})}{EoT_u}.$$

5 Empirical Study

As one can see in Algorithm 1, the decision to send a new request depends on the values of the personal and interactive satisfaction functions. Their update leads to an altruistic reaction from users to their partners.

For the evaluation, we have been working with an industry partner in the area of ride-sharing. Our model was proposed to them. They implemented it in full, except for the notion of *personal satisfaction*. They deployed the application over a number of days to measure the impact on system performance. The results were very good, significantly reducing the number of rejections and the number of requests sent to users. However, we explained to them the necessity in also using the personal satisfaction parameter to ensure their service could scale as the number of uses increases over time.

We do not present this evaluation in the paper because (a) of confidentiality issues, and (b) because the number of users using the apps were not large enough to see extreme cases and prove our model. This is why we rely on simulation on huge numbers of users and varying several parameters.

Our experiments focus on the specificities of these functions and on the objectives to control the risk of rejection while minimising the number of requests. In the following, we present the different parameters which will vary in the simulations.

5.1   Variation of Parameters

As seen in Definition 7, we define a specific function to describe the personal satisfaction. This function depends on three parameters k, T_{min} and EoT_u. By definition, the higher the parameters T_{min} and k are, the longer the time required to send a new request. We focus our simulation around these three parameters, in order to identify the cases where the reduction of the number of requests or rejections is the most visible. The variations are the following:

- k will vary between 2 and 6. It represents the number of time that the personal satisfaction is decreasing by -1 at rate $-1/T_{min}$ between t_u^0 and EoT_u;
- T_{min} will vary between 300 s and 600 s. In the greedy system, the time between each new request was 300 s. From this information, we put this value as a lower bound and to observe its impact on our model by increasing it;
- For each user, the horizon EoT_u will take a commun value of 3600 s, 7200 s or 10800 s. An analysis at 1 h can show the limit of our technique if there are not enough answers, whereas at 2/3 h the system should be more robust;
- The density of the graph will also vary to check the impact of a large number of partners. We will focus our simulation on a small density of 30% or a larger one of 60%.

5.2   Results and Discussion

In Fig. 5, we present the simulation results between the **Greedy system** where requests are sent every T_{min} minutes to a new partner and our **Multi-Agent model** (M-A). We considered randomly generated graphs and simulated our on-line protocol to manage the sending of requests, then compared the results according the different parameters of the functions introduced in this paper. We computed the percentage of benefit/loss between the two systems. All experiments were run on Intel quad-core $i7$ running Mac-OS X 10.9.5 with 16 GB of RAM. We present for each instance (depending on a small number of critical parameters) the number of rejections, requests sent, sum of weights for deals selected and the number of satisfied users.

One can see quickly that the number of requests sent decreases strongly. For each instance, there are in average a reduction of 700 requests sent in the new model. One can observe that in each instance where we increase the value of T_{min} or k, the number of rejections is reduced in the new model.

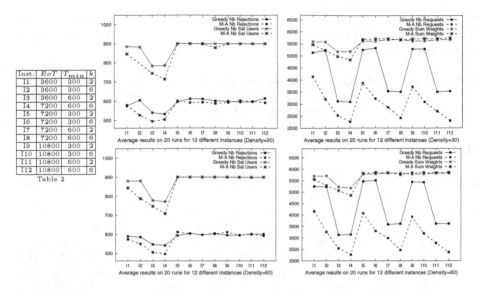

Inst.	EoT	T_{min}	k
I1	3600	300	2
I2	3600	300	6
I3	3600	600	2
I4	7200	600	6
I5	7200	300	2
I6	7200	300	6
I7	7200	600	2
I8	7200	600	6
I9	10800	300	2
I10	10800	300	6
I11	10800	600	2
I12	10800	600	6

Table 2

Fig. 5. Comparaison between the Greedy model and the Multi-Agent one using different values of k, T_{min}, EoT and density

The good news is that our approach seems robust in terms of its ability to reduce the number of requests sent in all cases. When the number of satisfied users is close between the two versions, the sum of weights is always better for our system. But for smaller values of EoT_u, the number of satisfied users is low for our system, since it does not have enough time to find a matching to everyone. This leads to a reduction of the number of deals obtained, and thereby of the sum of weights. By computing a ratio between the number of deals realised and the sum obtained, the matching obtained has a low level of risk as shown by the average value (weight) of each deal done for our system. Note that in real-world settings, users will start the application most often and the matching process three or four hours in advance. Finally, as the graph density increases the performance of our proposed method dramatically increases and describes well what we shown previously.

6 Conclusions

We have presented a new multi-agent approach to managing the requests made between users of applications for reaching bi-lateral agreements. The objective is to maximise the likelihood of acceptable matches while minimising the burden on the users due to unnecessary requests being sent. We presented a general model for this kind of smartphone-based matching problem, with two specific functions to describe the satisfaction/stress behaviours and to limit spamming. We also defined and experimented with a specific model for the Ride-Sharing Requests Problem. Our results showed that this system succeeded in significantly and robustly reducing the number of requests sent even with large variation between the parameters. Their increase leads to a reduction the number of rejections and an increase in the sum of weights in the solution.

Acknowledgement. This paper has emanated from research conducted with the financial support of Science Foundation Ireland (SFI) under Grant Number SFI/12/RC/2289 and from industry partner Carma (https://www.gocarma.com).

References

1. Balch, T., Arkin, R.: Communication in reactive multi-agent robotic systems. Autonom. Robots **1**, 27–52 (1994)
2. Chapelle, J., Simonin, O., Ferber, J.: How situated agents can learn to cooperate by monitoring their neighbors' satisfaction. In: ECAI, pp. 68–72 (2002)
3. Grinshpoun, T., Grubshtein, A., Zivan, R., Netzer, A., Meisels, A.: Asymmetric distributed constraint optimization problems. J.Artif. Intell. Res. **47**, 613–647 (2013)
4. Hilaire, V., Gruer, P., Koukam, A., Simonin, O.: Formal driven prototyping approach for multiagent systems. Int. J. Agent-Orient. Softw. Eng. **2**(2), 246–266 (2008)
5. Hirayama, K., Yokoo, M.: The distributed breakout algorithms. Artif. Intell. **161**(1–2), 89–115 (2005)
6. Maheswaran, R.T., Pearce, J.P., Tambe, M.: A family of graphical-game-based algorithms for distributed constraint optimization problems. In: Scerri, P., Vincent, R., Mailler, R. (eds.) Coordination of Large-Scale Multiagent Systems, pp. 127–146. Springer, Heidelberg (2006)
7. Mataric, M.J.: Behaviour-based control: examples from navigation, learning, and group behaviour. J. Exp. Theor. Artif. Intell. **9**(2–3), 323–336 (1997)
8. Lucidarme, P., O.S., Liégeois, A.: Implementation and evaluation of a satisfaction/altruism based architecture for multi-robot systems. In: IEEE International Conference on Robotics and Automation, vol. 1, pp. 1007–1012 (2002)
9. Simonin, G., O'Sullivan, B.: Optimisation for the ride-sharing problem: a complexity-based approach. In: ECAI, pp. 831–836 (2014)

Self-organizing Neural Network for Adaptive Operator Selection in Evolutionary Search

Teck-Hou Teng$^{(\boxtimes)}$, Stephanus Daniel Handoko, and Hoong Chuin Lau

School of Information Systems, Singapore Management University,
Singapore, Singapore
{thteng,dhandoko,hclau}@smu.edu.sg

Abstract. Evolutionary Algorithm is a well-known meta-heuristics para-digm capable of providing high-quality solutions to computationally hard problems. As with the other meta-heuristics, its performance is often attributed to appropriate design choices such as the choice of crossover operators and some other parameters. In this chapter, we propose a continuous state Markov Decision Process model to select crossover operators based on the states during evolutionary search. We propose to find the operator selection policy efficiently using a self-organizing neural network, which is trained offline using randomly selected training samples. The trained neural network is then verified on test instances not used for generating the training samples. We evaluate the efficacy and robustness of our proposed approach with benchmark instances of Quadratic Assignment Problem.

1 Introduction

Evolutionary algorithms (EAs) such as genetic algorithm (GA) and memetic algorithm (MA) have been widely used for solving NP-hard problems [8,10,21]. Using EA, a population of chromosomes representing the candidate solutions to an NP-hard problem are evolved over a number of generations. The aim of the search process is to optimize some objective value. These evolutionary operators constitute the algorithmic core of the evolutionary search (ES). Therefore, the quality of the evolutionary operators is critical for the performance of EAs [5].

Many genetic operators are known, and the applicability of these operators vary across problems. And this is also true even among different instances in the same problem [4]. The success of an evolutionary operator depends (among other things) on the characteristics of the fitness landscape of the problem. This information, however, is usually not readily available. Although the literature might provide comparisons between operators on certain problem instances, an evolutionary search algorithm designer is left with a difficult choice of operator selection when designing an evolutionary search algorithm for a new problem. These technical choices quite often lead to the design of ad-hoc methods to solve specific problem instances.

This work is funded by the National Research Foundation, Singapore under its Corp Lab @ University scheme.

© Springer International Publishing AG 2016
P. Festa et al. (Eds.): LION 2016, LNCS 10079, pp. 187–202, 2016.
DOI: 10.1007/978-3-319-50349-3_13

Automated tuning algorithms [3,15] can be used to adjust parameters of evolutionary search prior to testing it on new problems. This does not, however, exploit the fact that the usefulness of the operators often changes during the evolutionary search. Besides, dependencies between operators can exist and the interaction of multiple operators might lead to better results than when used alone. During the search for a solution to the problem, the application rates of the variation operators can be controlled using recent performance of the operators. Such methods are categorically referred to as Adaptive Operator Selection (AOS) [7]. At each iteration, AOS provides an adaptive mechanism for selecting suitable variation operators during the evolutionary search. Recent works [12,19] have proposed such adaptive mechanisms for general evolutionary search with possibly many variation operators whose behavior may be unknown, giving rise to uncertainty.

Against this backdrop, we propose a formulation of a Markov Decision Process (MDP) [23] for selecting crossover operators adaptively. Uncertainty over the outcomes of applying a certain crossover operator suggests that MDP can be an alternative approach for performing AOS. Furthermore, MDP uses concepts of states and transitions, which is in line with the fact that the best choice of the operator is very much dependent on, among others, the multi-modality of the fitness landscape and the diversity levels of the population.

We claim the following contributions for this chapter.

1. We formulate a MDP with continuous states and discrete actions for AOS with discrete action space. Decisions are made in discrete time. The state features are generic in that they are problem-independent, i.e., they represent common features found in most ES methods for solving optimization problems.
2. We proposed the use of a self-organizing neural network within such a MDP. In this work, our self-organizing neural network is trained offline using training samples of problem instances to discover action policies.
3. We compare and contrast the proposed neural network approach to several benchmark AOS methods on QAP instances. Results of our experiments show our approach have the best performance outcome for the optimization of QAP instances using evolutionary search.

This chapter continues with Sect. 2 where the related works are surveyed. Section 3 formulates AOS as MDP with continuous states and discrete actions. Components of the MDP and our proposed neural network approach are also described. The experiments and the results are presented in Sect. 5. Section 6 concludes and suggests some extensions to this work.

2   Related Works

Reactive Tabu Search (RTS) [2] is a state-based adaptive search method. The tabu list length is adjusted based on whether or not repeated solutions appear or a new best-so-far solution is found. Enhancement to the original RTS [1] features a Markov Decision Process (MDP) model with continuous states. The states include among others the current objective function value.

In evolutionary computing context, reinforcement learning (RL) has been used to adapt the step size in $(1 + 1)$ evolutionary strategy [22] while solving continuous optimization problem. Furthermore, reinforcement learning has also been used to adapt crossover probability, mutation probability, and population size based on some continuous states including best fitness, mean fitness, standard deviation, breeding success number, average distance from the best, number of evaluations, and fitness growth [9].

While these works deal with the adaptation of numerical parameters, more recently, a discrete-state RL has been proposed to deal with the adaptation of categorical parameters [14] in the context of memetic algorithm. Specifically, the crossover operator is the parameter being adapted.

This work differs from the above works in two ways. Firstly, we propose the use of continuous state features for learning action policies used for AOS. Secondly, we learn the action policies offline using a self-organizing neural network. This is in line with the spirit of [13] which found the advantage of offline-tuning the parameters of an algorithm for online adaption.

3   Adaptive Operator Selection as Markov Decision Process

Adaptive Operator Selection (AOS) is used in several context. In [18], AOS is employed in the context of multi-objective evolutionary algorithms. In [17] and [29], AOS is used to select local search operators and meta-heuristics algorithms, respectively. In this chapter, AOS seeks out crossover operators best suited to the current stage of evolutionary search.

At times, a crossover operator can degrade the overall quality of a population in the immediate steps but may lead to a global optimum eventually. Therefore, it is essential to choose the crossover operators strategically to optimize the cumulative reward over a sequence of heuristic search. To employ the exploratory operators at the right stage of the search, we propose to model AOS as a markov decision process (MDP) with continuous states and discrete actions.

We define such a MDP as a 5-tuple $< \mathbf{S}, \mathbf{A}, T, R, \gamma >$ [30] where

- $\mathbf{S}$ is the set of states, also known as the state space
- $\mathbf{A}$ is a finite set of action choices, also known as the action space
- $T : S \times A \times S$ is the transition function specifying the probability of going from state s_1 to state s_2.
- $R : S \times A \times S$ is the reward function giving a numerical reward value $r \in [0.0, 1.0]$ received at time t after transiting from state s_1 to state s_2 as a result of applying action choice a at time $t - 1$
- γ is the discount factor of the feedback signals recieved over time

The state space, action space and reward function of this AOS and the function approximator are described below.

3.1 The State Space

A total of 15 features is defined as the state features. The state features are dichotomized into fitness landscape features and parent-oriented features. The eight continuous and a binary fitness landscape features are described below.

1. Restart $rs \in \{0, 1\}$: At each generation, if $div(p) < div_t(p)$ where $div_t(p)$ is a threshold of population diversity or when the average fitness has remained unchanged for g_{es} generations, the population will be restarted. All the chromosomes except the best chromosomes are mutated to give $dist(c_o, p_{c1}) > \delta_{dist}$ and $dist(c_o, p_{c2}) > \delta_{dist}$ where δ_{dist} is the percentage of differences among the parent chromosomes.

2. Population Diversity $div(P) \in [0.0, 1.0]$: Different crossover operators may be preferred at different population diversity. The population diversity tracks the number of differences among the corresponding elements of a pair of chromosomes. It is normalized by the size of a problem instance n and the number of pairs of chromosomes n_{pair}.

3. Population Fitness Diversity $fdiv(P) \in [0.0, 1.0]$: The population fitness diversity tracks the fitness differences between a pair of chromosomes. It is normalized by the number of pairs of chromosomes.

4. Proportion of new best offspring $N_{best} \in [0.0, 1.0]$: Offsprings are created by crossover and mutation processes. An offspring is a new best offspring when its fitness level f_o is better than the best fitness level from the previous generation f_{best}. In this work, an offspring is considered to be better than the best offspring from the previous generations when $f_o < f_{best}$. N_{best} is normalized using the number of crossover operations N_{co} and the number of mutation operation N_{mu}.

5. Proportion of improving offspring $N_{imp} \in [0.0, 1.0]$: An offspring is improving when $f_o < f_{better}$ where f_{better} is the fitness level of the better parent chromosome c_{better}. N_{imp} tracks the number of improving offspring and is normalized using the number of crossover operations N_{co}.

6. Proportion of worsening chromosomes $N_{wrs} \in [0.0, 1.0]$: An offspring is worsening when $f_o > f_{worse}$ where f_{worse} is the fitness level of the worse parent chromosome c_{worse}. N_{wrs} tracks the number of worsening offsprings and is normalized using the number of crossover operations N_{co}.

7. Proportion of equal quality offspring $N_{eql} \in [0.0, 1.0]$: It may also be possible that $f_{better} < f_o < f_{worse}$. Such offspring is considered to be of equal quality to the parent chromosomes. N_{eql} tracks the number of equal quality offsprings and is normalized using the number of crossover operations N_{co}.

8. Amount of improvements $\Delta_{imp} \in [0.0, 1.0]$: The amount of improvement Δ_{imp} is recorded as $\frac{f_{better} - f_o}{f_{better}}$. It is aggregated over N_{co} crossover operations in a generation. Δ_{imp} is normalized using N_{co}.

9. Amount of worsening $\Delta_{wrs} \in [0.0, 1.0]$: The amount of worsening to the fitness level Δ_{wrs} is recorded as $\frac{f_o - f_{worse}}{f_{worse}}$. It is aggregated over N_{co} crossover operations in a generation and then normalized using N_{co}.

An offspring is a product of the parent chromosomes after crossover operation. Therefore, the following parent-oriented features are included as the observable features.

1. Normalized distance between parent chromosomes $dist(c_{p1}, c_{p2}) \in [0.0, 1.0]$: This is a measure of the number of differences between the corresponding elements of c_{better} and c_{worse}. $dist(c_{worse}, c_{better})$ is normalized using the size of the given QAP instance n.

2. Normalized fitness gap between parent chromosomes $fgap(c_{p1}, c_{p2}) \in [0.0, 1.0]$: This is a measure of the amount of difference between the fitness levels of c_{worse} and c_{better}, i.e., $\frac{f_{worse} - f_{better}}{f_{worse}}$.

3. Mean distance of c_{better} with population $dist(c_{better}, P) \in [0.0, 1.0]$: The number of differences between the corresponding elements of c_{better} and chromosome c_i where $c_i \neq c_{better}$ and $c_i \neq c_{worse}$ is tracked as part of the observable feature. $dist(c_{better}, P)$ is derived using $\frac{1}{p-2} \sum_{i=0}^{p} dist(c_{better}, c_i)$.

4. Mean distance of c_{worse} with population $dist(c_{worse}, P) \in [0.0, 1.0]$: The number of differences between the corresponding elements of c_{worse} and c_i where $c_i \neq c_{better}$ and $c_i \neq c_{worse}$ is tracked as part of the observable feature. $dist(c_{worse}, P)$ is derived using $\frac{1}{p-2} \sum_{i=0}^{p} dist(c_{better}, c_i)$.

5. Mean fitness gap of c_{better} with population $fgap(c_{better}, P) \in [0.0, 1.0]$: The amount of differences between the fitness levels of c_{better} with c_i where $c_i \neq c_{better}$ and $c_i \neq c_{worse}$ is also tracked as part of the observable feature. $fgap(c_{better}, P)$ is derived using $\frac{1}{p-2} \sum_{i=0}^{p} fgap(c_{better}, c_i)$.

6. Mean fitness gap of c_{worse} with population $fgap(c_{worse}, P) \in [0.0, 1.0]$: Similarly, the amount of differences between the fitness levels of c_{worse} and c_i where $c_i \neq c_{better}$ and $c_i \neq c_{worse}$ is also tracked as part of the observable feature. $fgap(c_{worse}, P)$ is derived using $\frac{1}{p-2} \sum_{i=0}^{p} fgap(c_{worse}, c_i)$.

3.2 The Action Space

The action choices are the crossover operators $c \in \Gamma$ where Γ is the set of all crossover operators. This implies Γ is equivalent to action space A. Four crossover operators from [13] are used here.

1. Cycle crossover (CX): This crossover operator first include the elements of the chromosome common to parent chromosomes c_{p1} and c_{p2} for creating offspring c_o. An unassigned element of c_o indexed at j is then chosen. $c_o(j)$ is then given the value at $c_{p1}(j)$, i.e., $c_o(j) = c_{p1}(j)$. A second element of c_o indexed at k is again picked. $c_o(k)$ is then given the value at $c_{p2}(j)$, i.e., $c_o(k) = c_{p2}(j)$.

2. Distance-preserving crossover (DPX): This crossover operator produces c_o equi-distance apart from c_{p1} and c_{p2}. Like CX, this crossover operator first include the elements of the chromosome common to c_{p1} and c_{p2} for creating c_o. The remaining unassigned elements of c_o are given values that is the permutation of c_{p1} and c_{p2}. In this way, c_o will have the same distance to c_{p1} and c_{p2}.

3. Partially-mapped crossover (PMX): This crossover operator chooses randomly indices j and j' where $j < j'$. $c_o(k)$ is then assigned the value at $c_{p1}(k) \forall k \notin [j, j']$ and $c_o(k))$ is assigned the value at $c_{p2}(k) \forall k \in [j, j']$. If c_o is an invalid permutation, then for each $c_o(k)$ and $c_o(z)$ where $j < z < j'$, $c_o(k) = c_{p1}(z)$.
4. Order crossover (OX): Like PMX, this crossover operator also chooses randomly indices j and j' where $j < j'$. $c_o(k)$ is assigned the value at $c_{p1}(k) \forall k \in [j, j']$. The kth unassigned elements of c_o is assigned the value of the kth element of c_{p2} such that $c_{p2}(k) \neq c_o(z)$ for $j \leq z \leq j'$.

3.3 The Reward Function

The feedback signal r_c is derived using a reward function R. Known reward functions differ mainly in the calculation of the credit and the measurement method [7,12,16,28]. Our reward function makes reference to the current best, and the reward is assigned only when the offspring improves over both its parents. The feedback signal r_c is derived using

$$r_c = \min \left\{ \frac{\mathcal{C}_{\text{best}}}{\mathcal{C}_{\text{offspring}}} \text{sign}(\mathcal{C}_{\text{parent}} - \mathcal{C}_{\text{offspring}}), \zeta \right\}$$

where $\zeta \in [0.5, 1.0]$. The $\text{sign}(\cdot)$ function returns 1 if the offspring is better than the fitter parent. Otherwise, 0 is returned. The feedback signal r_c is used in (1) to estimate the long-term value $Q(s, a)$ of choosing the crossover operators.

3.4 Function Approximator

The state space described in Sect. 3.1 is continuous. Other than discretizing the continuous state space, a function approximator can be used to generalize the states. The function approximator used here is a self-organizing neural known as FL-FALCON [26]. Based on the adaptive resonance theory (ART) [6], it can learn incrementally in real time while generalizing on the states without compromising on its prediction accuracy.

Structure and Operating Modes. Seen in Fig. 1, FL-FALCON has a two-layer architecture, comprising an input/output (IO) layer and a knowledge layer. The IO layer has a sensory field F_1^{c1} for accepting state vector $\mathbf{S}$, a motor field F_1^{c2} for accepting action vector $\mathbf{A}$, and a feedback field F_1^{c3} for accepting reward vector $\mathbf{R}$. The category field F_2^c at the knowledge layer stores the committed and uncommitted cognitive nodes. Each cognitive node j has template weights $\mathbf{w}^{ck}$ for $k = \{1, 2, 3\}$.

FL-FALCON operates in one of the following modes. In *PERFORM* mode, the Fusion-ART algorithm is used to select cognitive node J for deriving action choice a for state s. In *LEARN* mode, FL-FALCON learns the effect of action choice a on state s.

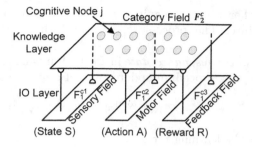

Fig. 1. The FL-FALCON architecture.

The Fusion-ART Algorithm. The Fusion-ART algorithm [25] is used for selecting winning cognitive node J from a collection of committed cognitive nodes. In *PERFORM* mode, cognitive node J is used to derive action choice a. In *LEARN* mode, the weights $\mathbf{w}_J^{ck}$ for $k = \{1, 2, 3\}$ of cognitive node J will be updated using template learning. The performance of FL-FALCON is dependent on the use of suitable vigilance parameters ρ^{ck} for the operating modes.

Using activity vector $\mathbf{x}^{ck}$ for $k = \{1, 2, 3\}$ as the inputs, the process of selecting winning cognitive node J begins with the *code activation* procedure. This procedure derives the choice function T_j^c using

$$T_j^c = \sum_{k=1}^{3} \gamma^{ck} \frac{|\mathbf{x}^{ck} \wedge \mathbf{w}_j^{ck}|}{\alpha^{ck} + |\mathbf{w}_j^{ck}|},$$

where the fuzzy AND operation $(\mathbf{p} \wedge \mathbf{q})_i \equiv min(p_i, q_i)$, the norm $\|.\|$ is defined by $|\mathbf{p}| \equiv \sum_i p_i$ for vectors $\mathbf{p}$ and $\mathbf{q}$, $\alpha^{ck} \in [0, 1]$ is the choice parameters, $\gamma^{ck} \in [0, 1]$ is the contribution parameters and $k = \{1, 2, 3\}$.

The choice function T_j^c is then used for selecting a winning cognitive node J during the *code competition* procedure. This procedure selects cognitive node J using

$$J = \arg \max_j \{T_j^c : \text{for all } F_2^c \text{ node } j\}.$$

The match function m_J^{ck} of cognitive node J is then derived in the *template matching* procedure using

$$m_J^{ck} = \frac{\|\mathbf{x}^{ck} \wedge \mathbf{w}_J^{ck}\|}{|\mathbf{x}^{ck}|} \geq \rho^{ck},$$

where $\rho^{ck} \in [0, 1]$ for $k = \{1, 2, 3\}$ are the vigilance parameters.

A *resonance state* is attained when the vigilance criterion, $\mathbf{m}^{ck} \geq \rho^{ck}$ for $k = \{1, 2, 3\}$, is satisfied. Otherwise, a reset is performed by $T_J^c = 0.0$ and the state vigilance ρ^{c1} is modified in a *match tracking* procedure using

$$\rho^{c1} = \min\{m_J^{c1} + \psi, 1.0\},$$

where ψ is a very small step increment to match function m_J^{c1}.

After that, another winning cognitive node J is determined using the *code competition* procedure. The process repeats until the *vigilance criterion* is satisifed.

The attainment of the *resonance state* in *LEARN* mode leads to the *template learning* procedure. This procedure updates $\mathbf{w}_J^{ck}$ of cognitive node J using

$$\mathbf{w}_J^{ck(\text{new})} = (1 - \beta^{ck})\mathbf{w}_J^{ck(\text{old})} + \beta^{ck}(\mathbf{x}^{ck} \wedge \mathbf{w}_J^{ck(\text{old})}),$$

where $\beta^{ck} \in [0, 1]$ is the learning rate.

The attainment of the *resonance state* in *PERFORM* mode leads to the *activity readout* procedure. The action choice a is obtained by decoding action vector $\mathbf{x}^{c2(new)}$ using

$$\mathbf{x}^{c2(new)} = \mathbf{x}^{c2(old)} \wedge \mathbf{w}_J^{c2}.$$

In this work, FL-FALCON operates in *LEARN* mode when it is trained offline. After that, the trained FL-FALCON operates in *PERFORM* mode to select the crossover operators during evolutionary search.

Offline Training. The action policies π for selecting crossover operators are discovered by presenting training samples to FL-FALCON operating in *LEARN* mode. The training samples are gathered from the experiments conducted based on the problem instances.

To train FL-FALCON, training samples comprising the state features, the selected crossover operator and the estimated value function $Q(s, a)$ are presented to the sensory, motor and feedback fields respectively as $\mathbf{x}^{ck} = \{S, A, Q(s, a)\}$. Depending on the degree of match $\mathbf{x}^{ck}$ has with the existing cognitive nodes, the presented training sample is either learned as a new cognitive node or used to update a matching cognitive node J. One-shot training of FL-FALCON is performed by presenting randomly selected training samples to it.

The trained FL-FALCON is used in *PERFORM* mode for AOS. The choice of crossover operators is made by selecting a winning cognitive node J using the Fusion-ART algorithm. FL-FALCON is used in this way because the scope of this work is limited to testing its generalization capability for AOS.

4   The Operator Selection Policies

This section briefly reviews various strategies for AOS. Based on a recent study [13], there are two probabilistic and one deterministic allocation strategies, namely probability matching (PM), adaptive pursuit (AP) and Multi-Armed Bandit (MAB). In addition, for the sake of comparing with an MDP-based approach, we will review a reinforcement learning (RL) approach [14] for finding the action policy.

4.1   Probabilistic Matching

As the feedback signal R_c is received only after applying crossover operator c, it is difficult to estimate the quality of c using a fixed strategy. A probabilistic strategy (PS) that biases towards the good-performing operators while still allowing unused operators to be chosen would be desirable. Such a PS tuner assigns a probability value proportional to the credits of crossover operators.

At each generation, c is then randomly chosen following this probability distribution. For this purpose, each crossover operator c is assigned a quality $Q_c \in [0.0, 1.0]$ which is updated using $Q_c^{old} + \alpha(R_c - Q_c^{old})$. Using a probability matching (PM) mechanism, the probability P_c of choosing a crossover operator c is then derived using $P_{min} + (1 - |\Gamma|P_{min})\frac{Q_c}{\sum_{c'} Q_{c'}}$ where Γ is the set of all crossover operators considered. The lower threshold P_{min} is included to guarantee that every operator has a chance to be chosen. A drawback of PM is that convergence of P_c can be slow in some cases.

4.2   Adaptive Pursuit

The adaptive pursuit mechanism [27] may be used to speed up convergence of P_c using

$$P_c = \begin{cases} P_c + \beta(P_{max} - P_c), & \text{if } Q_c = \max_{c'} Q_{c'} \\ P_c + \beta(P_{min} - P_c), & \text{otherwise,} \end{cases}$$

where $P_{max} = 1 - (|\Gamma| - 1)P_{min}$. Eventually, P_c of a promising operator converge to P_{max} while P_c of the less promising crossover operators is reduced to P_{min}.

4.3   Multi-armed Bandit

The problem of Multi-armed Bandit (MAB) [24] chooses a slot machine to minimize the regret level over a fixed time horizon. The players are initially unaware of the amount of payoff from the slot machines. The players would begin discovering the payoff of the slot machines. As the players gain more knowledge of the payoff of the slot machines, decisions have to be made on whether to explore further or to exploit the existing knowledge. Well-studied methods to solve MAB problem are widely applied in real world domains such as network routing, financial investment, machine controller and clinical trials.

In the context of AOS [11], each crossover operator can be modelled as a slot machine. The corresponding problem is then to choose crossover operator c that maximize reward over time using $\text{argmax}_{c \in \Gamma} \left\{ \overline{Q_c} + \gamma\sqrt{\frac{2\ln\sum_{c'} n_{c'}}{n_c}} \right\}$ where $\overline{Q_c}$ is the average reward computed since the beginning of the search and n_c is the number of times the crossover operator c is chosen. The second term in the above equation is an exploratory term with similar significance to P_{min} seen in the probabilistic strategies.

4.4   Reinforcement Learning

Following [14], the evolutionary search is modelled as a Markov Decision Process (MDP). This approach estimates the expected Q-value $Q(s,a)$ of each state-action (s,a) pair using

$$Q(s,a) = R(s,a) + \sum_{s'} P(s',a,s) \max_{a'} Q(s',a'), \tag{1}$$

where $P(s',a,s)$ is the probability of entering into new state s' and $R(s,a)$ is the expected reward from executing action a in state s.

Given that $R(s,a)$ and $P(s',a,s)$ are well estimated, $Q(s,a)$ can be estimated using (1). However, in the evolutionary search domain, $Q(s,a)$ is instance-specific value. Significant number of runs on several instances is needed to estimate the model. This is infeasible and has also gone out of the scope of this work. Therefore, a model-free RL approach such as the Q-learning is used to solve the MDP-based search progressively. Using Q-learning, the Q-value of using different action choices can be explored and learned as part of the action policies. Within the MDP framework, $Q(s,a)$ is estimated using Q-learning [24].

$$Q_{t+1}(s_t,a_t) = Q_t(s_t,a_t) + \alpha_t(s_t,a_t)TD_{err},$$

where $\alpha_t(s_t,a_t)$ is the learning rate and TD_{err} is derived using

$$TD_{err} = R_{t+1}(s_t,a_t) + \delta \max_a Q_t(s_{t+1},a) - Q_t(s_t,a_t),$$

where δ is the discount rate. It is used in our search process to reflect the fact that the chance of improvement is reduced as the search continues.

5   Performance Evaluation

This section presents experiments that evaluate and compare the efficacy of our proposed approach on QAP instances [20].

5.1   Experiment Setup

The AOS methods used as the benchmark methods are the Reinforcement Learning approach (RL), Probability Matching (PM), Adaptive Pursuit (AP), Multi-Armed Bandit (MAB), and a *naive* (N) allocation strategies. Due to similar context to [13], the following parameter settings from [13] were used.

PM: $\alpha_{pm} = 0.30, P_{min} = 0.05$; AP: $\alpha_{ap} = \beta = 0.30, P_{min} = 0.05$;
MAB: $\gamma_{mab} = 1.00$;   RL: $\alpha = 0.03, \delta = 0.90$

For the memetic algorithm, a population size p of 40 chromosomes was adopted. At each generation, as many as 20 offsprings were produced using a selected crossover operator c. Each offspring was then refined using the random-order first-improvement 2-opt local search. When the average distance over all pairs of chromosomes is below 10 or the average fitness has remained unchanged for 30 generations, restart was initiated. This was achieved by mutating all but the best chromosomes in the population until each resulting chromosome differed from its parent by as much as $\delta_{dist} = 0.30$ of problem size n. Each run is terminated after 100 generations.

There are two rounds of experiments. The first round of experiments described in Sect. 5.2 is the offline training of the neural network-based function approximator. The second round of experiments described in Sect. 5.3 compares our proposed approach to the benchmark AOS methods.

5.2   Offline Training of Neural Network

The training samples were prepared by experimenting each crossover operator $c \in \Gamma$ on each QAP instance p_i taken from the QAPLib (from bur26a to scr20). Experiments were performed using four crossover operators mentioned in Sect. 3.2. Each experiment based on the pairing of c and p_i was performed for 100 generations. A training sample is produced for each generation of the evolutionary process.

A training sample is assembled as state s comprising 15 state features, action choice a which a particular crossover operator $c \in \Gamma$ and feedback signal r_c. State s and action choice a are taken at time t while the feedback signal r_c is taken at time $t + 1$. This means a training sample $t_s = \{s, a, r_c\}$ is only fully assembled at time $t + 1$. For $t = 0$, state s comprise the initial value of the state features, action choice a comprise the crossover operator c used for that round of experiment based on problem instance p_i.

The raw form of the test results are the aggregated fitness levels of the offsprings. Comparisons are made among the neural networks by ranking the aggregated fitness levels in ascending order. Smaller rank value is given to train-test configuration with lower aggregated fitness level. The ranks of 100 train-test configurations are illustration in Fig. 2.

From Fig. 2, the neural network has the best test performance across the different problem instances when trained using Bur-based problem instances. The worst test performance across the different problem instances is observed when neural network is trained using the Rou-based problem instances. The mean rank of the train instances plotted using the dotted line illustrates the aggregated effect of using different problem instances to train the neural network.

Following the above observation, the performance of neural networks trained differently are ranked and compared directly. Seen in Fig. 3, the mean rank of NN-S is based on the mean fitness values of neural networks trained using the same train and test problem instances. The mean rank of NN-D is based on the mean fitness values of neural networks trained using training samples of problem instances not used for testing. Following this train-test criteria, for each test problem instance $p_i \in \mathcal{P}$, $|\mathcal{P} - p_i|$ other QAP instances can be used for training NN-D. The heterogeneity of training samples are ensured by picking it randomly from those QAP instances. Then, there is NN whose rank is based on

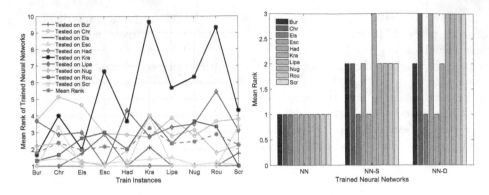

Fig. 2. The mean rank of different train-test configurations.

Fig. 3. The mean rank of the trained neural networks for the QAP instances.

the mean fitness values of the best performing neural network trained using a problem instance selected based on Table 1.

From Fig. 3, NN is observed having the best ranking of 1.0 when compared to NN-S and NN-D. NN-S has the next best ranking of the three trained neural network while NN-D is the trained neural network with the lowest rank. The rank of NN-S imply that the neural network may not necessarily have the best performance when trained using the problem instances used for testing it. The rank of NN-D implies that the neural network cannot be trained arbitrarily and expect it to generalize well. The rank of NN shows that the neural network can generalize well when trained selectively.

Table 1. Mean rank of train-test configurations

Test-train	Bur	Chr	Els	Esc	Had	Kra	Lipa	Nug	Rou	Scr
Bur	N.A.	1.00	1.75	2.88	1.00	2.13	1.00	1.00	1.00	1.75
Chr	3.71	N.A.	4.64	2.93	2.86	2.71	3.86	2.79	3.64	3.79
Els	1.00	1.00	N.A.	1.00	1.00	1.00	1.00	1.00	1.00	1.00
Esc	1.30	1.45	1.20	N.A.	1.30	1.40	1.50	1.05	1.20	1.05
Had	1.00	1.00	1.00	1.00	N.A.	1.00	1.00	1.00	1.00	1.00
Kra	1.67	4.00	2.00	6.67	3.67	N.A.	5.67	6.33	9.33	4.33
Lipa	3.69	2.88	3.00	1.00	4.31	2.75	N.A.	3.50	5.44	3.13
Nug	2.21	2.36	1.29	1.00	1.93	3.93	2.79	N.A.	1.79	2.21
Rou	1.33	1.67	2.67	3.00	2.00	4.00	2.33	3.67	N.A	1.00
Scr	1.00	3.33	1.00	1.00	1.00	4.00	1.00	1.00	1.00	N.A

5.3   Adaptive Operator Selection

Further experiments were conducted to compare the performance of the neural network-based approach with the several benchmark AOS methods. The benchmark AOS methods include Reinforcement Learning (RL), Probability Matching (PM), Adaptive Pursuit (AP), Multi-Armed Bandit (MAB), naïve (N) allocation strategy and the proposed neural network (NN) approach. The RL approach regards AOS as a finite MDP while the neural network-based approach regards AOS as a MDP. The test results using fixed crossover operators such as Cycle crossover (CX), Distance-preserving crossover (DPX), Partially-mapped crossover (PMX) and Order crossover (OX) are also compared in Fig. 5.

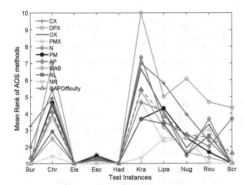

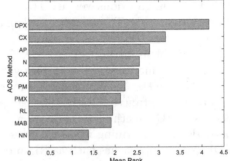

Fig. 4. Mean rank of the AOS methods for the QAP instances

Fig. 5. Aggregation of the mean rank of the AOS methods

The test results are the mean ranks on the mean fitness values of the AOS methods. The mean value is based on 10 runs of each experiment. Test results for the QAP instances mentioned in Sect. 5.2 are presented here. The neural network can be trained using any problem instance that has the lowest mean rank as seen in Table 1.

The illustration of the mean rank of the AOS methods for the QAP instances in Fig. 4 shows a broad spectrum of performance characteristics. The dotted-line plot shows the mean value of the mean rank of the AOS methods and it implies the level of difficulty of the QAP instances. It can be observed that many AOS methods perform quite well for Bur, Els, Had and Scr type of QAP instances, The more challenging QAP instances are Chr, Kra, Lipa, Nug and Rou.

Aggregation of the mean rank of the AOS methods over all QAP instances used here confirms NN to be the best performing approach while DistPres to be the worst performing approach. It turns out that the AP, Cycle and DistPres methods are performing worse than the *naïve* allocation strategy N. The best performance of NN implies the robustness of the neural network for selecting crossover operators during the evolutionary search aimed at optimizing a broad spectrum of QAP instances.

6   Conclusions

Many methods for AOS such as Probability Matching, Adaptive Pursuit, Multi-Armed Bandit and Reinforcement Learning (RL) are known to date. In this chapter, we propose the use of a self-organizing neural network for AOS. In contrast to the RL approach where AOS is formulated as discrete state MDP, our neural network approach is better suited for AOS modelled as MDP with continuous states and discrete actions.

The neural network used is a self-organizing neural network based on the Adaptive Resonance Theory (ART) known for addressing the generalization-specialization dilemma [6]. This work evaluates the generalization capability of such a self-organizing neural network for AOS. To do that, the neural network was trained offline using training samples from the problem instances. Several train-test configurations were used to study the performance of the trained neural network rigorously. To generalize well on the test instances, the results imply that the neural network has to be trained properly using training samples from the suitable problem instances.

The performance of the neural network is compared with the benchmark AOS methods. The test results reveal several characteristics of the AOS methods and the problem instances. Firstly, there are problem instances that can be optimized using any AOS methods. There are also problem instances where several AOS methods are performing worse than the naïve allocation strategy. In such cases, MDP-based approaches such as RL and our proposed neural network approach are among the better performing AOS methods. Such observations imply the feasibility of using MDP-based approaches for AOS in evolutionary search. Last but not least, aggregation of the mean rank of the AOS methods shows the proposed neural network to be sufficiently robust to give the best performance for all the QAP instances.

There are several directions to extend this work. First, we should determine the performance of the neural network in more challenging QAP instances as well as other permutation-based instances such as the flow-shop scheduling problem (FSP). Second, the incremental learning capability of the neural network can be tested using RL during evolutionary search. Doing so, allow the neural network to improve on the learned action policies while performing AOS during evolutionary search. The hypothesis here is that the neural network may able to be converge faster for simpler problem instances and still converge on the more challenging problem instances using available resources.

References

1. Battiti, R., Brunato, M., Campigotto, P.: Learning while optimizing an unknown fitness surface. In: Maniezzo, V., Battiti, R., Watson, J.-P. (eds.) LION 2007. LNCS, vol. 5313, pp. 25–40. Springer, Heidelberg (2008). doi:10.1007/978-3-540-92695-5_3
2. Battiti, R., Tecchiolli, G.: The reactive Tabu search. ORSA J. Comput. 6(2), 126–140 (1994)

3. Birattari, M., Yuan, Z., Balaprakash, P., Stützle, T.: F-Race, iterated F-Race: an overview. In: Bartz-Beielstein, T., Chiarandini, M., Paquete, L., Preuss, M. (eds.) Experimental Methods for the Analysis of Optimization Algorithms, pp. 311–336. Springer, Heidelberg (2010)
4. Boomsma, W.: A comparison of adaptive operator scheduling methods on the traveling salesman problem. In: Gottlieb, J., Raidl, G.R. (eds.) EvoCOP 2004. LNCS, vol. 3004, pp. 31–40. Springer, Heidelberg (2004). doi:10.1007/978-3-540-24652-7_4
5. Candan, C., Goeffon, A., Lardeux, F., Saubion, F.: A dynamic island model for adaptive operator selection. In: Proceedings of 14th GECCO, pp. 1253–1260 (2012)
6. Carpenter, G.A., Grossberg, S.: A massively parallel architecture for a self-organizing neural pattern recognition machine. Comput. Vis. Graph. Image Process. **37**(1), 54–115 (1987)
7. Davis, L.: Adapting operator probabilities in genetic algorithms. In: Proceedings of 3rd International Conference on Genetic Algorithms, pp. 61–69 (1989)
8. de Jong, K.A.: Evolutionary Computation - A Unified Approach. MIT Press, Cambridge (2006)
9. Eiben, A.E., Horvath, M., Kowalczyk, W., Schut, M.C.: Reinforcement learning for online control of evolutionary algorithms. In: Brueckner, S.A., Hassas, S., Jelasity, M., Yamins, D. (eds.) ESOA 2006. LNCS (LNAI), vol. 4335, pp. 151–160. Springer, Heidelberg (2007). doi:10.1007/978-3-540-69868-5_10
10. Eiben, A.E., Smith, J.E.: Introduction to Evolutionary Computing. Springer, Heidelberg (2003)
11. Fialho, Á., Costa, L., Schoenauer, M., Sebag, M.: Dynamic multi-armed bandits and extreme value-based rewards for adaptive operator selection in evolutionary algorithms. In: Stützle, T. (ed.) LION 2009. LNCS, vol. 5851, pp. 176–190. Springer, Heidelberg (2009). doi:10.1007/978-3-642-11169-3_13
12. Fialho, Á., Costa, L., Schoenauer, M., Sebag, M.: Extreme value based adaptive operator selection. In: Rudolph, G., Jansen, T., Beume, N., Lucas, S., Poloni, C. (eds.) PPSN 2008. LNCS, vol. 5199, pp. 175–184. Springer, Heidelberg (2008). doi:10.1007/978-3-540-87700-4_18
13. Francesca, G., Pellegrini, P., Stützle, T., Birattari, M.: Off-line and on-line tuning: a study on operator selection for a memetic algorithm applied to the QAP. In: Merz, P., Hao, J.-K. (eds.) EvoCOP 2011. LNCS, vol. 6622, pp. 203–214. Springer, Heidelberg (2011). doi:10.1007/978-3-642-20364-0_18
14. Handoko, S.D., Yuan, Z., Nguyen, D.T., Lau, H.C.: Reinforcement learning for adaptive operator selection in memetic search applied to quadratic assignment problem. In: Proceedings of GECCO, pp. 193–194 (2014)
15. Hutter, F., Hoos, H.H., Leyton-Brown, K., Stützle, T.: Paramils: an automatic algorithm configuration framework. J. Artif. Intell. Res. **36**(1), 267–306 (2009)
16. Julstrom, A.B.: What have you done for me lately? Adapting operator probabilities in a steady-state genetic algorithm. In: Proceedings of 6th International Conference on Genetic Algorithms, San Francisco, USA, pp. 81–87 (1995)
17. Krempser, E., Fialho, Á., Barbosa, H.J.C.: Adaptive operator selection at the hyper-level. In: Coello, C.A.C., Cutello, V., Deb, K., Forrest, S., Nicosia, G., Pavone, M. (eds.) PPSN 2012. LNCS, vol. 7492, pp. 378–387. Springer, Heidelberg (2012). doi:10.1007/978-3-642-32964-7_38
18. Li, K., Fialho, Á., Kwong, S., Zhang, Q.: Adaptive operator selection with bandits for multiobjective evolutionary algorithm based decomposition. IEEE Trans. Evol. Comput. **18**(1), 114–130 (2013)
19. Maturana, J., Lardeux, F., Saubion, F.: Autonomous operator management for evolutionary algorithms. J. Heuristics **16**(6), 881–909 (2010)

20. Merz, P., Freisleben, B.: Fitness landscape analysis and memetic algorithms for the quadratic assignment problem. IEEE Trans. Evol. Comput. **4**(4), 337–352 (2000)
21. Michalewicz, Z.: Genetic Algorithms + Data Structures = Evolution Programs, 3rd edn. Springer, London (1996)
22. Müller, S., Schraudolph, N.N., Koumoutsakos, P.D.: Step size adaptation in evolution strategies using reinforcement learning. In: Proceedings of IEEE Congress on Evolutionary Computation, pp. 151–156 (2002)
23. Puterman, M.L.: Markov Decision Processes: Discrete Stochastic Dynamic Programming. Wiley, Hoboken (1994)
24. Sutton, R.S., Barto, A.G.: Introduction to Reinforcement Learning, 1st edn. MIT Press, Cambridge (1998)
25. Tan, A.-H.: FALCON: a fusion architecture for learning, cognition, and navigation. In: Proceedings of IJCNN, pp. 3297–3302 (2004)
26. T.-H. Teng and A.-H. Tan. Fast reinforcement learning under uncertainties with self-organizing neural networks. In: Proceedings of IAT, pp. 51–58, December 2015
27. Thierens, D.: An adaptive pursuit strategy for allocating operator probabilities. In: Proceedings of IEEE Congress on Evolutionary Computation, pp. 1539–1546 (2005)
28. Tuson, A., Ross, P.: Adapting operator settings in genetic algorithms. Evol. Comput. **6**(2), 161–184 (1998)
29. Veerapen, N., Maturana, J., Saubion, F.: An exploration-exploitation compromise-based adaptive operator selection for local search. In: Proceedings of 14th GECCO, pp. 1277–1284 (2012)
30. Wiering, M., van Otterlo, M.: Reinforcement Learning: State-of-the-Art. Springer, Berlin (2012)

Quantifying the Similarity of Algorithm Configurations

Lin Xu[1], Ashiqur R. KhudaBukhsh[2], Holger H. Hoos[1(⊠)],
and Kevin Leyton-Brown[1(⊠)]

[1] University of British Columbia, Vancouver, BC, Canada
{xulin730,hhoos,kevinlb}@cs.ubc.ca
[2] Carnegie Mellon University, Pittsburgh, PA, USA
akhudabu@cs.cmu.edu

Abstract. A natural way of attacking a new, computationally challenging problem is to find a novel way of combining design elements
introduced in existing algorithms. For example, this approach was made
systematic in SATenstein [15], a highly parameterized stochastic local
search (SLS) framework for SAT that unifies techniques across a wide
range of well-known SLS solvers. The focus of such work so far has been
on building frameworks and identifying high-performing configurations.
Here, we focus on analyzing such frameworks, a problem that currently
requires considerable manual effort and domain expertise. We propose
a quantitative alternative: a new metric that measures the similarity
between a new configuration and previously known algorithm designs.
We first introduce concept DAGs, a data structure that preserves the
hierarchical structure of configurations induced by conditional parameter dependencies. We then quantify the degree of similarity between two
configurations as the transformation cost between the respective concept DAGs. In the context of analyzing SATenstein configurations, we
demonstrate that visualizations based on transformation costs can provide useful insights into the similarities and differences between existing
SLS-based SAT solvers and novel solver configurations.

Keywords: SAT · Stochastic local search · Algorithm configuration
similarity

1   Introduction

When faced with a new, computationally hard problem to solve, researchers do
not typically want to reinvent the wheel. Instead, it makes sense to draw on
design ideas from existing high-performance solvers. Such an approach can be
made systematic by designing a single, highly parameterized solver that incorporates these different ideas, and then identifying a parameter configuration that
achieves good performance via an automatic algorithm configuration method

---

Lin Xu and Ashiqur R. KhudaBukhsh contributed equally to this work.

© Springer International Publishing AG 2016
P. Festa et al. (Eds.): LION 2016, LNCS 10079, pp. 203–217, 2016.
DOI: 10.1007/978-3-319-50349-3_14

[9,15]. Indeed, many powerful configuration procedures have recently become available to meet this challenge [10,11,14,20]. The types of solvers configured in this way can range from simple heuristic switching [30] to a complex combination of multiple algorithms [28]. While the result is often an algorithm with excellent performance characteristics, it can be difficult to understand such an algorithm, e.g., in terms of how similar (or dissimilar) it is to the existing solvers from which design ideas were drawn—a problem that has received little attention to date by the research community. This work seeks to address this gap. We propose a new metric for quantitatively assessing the similarity between configurations for highly parametric solvers, which computes the distance between two algorithm configurations in two steps. In the first step, the hierarchical structure of algorithm parameters is represented by a novel data structure called a concept DAG. In the second step, we estimate the similarity of two configurations as the transformation cost from one configuration to another, using concept DAGs.

In order to demonstrate the effectiveness of our approach, we investigate the configurations of SATenstein, a well-known, highly parameterized SLS solver. SATenstein has a rich and complex design space with 43 parameters, drawing design ideas from several existing solvers, and is one of the most complex SLS solvers in the literature. We show that visualizations based on transformation costs can provide useful insights into similarities and differences between solver configurations. In addition, we argue that this metric can help to suggest potential links between algorithm structure and algorithm performance.

To our knowledge, there is little previous work directly relevant to the problem of quantifying the similarity of algorithm configurations. Visualization techniques have been used previously to characterize the structure of instances of the well-known propositional satisfiability problem (SAT) [26]; instead, we focus on algorithm design elements. Most similar to our work, Nikolić et al. [21] used the notion of edit distance to automatically quantify algorithm similarity. Our main innovation is to address hierarchies of conditional parameters by saying that edits to lower-level parameters are less significant than edits to higher-level parameters. Conditional parameters are increasingly important as algorithm development shifts to rely on algorithm configuration tools and hence parameter spaces become richer and more complex; see e.g., recent work on assessing parameter importance [13] and finding critical parameters [4].

The remainder of this paper is organized as follows. We present a high-level description of SATenstein in Sect. 2. Next, we describe concept DAGs (Sect. 3) and then present our experimental setup (Sect. 4). We describe our results on quantifying similarities between algorithm configurations in Sect. 5 and then conclude (Sect. 6).

2   SATenstein

In this section, we provide a short description of the overall design of SATenstein. A detailed description of SATenstein is given in [16]. As shown in the high-level algorithm outline, any instantiation of SATenstein proceeds as follows:

1. Optionally execute $B1$, which performs search diversification.
2. Execute either $B2$, $B3$ or $B4$, thus performing WalkSAT-based local search, dynamic local search or G^2WSAT-based local search, respectively.
3. Optionally execute $B5$ to update data structures such as promising list, clause penalties, dynamically adaptable parameters or tabu attributes.

SATenstein consists of five building blocks and eight components, some of which are shared across different building blocks. It has 43 parameters in total. The choice of building block is encoded by several high-level categorical parameters, while the design strategies within each component are determined by a larger number of low-level parameters.

3 Concept DAGs

We now introduce *concept DAGs*, a novel data structure for representing algorithm configurations that preserves the hierarchical structure of parameter dependencies. Our notion of a concept DAG is based on that of a concept tree [31]. We work with a DAG-based data structure because parameters may have more than one parent, where the child is only active if the parents take certain values (e.g., SATenstein's noise parameter *phi* is only activated when both *useAdaptiveMechanism* and *singleClauseAsNeighbor* are turned on). We then define four operators whose repeated application can be used to map between arbitrary concept DAGs, and assign each operator a cost. To compare two parameter configurations, we first represent them using concept DAGs and then define their similarity as the minimal total cost of transforming one DAG into the other.

A *concept DAG* is a six-tuple $G = (V, E, L^V, R, D, M)$, where V is a set of nodes, E is a set of directed edges between the nodes in V such that (V, E) is an acyclic graph, L^V is a set of lexicons (terms) for concepts used as node labels, R is a distinguished node called the root, D is the domain of discourse (i.e., the set of all possible node labels), and M is an injective mapping from V to L^V that assigns a unique label to every node. A parameter configuration can be expressed as a concept DAG in which each node in V represents a parameter, and each directed edge in E represents the conditional dependence relationship between two parameters. L^V is the set of parameter values used in a particular configuration (i.e., a set containing exactly one value from the domain of each parameter), D is the union of the domains of all parameters, and M specifies which value assigned to each parameter $v \in V$ in the given configuration. We add an artificial root node R, which connects to all parameter nodes that do not have any parent, and refer to these parameters as *top-level parameters*.

We can transform one concept DAG into another by a series of delete, insert, relabel and move operations, each of which has an associated cost. For measuring the degree of similarity between two algorithm configurations, we first express them as concept DAGs, DAG_1 and DAG_2. We define the distance between these DAGs as the minimal total cost required for transforming DAG_1 into DAG_2. Obviously, the distance between two identical configurations is 0.

Input: CNF formula ϕ; real number *cutoff*;
        Booleans *performDiversification, singleClauseAsNeighbor,*
             usePromisingList;
Output: Satisfying variable assignment
Start with random assignment A;
Initialize parameters;
while runtime < *cutoff* **do**
    if *A satisfies ϕ* **then**
        return A;
    end
    varFlipped ← FALSE;
    if *performDiversification* **then**
        with *probability diversificationProbability()* **do**   **[B1]**
            c ← selectClause();   **[B1]**
            y ← diversificationStrategy(c) ;   **[B1]**
            varFlipped ← TRUE;   **[B1]**
    end
    if *not varFlipped* **then**
        if *not usePromisingList* **then**
            if *singleClauseAsNeighbor* **then**
                c ← selectClause();   **[B2]**
                y ← selectHeuristic(c) ;   **[B2]**
            else
                sety ← selectSet();   **[B3]**
                y ← tieBreaking(sety);   **[B3]**
            end
        else
            if *promisingList is not empty* **then**
                y ← selectFromPromisingList() ;   **[B4]**
            else
                c ← selectClause();   **[B4]**
                y ← selectHeuristic(c) ;   **[B4]**
            end
        end
        flip y ;
        update();   **[B5]**
    end
end

Procedure. SATenstein(...)

The parameters with the biggest impact on an algorithm's execution path are likely to appear high in the DAG (i.e., to be conditional upon few or no other parameters) and/or to turn on a complex mechanism (i.e., to have many parameters conditional upon them). Therefore, we say that the importance of a parameter v is a function of its depth (the length of the longest path from the root R of the given concept DAG to v) and the total number of other parameters conditional on it. To capture this definition of importance, we define the cost of each of the four DAG-transforming operations as follows.

Deletion cost $C(\text{delete}(v)) = \frac{1}{|V|} \cdot (height(DAG) - depth(v) + 1 + |DE(v)|)$, where $height(DAG)$ is the height of the DAG, $depth(v)$ is the depth of node v and $DE(v)$ is the set of descendants of node v. This captures the idea that it is more costly to delete top-level parameters and parameters that (de-)activate complex mechanisms.

Insertion cost $C(\text{insert}(u, v)) = \frac{1}{|V|} \cdot (height(DAG) - depth(u) + 1 + |DE(v)|)$, where $DE(v)$ is the set of descendants of v after the insertion, and u is the node under which v is inserted.

Moving cost $C(\text{move}(u, v)) = \frac{|V|-2}{2 \cdot |V|} \cdot [C(\text{delete}(v)) + C(\text{insert}(u, v))]$, where $|V| > 2$.

Relabelling cost $C(\text{relabel}(v, l^v, l^{v^*})) = [C(\text{delete}(v)) + C(\text{insert}(u, v))] \cdot s(l^v, l^{v^*})$, where u is the parent node of v and $s(l^v, l^{v^*})$ is a measure of the distance between the old label, l^v, and the new label, l^{v^*}, of node v. For parameters with continuous domains, $s(l^v, l^{v^*}) = |l^v - l^{v^*}|$. For parameters whose domains are some finite, ordinal and discrete set $\{l^{v_1}, l^{v_2}, \ldots, l^{v_k}\}$, $s(l^v, l^{v^*}) = \text{abs}(v - v^*)/(k - 1)$, where $\text{abs}(v - v^*)$ measures the number of intermediate values between v and v^*. For categorical parameters, $s(l^v, l^{v^*}) = 0$ if $l^v = l^{v^*}$ and 1 otherwise. Since ParamILS, the algorithm configurator we used, requires discrete parameter domains, all our parameters are either categorical or have finite, ordinal and discrete domains; therefore, $s(l^v, l^{v^*})$ is always bounded between $[0, 1]$.

In our implementation, we did not use the move operator, because the structure of SATenstein's parameter space does not provide much scope for its application. Also, instead of finding the minimal transformation cost over all sequences of delete and insert operations (a potentially expensive computation), we used an easily implemented, yet effective stochastic local search procedure to produce upper bounds. This procedure is based on randomised iterative first improvement (with a random walk probability of 0.01); starting from a permutation of the delete and insert operations required to transform the first concept DAG into the second that is chosen uniformly at random, it swaps two operations in each step. The search process is restarted whenever no improvement in the best transformation cost seen so far has been obtained within 200 iterations and terminates upon the 10th such restart. For each sequence of inserts and deletes, it is straightforward to compute the transformation cost that also takes into account the corresponding relabelling operations.

4   Experimental Setup

Our quantitative analysis of SATenstein configurations is based on performance comparisons with eleven high-performance SLS solvers on six well-known SAT distributions, listed in Table 1 (we call each of these solvers a challenger) and Table 2, respectively.

Table 1. Our eleven challenger algorithms.

Algorithm	Abbrev	Reason for inclusion	Parameters
Ranov [22]	Ranov	gold 2005 SAT Competition (random)	wp
G^{2}WSAT [17]	G2	silver 2005 SAT Competition (random)	novNoise, dp
VW [24]	VW	bronze 2005 SAT Competition (random)	c, s, wpWalk
gNovelty$^+$ [23]	GNOV	gold 2007 SAT Competition (random)	novNoise, wpWalk, ps
adaptG^{2}WSAT$_0$ [18]	AG20	silver 2007 SAT Competition (random)	NA
adaptG^{2}WSAT$_+$ [19]	AG2+	bronze 2007 SAT Competition (random)	NA
adaptNovelty$^+$ [8]	ANOV	gold 2004 SAT Competition (random)	wp
adaptG^{2}WSAT$_p$ [19]	AG2p	performance comparable to G^{2}WSAT [17], Ranov, and adaptG^{2}WSAT$_+$; see [18]	NA
SAPS [12]	SAPS	prominent DLS algorithm	alpha, ps, rho, sapsthresh, wp
RSAPS [12]	RSAPS	prominent DLS algorithm	alpha, ps, rho, sapsthresh, wp
PAWS [27]	PAWS	prominent DLS algorithm	maxinc, pflat

Table 2. Our six benchmark distributions.

Distribution	Description	Generator parameters	Train/Test size
QCP	SAT-encoded quasi-group completion problems [6]	order $O \in [10, 30]$; holes $H = h * O^{1.55}$, $h \in [1.2, 2.2]$	1000/1000
SW-GCP	SAT-encoded small-world graph-colouring problems [5]	ring lattice size $S \in [100, 400]$; nearest neighbors connected: 10; rewiring probability: 2^{-7}; chromatic numbers: 6	1000/1000
R3SAT	uniform-random 3-SAT instances [25]	variable: 600; clauses-to-variables ratio: 4.26	250/250
HGEN	random instances generated by HGEN2 [7]	variable $n \in [200, 400]$	1000/1000
FAC	SAT-encoded factoring problems [29]	prime number $\in [3000, 4000]$	1000/1000
CBMC(SE)	SAT-encoded bounded model checking [1], preprocessed by SatELite [3]	array size $s \in [1, 2000]$; loop unwinding $n \in 4, 5, 6$	302/302

We performed algorithm configuration using ParamILS [11], a well-known automatic algorithm configurator. On each benchmark distribution, we configured SATenstein on the training set, and evaluated its performance of the configuration on the test set. For each test set instance, we ran each solver 25 times with a per-run cutoff of 600 CPU seconds. Following [11], we evaluate performance in terms of penalized average run time (PAR), which is defined as average run time with each timed out run counted as having completed in 10 times the cutoff time (in this case, 6000 CPU seconds). For a particular solver, we consider an instance solved if a majority of runs found a satisfying assignment. In practice, PAR can be sensitive to the choice of cutoff; however, in past work [15], we showed that PAR did not affect the qualitative evaluation of SATenstein's performance in all six distributions we considered. We conducted all of our experiments on a cluster of 55 machines each equipped with dual 3.2 GHz Intel Xeon CPUs with 2 MB cache and 2 GB RAM, running OpenSuSE Linux 11.1 and managed by Sun Grid Engine (version 6.0). Code for SATenstein and our transformation cost computation can be found at http://www.cs.ubc. ca/labs/beta/Projects/SATenstein/.

5   Quantitative Comparison of Algorithm Configurations

In previous work, we performed an extensive performance evaluation on six well-known benchmark distributions, finding that SATenstein outperformed all challengers in every distribution [16]. Moreover, we found that SATenstein outperformed tuned challengers as well, albeit to a reduced extent. In order to refer to them in what follows, we summarize these results in Tables 4 and 5.

Table 3 gives a high-level description of SATenstein solvers in terms of building blocks used and overall SLS category. Recall that SATenstein draws components from three major SLS solver categories: WalkSAT, dynamic local search and G^2WSAT-based algorithms.

5.1   Comparison of SATenstein Configurations

We now compare our automatically identified SATenstein solver designs to all of the challengers. As shown in Table 1, 3 of our 11 challengers (AG2p, AG2+, and AG20) are parameter-less. Furthermore, RANOV only differs from ANOV by the addition of a preprocessing step, and so can be understood as a variant of the

Table 3. High-level summary of SATenstein solvers.

Solver	Uses building blocks	Broad category
SATenstein[QCP]	1, 2 and 5	WalkSAT
SATenstein[SW-GCP]	2 and 5	WalkSAT
SATenstein[R3SAT]	1, 3 and 5	Dynamic local search
SATenstein[HGEN]	1, 2 and 5	WalkSAT
SATenstein[FAC]	3 and 5	Dynamic local search
SATenstein[CBMC(SE)]	1, 3 and 5	Dynamic local search

Table 4. Performance of SATenstein and the 11 challengers. Every algorithm was run 25 times on each instance with a cutoff of 600 CPU seconds per run. Each cell $\langle i, j \rangle$ summarizes the test-set performance of algorithm i on distribution j as a/b, where a (top) is the penalized average runtime; b (bottom) is the percentage of instances solved (i.e., those with median runtime < cutoff). The best-scoring algorithm(s) in each column are indicated in bold, and the best-scoring challenger(s) are underlined.

SATenstein[D] [15]	0.08 100%	0.03 100%	1.11 100%	0.02 100%	10.89 100%	4.75 100%
Solvers	QCP	SW-GCP	R3SAT	HGEN	FAC	CBMC(SE)
AG20 [18]	1054.99	0.64	2.14	137.02	3594.40	2169.77
	81.2%	**100%**	**100%**	98.1%	35.9%	61.1%
AG2p [19]	1119.96	0.43	2.35	105.30	1954.83	2294.24
	80.1%	**100%**	**100%**	98.4%	80.6%	61.1%
AG2+ [19]	1091.37	0.67	3.04	148.28	1450.89	2181.92
	80.3%	**100%**	**100%**	98.0%	91.0%	61.1%
ANOV [8]	25.42	4.86	11.17	109.94	2897.52	2021.22
	99.6%	**100%**	**100%**	98.6%	51.4%	61.1%
G2 [17]	2942.13	4092.29	3.69	104.55	5947.80	2139.12
	50.9%	31.0%	**100%**	98.7%	0%	65.4%
GNOV [23]	414.69	1.20	11.14	52.58	5935.39	2236.85
	93.3%	**100%**	**100%**	99.4%	0%	61.5%
PAWS [27]	1127.84	4495.50	1.77	62.18	22.05	1693.82
	81.0%	24.3%	**100%**	99.4%	**100%**	70.8%
RANOV [22]	73.38	0.15	18.29	151.11	887.33	1227.07
	99.1%	**100%**	**100%**	98.2%	96.8%	79.7%
RSAPS [12]	1255.94	5635.54	18.42	33.28	17.86	827.81
	79.2%	5.4%	**100%**	99.7%	**100%**	85.0%
SAPS [12]	1248.34	3864.74	22.93	40.17	16.41	646.89
	79.4%	34.2%	**100%**	99.5%	**100%**	89.7%
VW [24]	1022.69	161.74	12.45	176.18	3382.02	385.12
	81.9%	99.4%	**100%**	97.8%	35.3%	93.4%

same algorithm. This leaves us with 7 parameterized challengers to consider. For each, we sampled 50 configurations (consisting of the default configuration, one configuration optimized for each of our 6 benchmark distributions, and 43 random configurations). We then computed the pairwise transformation cost between the resulting 359 configurations (7 × 50 challengers' configurations + 6 SATenstein solvers + AG2p + AG2+ + AG20). The result can be understood as a graph with 359 nodes and 128 522 edges, with nodes corresponding to concept DAGs, and edges labeled by the minimum transformation cost between them. To visualize this graph, we used a dimensionality reduction method to map it onto a plane, with the aim of positioning points so that the Euclidean distance between every pair of points approximates their transformation cost as accurately as possible; in particular, we used the MDS algorithm [2] as implemented in MATLAB's *mdscale* function, with option *sstress*.

Table 5. Performance summary of the automatically configured versions of 8 challengers (three challengers have no parameters). Every algorithm was run 25 times on each problem instance with a cutoff of 600 CPU seconds per run. Each cell $\langle i, j \rangle$ summarizes the test-set performance of algorithm i on distribution j as a/b, where a (top) is the penalized average runtime; b (bottom) is the percentage of instances solved (i.e., having median runtime < cutoff). The best-scoring algorithm(s) in each column are indicated in bold.

Solvers	QCP	SW-GCP	R3SAT	HGEN	FAC	CBMC(SE)
ANOV[D] [8]	26.13	0.06	2.68	119.75	1731.16	994.94
	99.6%	**100%**	**100%**	98.2%	90.1%	83.4%
G2[D] [17]	514.29	**0.05**	3.64	98.70	617.83	1084.60
	91.4%	**100%**	**100%**	99.1%	97.8%	81.4%
GNOV[D] [23]	417.33	0.22	8.87	68.24	5478.75	2195.76
	92.9%	**100%**	**100%**	99.4%	0.3%	61.8%
PAWS[D] [27]	68.06	0.70	1.91	64.48	22.01	1925.56
	99.2%	**100%**	**100%**	99.4%	**100%**	67.7%
RANOV[D] [22]	75.06	0.15	13.85	141.61	336.27	1223.83
	98.9%	**100%**	**100%**	98.1%	**100%**	80.4%
RSAPS[D] [12]	868.37	0.19	1.32	42.99	12.17	67.59
	85.2%	**100%**	**100%**	99.5%	**100%**	99.0%
SAPS[D] [12]	27.69	0.31	1.54	**31.77**	**10.68**	62.63
	99.8%	**100%**	**100%**	99.6%	**100%**	99.0%
VW[D][24]	**0.33**	417.71	**1.26**	57.44	32.38	**16.45**
	100%	94.8%	**100%**	99.6%	**100%**	**100%**

The final layout of similarities among 359 configurations (16 algorithms) is shown in Fig. 1. Observe that in most cases the 50 different configurations for a given challenger solver were so similar that they mapped to virtually the same point in the graph.

As noted earlier, the distance between any two configurations shown in Fig. 1 only approximates their true distance. In addition, the result of the visualization also depends on the number of configurations considered: adding an additional configuration may affect the position of many or all other configurations. Thus, before drawing further conclusions about the results illustrated in Fig. 1, we validated the fidelity of the visualization to the original distance data. As can be seen from Fig. 2, there was a strong correlation between the computed and mapped distances (Pearson correlation coefficient: 0.96). Also, the mapping preserved the relative ordering of the true distances between configurations quite well (Spearman correlation coefficient 0.96)—in other words, distances that appear similar in the 2D plot tend to correspond to similar true distances (and vice versa). Digging deeper, we confirmed that the top two closest challengers in Fig. 1 to each given SATenstein were always the ones having the lowest true

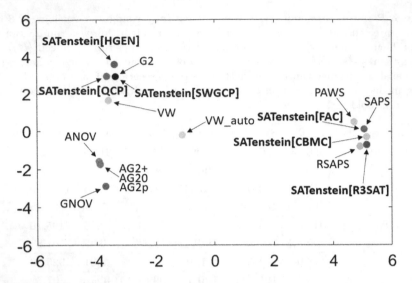

Fig. 1. Visualization of the transformation costs in the design of 16 high-performance solvers (359 configurations) obtained via MDS.

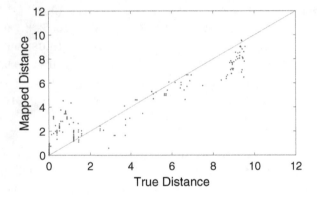

Fig. 2. True *vs* mapped distances in Fig. 1. The data points correspond to the complete set of SATenstein[D] for all domains and all challengers with their default and domain-specific, optimized configurations.

transformation costs. For distant challengers, relative distance in the visualization did not always reflect true relative transformation costs; however, we find this acceptable, since we are mainly interested in examining which configurations are similar to each other.

Having confirmed that our dimensionality reduction method is performing reliably, let us examine Fig. 1 in more detail. Overall, and unsurprisingly, we first note that the transformation cost between two configurations in the design space is very weakly related to their performance difference (quantitatively, the Spearman correlation coefficient between performance difference (PAR ratio) and configuration difference (transformation cost) was 0.25). Examining algorithms by type, we note that all dynamic local search algorithms are grouped together, on the right side of Fig. 1; likewise, the algorithms using adaptive mechanisms are grouped together at the bottom-left of Fig. 1. SATenstein() solvers were typically more similar to each other than to challengers, and fell into two broad clusters. The first cluster also includes the SAPS variants (SAPS, RSAPS), while the second also includes G2 and VW. None of the SATenstein solvers uses an adaptive mechanism to automatically adjust other parameters. In fact, as shown in Table 5, the same is true of the best performance-optimized challengers as neither SAPS, G2, or VW use adaptive mechanism. This suggests that in many cases, contrary to common belief (see, e.g., [8,19]) it may be preferable to expose parameters so they can be instantiated by sophisticated configurators rather than automatically adjusting them at running time using a simple adaptive mechanism.

We now consider benchmarks individually. For the FAC benchmark, SATenstein[FAC] had similar performance to SAPS[FAC]; as seen in Fig. 1, both solvers are structurally very similar as well. Overall, for the 'industrial' distributions, CBMC(SE) and FAC, dynamic local search algorithms often yielded the best performance amongst all challengers. Our automatically-constructed SATenstein solvers for these two distributions are also dynamic local search algorithms. Due to the larger search neighbourhood and the use of clause penalties, dynamic local search algorithms are more suitable for solving industrial SAT instances, which often have some special global structure.

For R3SAT, a well-studied distribution, many challengers showed good performance (the top three challengers were VW, RSAPS, and SAPS). The performance of SATenstein[R3SAT] is only slightly better than that of VW[R3SAT]. Figure 1 shows that SATenstein[R3SAT] is a dynamic local search algorithm similar to RSAPS and SAPS.

For HGEN, even the best performance-optimized challengers, RSAPS[HGEN] and SAPS[HGEN], performed poorly. SATenstein[HGEN] achieves more than 1 000-fold speedups against all challengers. Its configuration is far away from any dynamic local search algorithm (the best challengers), and closest to VW, a WalkSAT algorithm, and G2.

For QCP, VW[QCP] does not reach the performance of SATenstein[QCP], but significantly outperforms all other challengers. Our transformation cost analysis shows that VW is the closest neighbour to SATenstein[QCP]. For SWGCP, many challengers achieve similar performance to SATenstein[SWGCP]. Figure 1 shows that SATenstein[SWGCP] is close to G2[SWGCP], which is the best performing challenger on SWGCP.

5.2   Comparison to Configured Challengers

Since there were large performance gaps between default and configured challengers, we were also interested in the transformation cost between the configurations of individual challenger solvers. Table 5 shows that after configuring each challenger for each distribution, we found that SAPS was best on HGEN and FAC;

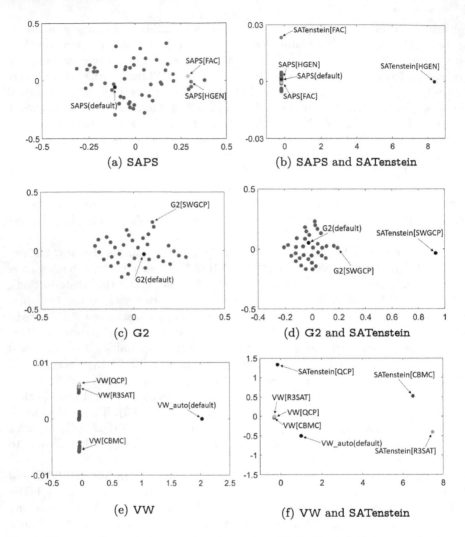

Fig. 3. The transformation costs of configuration of individual challengers and selected SATenstein solvers. (a): SAPS (best on HGEN and FAC); (b): SAPS and SATenstein[HGEN, FAC]; (c): G2 (best on SWGCP); (d): G2 and SATenstein[SWGCP]; (e): VW (best on CBMC(SE), QCP, and R3SAT); (f): VW and SATenstein[CBMC, QCP, R3SAT].

G2 was best on SWGCP, and VW was best on CBMC(SE), QCP, and R3FIX. Figure 3 (left) visualizes the parameter spaces for each of these three solvers (43 random configurations + default configuration + 6 optimized configurations). Figure 3 (right) shows the same thing, but also adds the best SATenstein() configurations for each benchmark on which the challenger exhibited top performance.

Examining these figures in the left column of Fig. 3, we first note that the SAPS configurations optimized for FAC and HGEN are very similar but differ substantially from SAPS's default configuration. On SWGCP, the optimized configuration of G2 not only performs much better than the default but, as seen in Fig. 3(c), is also quite different. All three top-performing VW configurations are rather different from VW's default, and none of them uses the adaptive mechanism for choosing parameter *wpWalk*, *s*, and *c*. Since the parameter useAdaptiveMechanism is a top-level parameter and many other parameters are conditionally dependent on it, the transformation costs between VW default and optimized configurations of VW are very large, due to the high relabelling cost for these nodes in our concept DAGs.

The right column of Fig. 3 illustrates the similarity between optimized SATenstein() solvers and the best performing challenger for each benchmark. As previously noted, SATenstein[FAC] and SAPS[FAC] are not only very similar in performance, but also structurally similar. Likewise, SATenstein[SWGCP] is similar to G2[SWGCP]. On R3SAT, many challengers had similar performance. SATenstein[R3SAT] (PAR = 1.11) was quite different from the best challenger VW[R3SAT] (PAR = 1.26), but resembled SAPS[R3SAT] (PAR = 1.53). For the three remaining benchmarks, SATenstein() solvers exhibited much better performance than the best optimized challengers, and their configurations likewise differed substantially from the challengers' configurations.

As an aside, it might initially be surprising that qualitative features of the visualizations in Fig. 3 appear to be absent from Fig. 1. In particular, the sets of randomly sampled challenger configurations that are quite well-separated in Fig. 3 are nearly collapsed into single points in Fig. 1. The reason for this lies in the fact that the 2D-mapping of the highly non-planar pairwise distance data performed by MDS focuses on minimal overall distortion. For example, when visualizing the differences within a set of randomly sampled SAPS configurations (Fig. 3(a)), MDS spreads these out into a cloud of points to represent their differences. However, the presence of a single SATenstein configuration that has large transformation costs from all of these SAPS configurations forces MDS to use one dimension to capture those differences, leaving essentially only one dimension to represent the much smaller differences between the SAPS configurations (Fig. 3(b)). Adding further very different configurations (as present in Fig. 1) leads to mappings in which the smaller differences between configurations of the same challenger become insignificant.

6 Conclusion

We have proposed a new metric for quantitatively assessing the similarity between configurations of highly parametric solvers. Our metric is based on a data structure, concept DAGs, that preserves the internal hierarchical structure of parameters. We estimate the similarity of two configurations as the transformation cost from one configuration to another. In the context of SATenstein, a highly parameterized SLS-based SAT solver, we have demonstrated that visualizations based on transformation cost can provide useful insights into similarities and differences between solver configurations. In addition, we believe that this metric could be useful for suggesting potential links between algorithm structure and algorithm performance further exploration of which could be an interesting future research direction.

References

1. Clarke, E., Kroening, D., Lerda, F.: A tool for checking ANSI-C programs. In: Jensen, K., Podelski, A. (eds.) TACAS 2004. LNCS, vol. 2988, pp. 168–176. Springer, Heidelberg (2004). doi:10.1007/978-3-540-24730-2_15
2. Cox, T.F., Cox, M.A.: Multidimensional scaling. CRC Press, Boca Raton (2000)
3. Eén, N., Biere, A.: Effective preprocessing in SAT through variable and clause elimination. In: Bacchus, F., Walsh, T. (eds.) SAT 2005. LNCS, vol. 3569, pp. 61–75. Springer, Heidelberg (2005). doi:10.1007/11499107_5
4. Fawcett, C., Hoos, H.H.: Analysing differences between algorithm configurations through ablation. J. Heuristics **22**, 1–28 (2013)
5. Gent, I.P., Hoos, H.H., Prosser, P., Walsh, T.: Morphing: combining structure and randomness. In: Proceedings of the Sixteenth National Conference on Artificial Intelligence (AAAI 1999), pp. 654–660 (1999)
6. Gomes, C.P., Selman, B.: Problem structure in the presence of perturbations. In: Proceedings of the Fourteenth National Conference on Artificial Intelligence (AAAI 1997), pp. 221–226 (1997)
7. Hirsch, E.A.: Random generator hgen2 of satisfiable formulas in 3-CNF (2002). http://logic.pdmi.ras.ru/~hirsch/benchmarks/hgen2-1.01.tar.gz. Accessed 18 December 2015
8. Hoos, H.H.: An adaptive noise mechanism for WalkSAT. In: Proceedings of the Eighteenth National Conference on Artificial Intelligence (AAAI 2002), pp. 655–660 (2002)
9. Hoos, H.H.: Programming by optimization. Commun. ACM **55**(2), 70–80 (2012)
10. Hutter, F., Hoos, H.H., Leyton-Brown, K.: Sequential model-based optimization for general algorithm configuration. In: Coello, C.A.C. (ed.) LION 2011. LNCS, vol. 6683, pp. 507–523. Springer, Heidelberg (2011). doi:10.1007/978-3-642-25566-3_40
11. Hutter, F., Hoos, H.H., Leyton-Brown, K., Stützle, T.: ParamILS: an automatic algorithm configuration framework. J. Artif. Intell. Res. (JAIR) **36**(1), 267–306 (2009)
12. Hutter, F., Tompkins, D.A.D., Hoos, H.H.: Scaling and probabilistic smoothing: efficient dynamic local search for SAT. In: Hentenryck, P. (ed.) CP 2002. LNCS, vol. 2470, pp. 233–248. Springer, Heidelberg (2002). doi:10.1007/3-540-46135-3_16
13. Hutter, F., Hoos, H., Leyton-Brown, K.: An efficient approach for assessing hyperparameter importance. In: Proceedings of the 31st International Conference on Machine Learning (ICML 2014), pp. 754–762 (2014)
14. Kadioglu, S., Malitsky, Y., Sellmann, M., Tierney, K.: ISAC - instance specific algorithm configuration. Eur. Conf. Artif. Intell. (ECAI) **2010**, 751–756 (2010)
15. KhudaBukhsh, A.R., Xu, L., Hoos, H.H., Leyton-Brown, K.: SATenstein: automatically building local search SAT solvers from components. In: Proceedings of the Twenty-First International Joint Conference on Artificial Intelligence (IJCAI 2009), pp. 517–524 (2009)
16. KhudaBukhsh, A.R., Xu, L., Hoos, H.H., Leyton-Brown, K.: Satenstein: automatically building local search sat solvers from components. Artif. Intell. **232**, 20–42 (2016)

17. Li, C.M., Huang, W.Q.: Diversification and determinism in local search for satisfiability. In: Bacchus, F., Walsh, T. (eds.) SAT 2005. LNCS, vol. 3569, pp. 158–172. Springer, Heidelberg (2005). doi:10.1007/11499107_12

18. Li, C.M., Wei, W., Zhang, H.: Combining adaptive noise and promising decreasing variables in local search for SAT (2007). Solver description, SAT competition 2007

19. Li, C.M., Wei, W., Zhang, H.: Combining adaptive noise and look-ahead in local search for SAT. In: Marques-Silva, J., Sakallah, K.A. (eds.) SAT 2007. LNCS, vol. 4501, pp. 121–133. Springer, Heidelberg (2007). doi:10.1007/978-3-540-72788-0_15

20. Malitsky, Y., Sabharwal, A., Samulowitz, H., Sellmann, M.: Algorithm portfolios based on cost-sensitive hierarchical clustering. In: Proceedings of the Twenty-Third International Joint Conference on Artificial Intelligence, pp. 608–614. AAAI Press (2013)

21. Nikolić, M., Marić, F., Janičić, P.: Instance-based selection of policies for SAT solvers. In: Kullmann, O. (ed.) SAT 2009. LNCS, vol. 5584, pp. 326–340. Springer, Heidelberg (2009). doi:10.1007/978-3-642-02777-2_31

22. Pham, D.N., Anbulagan, A.: Resolution enhanced SLS solver: R+AdaptNovelty+ (2007). Solver description, SAT competition 2007

23. Pham, D.N., Thornton, J., Gretton, C., Sattar, A.: Combining adaptive and dynamic local search for satisfiability. J. Satisf. Boolean Model. Comput. (JSAT) **4**, 149–172 (2008)

24. Prestwich, S.: Random walk with continuously smoothed variable weights. In: Bacchus, F., Walsh, T. (eds.) SAT 2005. LNCS, vol. 3569, pp. 203–215. Springer, Heidelberg (2005). doi:10.1007/11499107_15

25. Simon, L.: SAT competition random 3CNF generator (2002). www.satcompetition.org/2003/TOOLBOX/genAlea.c. Accessed 18 December 2015

26. Sinz, C.: Visualizing the internal structure of sat instances (preliminary report). In: SAT. Citeseer (2004)

27. Thornton, J., Pham, D.N., Bain, S., Ferreira, V.: Additive versus multiplicative clause weighting for SAT. In: Proceedings of the Nineteenth National Conference on Artificial Intelligence (AAAI 2004), pp. 191–196 (2004)

28. Tompkins, D.A.D., Balint, A., Hoos, H.H.: Captain Jack: new variable selection heuristics in local search for SAT. In: Sakallah, K.A., Simon, L. (eds.) SAT 2011. LNCS, vol. 6695, pp. 302–316. Springer, Heidelberg (2011). doi:10.1007/978-3-642-21581-0_24

29. Uchida, T., Watanabe, O.: Hard SAT instance generation based on the factorization problem (1999). http://www.is.titech.ac.jp/~watanabe/gensat/a2/GenAll.tar.gz

30. Wei, W., Li, C.M., Zhang, H.: A switching criterion for intensification, and diversification in local search for sat. J. Satisf. Boolean Model. Comput. **4**, 219–237 (2008)

31. Xue, Y., Wang, C., Ghenniwa, H., Shen, W.: A tree similarity measuring method and its application to ontology. J. Univ. Comput. Sci. **15**(9), 1766–1781 (2001)

Short Papers

Neighborhood Synthesis from an Ensemble of MIP and CP Models

Tommaso Adamo, Tobia Calogiuri, Gianpaolo Ghiani, Antonio Grieco,
Emanuela Guerriero, and Emanuele Manni$^{(\boxtimes)}$

Dipartimento di Ingegneria Dell'Innovazione, Università del Salento,
Via per Monteroni, 73100 Lecce, Italy
{tommaso.adamo,tobia.calogiuri,gianpaolo.ghiani,antonio.grieco,
emanuela.guerriero,emanuele.manni}@unisalento.it

Abstract. In this paper we describe a procedure that automatically synthesizes a neighborhood from an *ensemble* of Mixed Integer Programming (MIP) and/or Constraint Programming (CP) models. We move on from a recent paper by Adamo et al. (2015) in which a neighborhood structure is automatically designed from a (single) MIP model through a three-step approach: (1) a semantic feature extraction from the MIP model; (2) the derivation of neighborhood design mechanisms based on these features; (3) an automatic configuration phase to find the "proper mix" of such mechanisms taking into account the instance distribution. Here, we extend the previous work in order to generate a suitable neighborhood from an ensemble of MIP and/or CP models of a given combinatorial optimization problem. Computational results show relevant improvements over the approach considering a single model.

Keywords: Combinatorial optimization · Neighborhood search · Automatic neighborhood design · Feature extraction

1 Introduction

The definition of "good" neighborhood structures on the solution space is a key step when designing neighborhood search heuristics for combinatorial optimization problems. In order to make the search efficient, it is fundamental to tailor the neighborhood structures *not only* to the specific problem *but also* to the *reference instance population* (i.e., the particular distribution of the instances to be solved). Our aim is to develop mechanisms that may derive automatically suitable neighborhoods by exploiting some features of an ensemble of Mixed Integer Programming (MIP) and/or Constraint Programming (CP) models of the problem. In particular, we extend a recent work by Adamo et al. (2015) in which a neighborhood structure is automatically designed from a (single) MIP model with a three-step procedure: (1) extraction of semantic features from the MIP model; (2) derivation of neighborhood design mechanisms (based on these features), obtained by freeing a subset of the variables and then fixing them again;

P. Festa et al. (Eds.): LION 2016, LNCS 10079, pp. 221–226, 2016.
DOI: 10.1007/978-3-319-50349-3_15

(3) search of a good mix of such mechanisms in an automatic configuration phase. Since multiple models (taken from the existing literature or written from scratch by the analyst/researcher) may be known for a given combinatorial optimization problem, we extend this approach in order to feed the Automatic Neighborhood Design (AND) algorithm with an ensemble of MIP and/or CP models, and show the advantages of this approach.

We now review the most relevant literature related to our work. Along the line of research aiming to improve a given feasible solution by machine-generated neighborhood structures, it is relevant the work on the *Local Branching* algorithm (Fischetti and Lodi 2003) that identifies suitable *spherical neighborhoods* by means of branching conditions (*local branching cuts*) and explores them by using a generic black-box MIP solver. Other remarkable contributions are due to Danna et al. (2005) and Parisini and Milano (2012). More recently, Ghiani et al. (2015) move on from a MIP compact formulation and show how to exploit its features to automatically design efficient neighborhoods. They use an unsupervised learning approach to automatically identify good regions of the search space around a given feasible solution. Similarly, Van Hentenryck and co-authors (see, e.g., Mouthuy et al. 2012) show how to synthesize local search algorithms from high-level constraint-based local search models. In addition, our work is also related to *Automatic Algorithm Configuration* (AAC), in which the most appropriate parameter setting is found by an automatic procedure (Hutter et al. 2007).

2   Automatically Designed Neighborhoods

Given a combinatorial optimization problem P, a *neighborhood structure* associates a set of feasible solutions $N(s)$ to each feasible solution s of P. In this paper, we assume the user supplies an *ensemble* $M_1, \ldots, M_L$ of MIP and/or CP models, each of which is characterized by its own vector $x^{(l)} \in R^{n_l}$ of n_l decision variables $(l = 1, \ldots, L)$. Model M_1 (which is referred to as the *leading model* in the following) plays a special role since it is used by the AND procedure to extract the main features of both the problem P and its feasible solution s. We also assume the user describes in a high-level language how to convert each $x^{(l)}$ vector $(l = 2, \ldots, L)$ from/to the corresponding $x^{(1)}$ vector. Since a good neighborhood structure must be adapted to the specific problem as well as to the particular distribution of the instances to be solved, the user also provides a *training set* representative of the reference instance population. Moreover, we require the MIP and the CP models are written through an algebraic modeling language (e.g., AMPL, GAMS or OPL). Such languages are characterized by sets of entities (e.g., vehicles, commodities or customers) and parameters, variables and constraints of the model are then "tagged" (or indexed) by one or more entities. For example, the constraint $\sum_{i \in A} x_{ij} = d_j, j \in B$ is tagged by the entities in set B. Moreover, variables x are tagged by entity sets A and B, while parameters d are tagged by entity set B. This tagging defines implicitly relationships between entities. In this paper we require that the MIP and the CP models share the *same* set of entities E that can be subdivided into one or

---

Algorithm 1. Sketch of the Automatic Neighborhood Design procedure using an ensemble of models. The output is the (possibly) improving solution s'

 1: **procedure** AND($M_1, \ldots, M_L, TS, s$) //TS is the training set
           //compute the vector w used to weight the parameters of the model and
           //identify among $M_1, \ldots, M_L$ the model M_{l^*} to use in the repair phase
 2:       $w, l^* \leftarrow$ AAC($M_1, \ldots, M_L, TS$)
           //extract the set S of semantic features from M_1 and the current solution s
 3:       $S \leftarrow$ ExtractSemanticFeatures(M_1, w, s)
           //select the entities that make up set F
 4:       $F \leftarrow$ SelectEntities(S, s)
           //explore the automatically-designed neighborhood
 5:       $s' \leftarrow$ DestroyAndRepair(M_{l^*}, s, F)
 6: **end procedure**

---

more sets E_k ($k = 1, \ldots, K$) of *homogeneous* entities (e.g., a set of facilities, a set of commodities and a set of orders).

To measure the strength of the relationships between the entities, our procedure extracts some *semantic features* (that can be model- or solution-based) as follows. Given two entities $e, e' \in E_k$, we first identify the P_k parameters (i.e., the data of the problem) tagged by E_k and then we build a dataset in which the rows are associated with the entities in E_k, while each column reports the values of a parameter. Each parameter is given a weight w_k and the most appropriate values of the weight vector w are determined through an AAC phase performed on a training set. The similarity between e and e' is obtained by first normalizing the weighted columns of the dataset and then computing the inner product of the corresponding rows. In addition, we also consider some solution-based features that take into account the similarity between entities with respect to the current solution. More specifically, two entities $e, e' \in E$ are considered similar (*adjacent*) with respect to s if: (i) they are both tagged by a variable with a non-zero value in s or (ii) there is a constraint tagged by e that is active in s, in which a variable, tagged by e' and having a non-zero value in s, appears with a coefficient different from zero (or vice versa).

We define a neighborhood structure $N(s)$ as the set of feasible solutions that can be obtained by first destroying and then repairing the fragment of the current solution s tagged by a selected subset $F \subseteq E$ of entities. The destroy phase is obtained by freeing the decision variables tagged by F. On the other hand, the repair step fixes such variables again by means of an off-the-shelf MIP or CP solver (e.g., CPLEX or CP Optimizer with a given parameter setting). The model employed in the repair phase is chosen among $M_1, \ldots, M_L$ by the AAC procedure as the one providing the best performance across the training set. In our algorithm, a key aspect is how to choose F. In this paper, this is done on the basis of the semantic features previously extracted from the leading model M_1 and the current solution s (the details of the procedure can be found in Adamo et al. 2015). In particular, the semantic features are used to identify entities that are "similar" and may have a strong impact on the objective function. The main steps of our approach are summarized in Algorithm 1. First, the training set TS and the ensemble of models $M_1, \ldots, M_L$ are used to determine the values of the weights to be assigned to the parameters of the model as well as the most appropriate model M_{l^*} for the repair phase (line 2). Then, the EXTRACTSEMANTICFEATURES procedure utilizes the weight vector w, the leading model M_1 and the current solution s to extract a set S of semantic features (line 3). At line 4, the semantic features S and the current solution s are inputs for the SELECTENTITIES procedure that selects the entities constituting the set F. Finally, the model M_{l^*}, the set F and the current solution s are used to synthesize and explore the neighborhood of s (line 5).

3   Computational Results

We test our approach on several classical combinatorial optimization problems, as in Adamo et al. (2015). Here, we report the results on the Capacitated Vehicle Routing Problem with Time Windows (VRPTW) and the Generalized Traveling Salesman Problem (GTSP). For each test problem, we consider a MIP formulation for which a "repair" is made through CPLEX 12.5, and a CP formulation for which a "repair" is made through CP Optimizer 12.5. The initial solutions are the first feasible solutions provided by CP Optimizer. A time limit of 60 s is imposed on the exploration of each neighborhood (repair phase). Finally, we use ParamILS (Hutter et al. 2007) as an AAC tool. Given a feasible solution and a time limit, our computational experiments aim to compare the average percentage objective function improvements from the initial solution of: (1) our approach using an ensemble of models (ENS in the tables); (2) the single model approach of Adamo et al. (2015) (MIP in the tables).

3.1   Capacitated Vehicle Routing Problem with Time Windows

We consider two models: a MIP formulation (Cordeau et al. 2006, Sect. 3.1) and a CP formulation (CP Optimizer Forum 2009). Our tests are made on the classical Solomon (1987) instances which are made up of six classes: C1, C2, R1, R2, RC1, RC2. To assess the AND algorithm for different reference instance populations, we generate additional classes of instances by tightening the customers' time windows. In particular, if $[a_i, b_i]$ is the original time window of a customer i, we tighten it as: $[a_i + \alpha(b_i - a_i), b_i - \alpha(b_i - a_i)]$ for $\alpha = 0.1, 0.3$ and 0.5. We also completely relax the time windows. The most relevant feature to tune is the number of variables freed in the destroy phase. After a training step, the best results are achieved when such variables amount to 20% of the overall number. The results are reported in Table 1 and show that the approach considering an ensemble of models provides the largest objective function improvements when the time windows are removed (No TW) and for the original instances ($\alpha = 0$). In the former case, the average improvement of the ENS neighborhood is around 7% on average while the MIP-based neighborhood provides an improvement equal to 5.30%. In the latter case, the average ENS improvement is around 5% on average while the MIP-based neighborhood provides an improvement around 3.50%. On

Table 1. Results for the VRPTW (**bold** numbers represent the best results)

Dataset	No TW		$\alpha = 0$		$\alpha = 0.1$		$\alpha = 0.3$		$\alpha = 0.5$	
	ENS	MIP	ENS	MIP	ENS	MIP	ENS	MIP	ENS	MIP
C1	**6.31**	5.29	**4.95**	3.76	**3.87**	3.78	**3.84**	**3.84**	**3.75**	**3.75**
C2	**7.21**	5.75	**4.67**	3.95	**6.39**	**6.39**	**5.55**	**5.55**	**6.37**	**6.37**
R1	**5.87**	4.46	**4.37**	2.05	**3.69**	3.07	**4.10**	2.96	**3.66**	**3.66**
R2	**8.23**	5.60	**5.81**	4.64	**6.08**	**6.08**	**6.31**	**6.31**	**5.49**	**5.49**
RC1	**6.98**	5.90	**5.43**	2.59	**4.56**	4.49	**4.74**	**4.74**	**2.84**	**2.84**
RC2	**7.27**	4.82	**5.21**	4.29	**6.03**	**6.03**	**6.38**	**6.38**	**5.77**	**5.77**
AVERAGE	**6.98**	5.30	**5.07**	3.55	**5.10**	4.97	**5.15**	4.96	**4.65**	**4.65**

the other hand, for the tighter instances ($\alpha = 0.1$, $\alpha = 0.3$ and $\alpha = 0.5$) the two approaches are comparable (indeed, the results are exactly the same for $\alpha = 0.5$, meaning that the AAC phase identified the MIP model as the best model to perform the repair step). Here, the difference between the two classes of neighborhoods is rather small (around 0.1–0.2%).

3.2 Generalized Traveling Salesman Problem

The GTSP MIP model we consider is the same described in Adamo et al. (2015), whereas the CP model is implemented from scratch. The instances used to perform our experiments are taken from Fischetti et al. (1997), who derive test problems by modifying standard instances of the Traveling Salesman Problem. For the GTSP, after the training step it turns out that the best results are obtained when the maximum number of variables made free in the destroy phase amounts to 5% of the overall number. The results are reported in Table 2 and show that the ENS neighborhoods allow to achieve considerably higher improvements when

Table 2. Results for the GTSP (**bold** numbers represent the best results)

Instance	ENS	MIP	Instance	ENS	MIP
64lin318	**70.65**	24.88	131p654	**59.31**	25.53
80rd400	**63.67**	27.18	132d657	**63.69**	0.00
84fl417	**68.29**	26.42	145u724	**62.50**	0.00
88pr439	**62.31**	30.68	157rat783	**47.80**	0.00
89pcb442	**63.54**	34.34	201pr1002	**69.73**	0.00
99d493	**60.41**	0.00	212u1060	**74.48**	0.00
115u574	**69.15**	28.84	217vm1084	**29.17**	0.00
115rat575	**68.61**	32.58			
AVERAGE				**62.22**	15.36

compared to the neighborhoods based on a single MIP model. In particular, the average ENS objective function improvement is about 62%, while the MIP-based neighborhoods provide an average improvement of about 15%. This latter result is influenced by several instances (namely, 99d493, 132d657, 145u724, 157rat783, 201pr1002, 212u1060 and 217vm1084) for which the procedure is not able to improve the initial solution within the time limit.

4   Conclusions

In this paper, we propose a procedure to synthesize automatically "good" neighborhood structures from an ensemble of MIP and CP models, by extending a previous approach for the case of a single MIP model. The algorithm takes into account the characteristics of the problem as well as the reference instance population. In particular, the procedure employs a leading model to extract some semantic features used to derive automatically some neighborhood design mechanisms. Then, an automatic algorithm configuration phase allows to find a good mix of such mechanisms, as well as the best model to be used during the repair phase. Computational experiments show that the approach considering an ensemble of models outperforms the neighborhoods based on a single MIP model.

References

Adamo, T., Ghiani, G., Grieco, A., Guerriero, E., Manni, E.: MIP neighborhood synthesis through semantic feature extraction and automatic algorithm configuration. Comput. Oper. Res. (2015, submitted). http://www.emanuelemanni.unisalento.it/Sito/Papers/MIPneighSynth.pdf

Cordeau, J.-F., Laporte, G., Savelsbergh, M.W., Vigo, D.: Vehicle routing. In: Barnhart C., Laporte, G. (eds.) Transportation. Handbook in OR & MS, vol. 14, pp. 367–428 (2006)

CP   Optimizer   Forum.   https://www.ibm.com/developerworks/community/forums/ajax/download/77777777-0000-0000-0000-000014932402/b9379ac1-721e-4d9f-aa4f-e768de6502b4/attachment_14932402_cvrptw.mod

Danna, E., Rothberg, E., Le Pape, C.: Exploring relaxation induced neighborhoods to improve MIP solutions. Math. Prog. Ser. A **102**, 71–90 (2005)

Fischetti, M., Lodi, A.: Local branching. Math. Prog. Ser. B **98**, 23–47 (2003)

Fischetti, M., Salazar-González, J.J., Toth, P.: A branch-and-cut algorithm for the symmetric generalized traveling salesman problem. Oper. Res. **45**, 378–394 (1997)

Ghiani, G., Laporte, G., Manni, E.: Model-based automatic neighborhood design by unsupervised learning. Comput. Oper. Res. **54**, 108–116 (2015)

Hutter, F., Hoos, H., H., Stützle, T.: Automatic algorithm configuration based on local search. In: Proceedings of AAAI-07, vol. 22, pp. 1152–1157 (2007)

Mouthuy, S., Van Hentenryck, P., Deville, Y.: Constraint-based very large-scale neighborhood search. Constraints **17**, 87–122 (2012)

Parisini, F., Milano, M.: Sliced neighborhood search. Expert Syst. Appl. **39**, 5739–5747 (2012)

Solomon, M.M.: Algorithms for the vehicle routing and scheduling problems with time window constraints. Oper. Res. **35**, 254–265 (1987)

Parallelizing Constraint Solvers for Hard RCPSP Instances

Roberto Amadini[1], Maurizio Gabbrielli[2], and Jacopo Mauro[3(✉)]

[1] Department of Computing and Information Systems,
University of Melbourne, Melbourne, Australia
[2] DISI, University of Bologna, Italy/FOCUS Research Team, INRIA,
Rocquencourt, France
[3] Department of Informatics, University of Oslo, Oslo, Norway
mauro.jacopo@gmail.com

Abstract. The Resource-Constrained Project Scheduling Problem (RCPSP) is a well-known scheduling problem aimed at minimizing the makespan of a project subject to temporal and resource constraints. In this paper we show that hard RCPSPs can be efficiently tackled by a portfolio approach that combines the strengths of different constraint solvers Our approach seeks to predict and run in parallel the best solvers for a new, unseen RCPSP instance by enabling the bound communication between them. This on-average allows to outperform the oracle solver that always chooses the best available solver for any given instance.

1 Introduction

The *Resource-Constrained Project Scheduling Problem* (RCPSP) [10] is the problem of minimizing the makespan (i.e., the total duration) of a project, defined as a collection of tasks subject to precedence relations between the activities and constrained by resource availabilities. This well-known NP-hard problem [9] has countless industrial applications and it is probably one of the most studied scheduling benchmark. The *Constraint Programming* (CP) [18] paradigm allows to model and solve hard combinatorial problems, and in particular the RCPSP can be naturally and elegantly encoded into a *Constraint Optimization Problem* (COP) where: *(i)* integer variables are used to track the start time of each task; *(ii)* constraints over such variables ensure compliance with the precedence relations and resource capabilities; *(iii)* a special integer variable keeps track of the makespan. The goal is to find a consistent assignment of the variables which minimizes the makespan. An effective solving technique to do so is the *Lazy Clause Generation* (LCG) [17] approach. The key idea of LCG is to mimic the Finite Domain propagation by properly generating corresponding SAT clauses during the search.

To exploit the diverse nature and performance of different solving techniques, a fairly recent trend consists in using a portfolio approach [11]. Basically, given

---

Supported by the EU project FP7-644298 *HyVar: Scalable Hybrid Variability for Distributed, Evolving Software Systems.*

a portfolio $\{s_1, \ldots, s_m\}$ of $m > 1$ solvers, a *portfolio solver* seeks to predict the best solver(s) $s_{i_1}, \ldots, s_{i_k}$ (with $1 \leq i_j \leq m$ for $j = 1, \ldots, k$) for solving a new, unseen problem p and then runs such solver(s) on p. Scheduling $k > 1$ solvers can reduce the risk of selecting only one solver and especially enables the knowledge sharing between solvers, as well as their parallel execution. Surprisingly, despite their effectiveness, portfolio solvers have been poorly adopted in real-life applications [5].

In this work we show how CP can be successfully applied for solving hard RCPSP instances by means of a parallel portfolio approach. We retrieved a fairly large number of non-trivial RCPSPs encoded in *MiniZinc* [16] and we defined and test some variants of the parallel portfolio solver sunny-cp [3] to boost the resolution of the RCPSP instances. Experimental results show that state-of-the-art LCG solvers can be significantly overcome thanks to other constraint solvers not employing LCG. The message of the paper is twofold: *(i)* we prove that the belief that portfolios can not be applied in real-life scenarios characterized by a dominating solver is false, and *(ii)* we show that by parallelizing the solvers execution and by enabling the bounds communication between the scheduled solvers we can get an overall better solver which is even greater than the sum of its parts. To the best of our knowledge, we are not aware of similar approaches for efficiently solving hard RCPSP instances.

2 Background

The RCPSP resolution has attracted a lot of attention over the last decades, since this problem emerges from many real-life scenarios [12–14]. To our knowledge, the LCG approach gives the best results for RCPSP [19] and variants like RCPSP/max [20] and RCPSP/max-cal [15]. In this paper we examine a possible CP formulation of RCPSP, as it appears in MiniZinc 1.6 benchmarks.[1] The CP model is annotated with a default search strategy imposing to select the variable having smaller domain (*min-dom* heuristic) and trying to assign to such variable the smaller value of its domain (*min-value* heuristic). The study of alternative heuristics is outside the scope of this work.

sunny-cp is an open-source portfolio solver [3,4]. It enables to run more solvers simultaneously by exploiting their cooperation via bound sharing [6] and restarting policies. sunny-cp won the gold medal in the open category of MiniZinc Challenge 2015 and it is currently the only parallel portfolio solver able to solve generic COPs [1]. sunny-cp is built on top of SUNNY algorithm [2], which exploits the k-Nearest Neighbors algorithm to produce a sequential schedule of solvers for solving a given problem. In a multicore setting, the schedule is parallelized on the c available cores: the first and most promising $c - 1$ solvers are allocated to the first $c - 1$ cores, while the remaining ones are assigned to the last available core. For RCPSPs, the most promising solvers are those that are faster in finding the minimal makespan values in the k-neighborhood.

---

[1] Available at https://github.com/MiniZinc/minizinc-benchmarks/blob/master/rcpsp /rcpsp.mzn.

A *"bound-and-restart"* mechanism is used for enabling the bound sharing between the running solvers. Given a restarting threshold T_r, a running solver is stopped and restarted if it has not found a solution in the last T_r seconds and its current best bound is obsolete w.r.t. the overall best bound found by another scheduled solver. sunny-cp uses a portfolio of solvers disparate in their nature. Some of them are provided in two variants: *fixed* and *free*. The fixed variant is optional, and forces a solver to use the search strategy possibly defined in the MiniZinc input model. The free variant instead allows a solver to use its preferred search strategy. By default, sunny-cp uses the fixed variant. However, as later detailed, the free variant may significantly outperform the fixed one.[2]

3 Methodology

The RCPSP model mentioned in Sect. 2 is the most represented problem class in the MiniZinc 1.6 benchmarks with 2904 RCPSP instances coming from different scenarios. However, most of these instances are not challenging: often the best solvers of sunny-cp can solve them instantaneously. Therefore, we decided to consider a narrowed dataset Δ of 647 RCPSPs for each of which no solver can find an optimal solution in 90 s and at least one solver can find a feasible solution.

Fixed a universe of solvers $\mathcal{U}$ and a solving timeout T, we measure the performance of solver $s \in \mathcal{U}$ on a problem instance $p \in \Delta$ within T seconds in terms of:

- OPT: measures the optima proven. If s proves the optimality of a solution for p, then $\mathsf{OPT}(s, p) = 1$. Otherwise, $\mathsf{OPT}(s, p) = 0$;
- TIME: measures the optimization time. If s proves the optimality of a solution for p in $t < T$ seconds, then $\mathsf{TIME}(s, p) = t$. Otherwise, $\mathsf{TIME}(s, p) = T$;
- OBJ: measures the quality of a solution, by normalizing its makespan value in the range $[0, 1]$. If s finds no solution, then $\mathsf{OBJ}(s, p) = 0$. Otherwise, if $\mathrm{mkspan}(s, p)$ is the best makespan found by s for problem p and said $\mathcal{M}_p = \{\mathrm{mkspan}(s, p) \mid s \in \mathcal{U}\}$, we have: $\mathsf{OBJ}(s, p) = 1 - \dfrac{\mathrm{mkspan}(s, p) - \min \mathcal{M}_p}{\max \mathcal{M}_p - \min \mathcal{M}_p}$.

Table 1 shows the average performance of the individual solvers of sunny-cp with $T = 900$ s. We added as baseline the Virtual Best Solver (*VBS*), the oracle portfolio solver that —for a given problem and performance metric— always chooses the best solver in the portfolio. As can be seen, for almost 90% of the dataset Δ (578 instances) no solver is able to prove the optimality in 900 s. Chuffed clearly dominates all the other solvers, almost reaching the *VBS* performance. The effectiveness of LCG is also confirmed by the performance of the others LCG-based solvers, namely CPX and LazyFD. While using the free search is often effective, it is not always the best choice. For this reason we decided to test three different variants of sunny-cp:

---

[2] For more details about sunny-cp, we refer the reader to [3].

Table 1. Average performance. Fixed version is available only for the solvers marked with *.

Solver	OPT (%)		TIME (sec.)		OBJ × 100	
	Fixed	Free	Fixed	Free	Fixed	Free
Chuffed*	2.63	**7.42**	887.77	**858.64**	88.96	**96.50**
G12/CPX*	1.24	1.08	894.66	894.26	89.03	75.66
G12/LazyFD*	0.15	2.62	899.90	888.48	72.76	75.38
HaifaCSP	-	0.62	-	896.97	-	74.73
Choco*	0	0	900	900	66.75	72.76
OR-Tools*	0.31	0.16	899.05	899.14	65.00	67.11
G12/FD*	0	0	900	900	65.92	15.78
Gecode*	0	0	900	900	64.25	64.20
MinisatID	-	0.31	-	898.32	-	63.64
iZplus*	0	0	900	900	43.43	33.74
G12/Gurobi	-	2.94	-	885.90	-	2.94
VBS	10.67		841.33		100	

- `sunny-def`: the default version of `sunny-cp`. It uses the portfolio Π_{def} of the solvers listed in Table 1 and always chooses the fixed version when available;
- `sunny-all`: uses a portfolio Π_{all} of $11 + 8 = 19$ solvers which extends Π_{def} by including all the versions of all the available solvers;
- `sunny-stc`: uses a variable sized portfolio $\Pi_{c,\mu}$ of c solvers, where c is the number of available cores and $\mu \in \{\text{OPT}, \text{TIME}, \text{OBJ}\}$ is a performance measure. Specifically, $\Pi_{c,\mu}$ is the subset of the best c solvers of Π_{all} according to the average value of μ over dataset Δ.

Note that both `sunny-all` and `sunny-def` are *dynamic* approaches, since they select the solvers to run on-line according to the instance to be solved. `sunny-stc` follows instead a *static* approach. Indeed, since for each number of cores c its portfolio $\Pi_{c,\mu}$ contains exactly c solvers, no prediction is performed and all its solvers are launched simultaneously regardless of the instance to be solved.

4   Results

This Section presents the performance of `sunny-def`, `sunny-all`, and `sunny-stc` in terms of OBJ, OPT, and TIME metrics by considering $1, 2, 4$, and 8 cores. For all the `sunny-cp` variants, we used the default value $T_r = 5\,\text{s}$ for the restarting threshold, and we validated the predictions with a 10-fold cross-validation [7]. In addition to the *VBS* we introduce the *Virtual Parallel Solver* ($VPS_{c,\mu}$), an oracle portfolio solver that for $c \in \{1, 2, 4, 8\}$ and $\mu \in \{\text{OPT}, \text{TIME}, \text{OBJ}\}$ simulates the parallel and independent execution of the

Table 2. OBJ performance.

OBJ × 100	1 core	2 cores	4 cores	8 cores
sunny-def	92.14	94.25	95.83	96.04
sunny-all	94.67	96.36	**98.25**	98.45
sunny-stc	**94.97**	95.73	97.57	**99.05**
VPS	*94.97*	*96.92*	*97.43*	*98.21*
VBS	*98.38*			

solvers of the portfolio $\Pi_{c,\mu}$ introduced in Sect. 3. With this definition, the *Single Best Solver* (*SBS*) of the portfolio (i.e., the free version of Chuffed) is equivalent to the $VPS_{1,\mu}$, while $VBS = VPS_{|\Pi_{all}|,\mu}$. Where there is no ambiguity, we will use the notation *VPS* or VPS_c instead of $VPS_{c,\mu}$.

Table 2 shows OBJ results. We can say that most of the sunny-cp approaches provide high quality solutions ($0.95 < OBJ < 1$) even when optimality is not proven. The only approach that performs rather poorly is sunny-def with one core. For $c \geq 4$ the effectiveness of bounds communication becomes clear. sunny-stc with 4 cores is better than VPS_4, i.e., its corresponding version without bounds communication and synchronization issues. With 8 cores sunny-stc outperforms not only VPS_8, but also the *VBS*. In other terms, 8 cores are enough for providing better solution than a "magic solver" that runs simultaneously —without synchronization issues— all the 19 solvers of Π_{all}. The peak performance is reached by sunny-stc with 8 cores: the *VBS* is outperformed 158 times (24.42% of Δ), meaning that almost one time out of four it finds a better solution than *VBS*.

The OPT performance is depicted in Table 3. This metric is challenging in our context: we are dealing with hard RCPSP instances for which no solver is able to prove the optimality in less than 90 s and the *VBS* can prove only 69 optimum (10.66% of Δ). For $c \geq 2$ the best approach is sunny-all, which is able to outperform the *VBS* whith 4 or more cores. In particular, the gain with 8 cores is somewhat impressive: 4.8% optima proven more than *VBS*. Here the performance difference is not only due to parallelism and bounds communication, but especially due to the solver selection. Indeed, the gap with sunny-stc becomes

Table 3. OPT performance.

OPT (%)	1 core	2 cores	4 cores	8 cores
sunny-def	2.01	4.64	9.58	10.36
sunny-all	6.65	**9.58**	**12.52**	**15.46**
sunny-stc	**7.42**	7.88	9.12	9.58
VPS	*7.42*	*8.04*	*8.35*	*8.50*
VBS	*10.66*			

Table 4. TIME performance.

TIME (sec.)	1 core	2 cores	4 cores	8 cores
sunny-def	889.33	869.42	830.71	825.55
sunny-all	**858.20**	**838.41**	**823.18**	**797.53**
sunny-stc	858.63	853.55	849.45	844.69
VPS	*858.63*	*854.13*	*852.16*	*851.17*
VBS	*841.33*			

larger as the number of cores increases. Table 4 shows instead the average TIME performances. Being the majority of the instances of Δ very hard to solve, it is not surprising that the TIME values are very close to the timeout $T = 900$. All the tested approaches perform well, since they are very close to, or better than, the *VBS*. The effectiveness in reducing the optimization time is also corroborated by the fact that for 41 instances (6.33% of Δ) sunny-all can prove the optimality of a solution in less than 90 s.

Summarizing, we can say that all the sunny-cp variants we tested can be effective on the RCPSP instances of Δ, especially when more than one core is used. The solver's parallelization is not the only key for the success of such approaches. The use of the free search and the bounds communication between the scheduled solvers enable to outperform the *VBS*. Furthermore, it is also important —especially for OPT and TIME metrics— to properly schedule a subset of solvers dynamically, i.e., according to the instance to be solved. For a more in depth discussion of the results and more data we invite the interested reader to the companion technical report available at https://hal.inria.fr/hal-01295061.

5   Conclusions

The Resource-Constrained Project Scheduling Problem (RCPSP) is a well-known scheduling problem applicable in many real-life scenarios. In this paper we show how it is possible to boost its resolution by using a portfolio of different constraint solvers for selecting and running a subset of them on multiple cores. Improvements are manifold in terms of both solution quality, optima proven and optimization time. We noticed in particular significant performance gains in quickly proving more optima.

We believe that this work may open the way to several extensions. An interesting one concerns the analysis of how to properly integrate and combine the concurrent execution of different constraint solvers. It is also certainly worth to evaluate and combine other search heuristics for the RCPSP resolution (e.g., precedence-setting searches, texture-based heuristics [8]). From the portfolio perspective, we hope that this work can stimulate the utilization of portfolio solvers also in real-life scenarios where typically a single, dominant solver is used for solving different instances of the same problem.

References

1. Amadini, R., Gabbrielli, M., Mauro, J.: Portfolio approaches for constraint optimization problems. In: Pardalos, P.M., Resende, M.G.C., Vogiatzis, C., Walteros, J.L. (eds.) LION 2014. LNCS, vol. 8426, pp. 21–35. Springer, Heidelberg (2014). doi:10.1007/978-3-319-09584-4_3
2. Amadini, R., Gabbrielli, M., Mauro, J.: SUNNY: a Lazy portfolio approach for constraint solving. TPLP **4–5**, 509–524 (2014)
3. Amadini, R., Gabbrielli, M., Mauro, J.: A multicore tool for constraint solving. In: IJCAI, pp. 232–238 (2015)
4. Amadini, R., Gabbrielli, M., Mauro, J.: SUNNY-CP: a sequential CP portfolio solver. In: SAC, pp. 1861–1867 (2015)
5. Amadini, R., Gabbrielli, M., Mauro, J.: Why CP portfolio solvers are (under)utilized? Issues and challenges. In: Falaschi, M. (ed.) LOPSTR 2015. LNCS, vol. 9527, pp. 349–364. Springer, Heidelberg (2015). doi:10.1007/978-3-319-27436-2_21
6. Amadini, R., Stuckey, P.J.: Sequential time splitting and bounds communication for a portfolio of optimization solvers. In: O'Sullivan, B. (ed.) CP 2014. LNCS, vol. 8656, pp. 108–124. Springer, Heidelberg (2014). doi:10.1007/978-3-319-10428-7_11
7. Arlot, S., Celisse, A.: A survey of cross-validation procedures for model selection. Statist. Surv. **4**, 40–79 (2010)
8. Christopher Beck, J., Davenport, A.J., Sitarski, E.M., Fox, M.S.: Texture-based heuristics for scheduling revisited. In: AAAI, pp. 241–248 (1997)
9. Blazewicz, J., Lenstra, J.K., Rinnooy Kan, A.H.G.: Scheduling subject to resource constraints: classification and complexity. Discret. Appl. Math. **5**(1), 11–24 (1983)
10. Brucker, P., Drexl, A., Mohring, R.H., Neumann, K., Pesch, E.: Resource-constrained project scheduling: notation, classification, models, and methods. Eur. J. Oper. Res. **1**, 3–41 (1999)
11. Gomes, C.P., Selman, B.: Algorithm portfolios. Artif. Intell. **1–2**, 43–62 (2001)
12. Hartmann, S., Briskorn, D.: A survey of variants and extensions of the resource-constrained project scheduling problem. Eur. J. Oper. Res. **1**, 1–14 (2010)
13. Herroelen, W., De Reyck, B., Demeulemeester, E.: Resource-constrained project scheduling: a survey of recent developments. Comput. OR **4**, 279–302 (1998)
14. Kolisch, R., Hartmann, S.: Experimental investigation of heuristics for resource-constrained project scheduling: an update. Eur. J. Oper. Res. **1**, 23–37 (2006)
15. Kreter, S., Schutt, A., Stuckey, P.J.: Modeling and solving project scheduling with calendars. In: Pesant, G. (ed.) CP 2015. LNCS, vol. 9255, pp. 262–278. Springer, Heidelberg (2015). doi:10.1007/978-3-319-23219-5_19
16. Nethercote, N., Stuckey, P.J., Becket, R., Brand, S., Duck, G.J., Tack, G.: MiniZinc: towards a standard CP modelling language. In: CP, pp. 529–543 (2007)
17. Ohrimenko, O., Stuckey, P.J., Michael, C.: Propagation via lazy clause generation. Constraints **3**, 357–391 (2009)
18. Rossi, F., van Beek, P., Walsh, T. (eds.): Handbook of Constraint Programming (2006)
19. Schutt, A., Feydy, T., Stuckey, P.J., Wallace, M.G.: Explaining the cumulative propagator. Constraints **3**, 250–282 (2011)
20. Schutt, A., Feydy, T., Stuckey, P.J., Wallace, M.G.: Solving RCPSP/max by lazy clause generation. J. Sched. **3**, 273–289 (2013)

Characterization of Neighborhood Behaviours in a Multi-neighborhood Local Search Algorithm

Nguyen Thi Thanh Dang$^{(\boxtimes)}$ and Patrick De Causmaecker

KU Leuven KULAK, CODeS, iMinds-ITEC, Kortrijk, Belgium
{nguyenthithanh.dang,patrick.decausmaecker}@kuleuven-kulak.be

Abstract. We consider a multi-neighborhood local search framework with a large number of possible neighborhoods. Each neighborhood is accompanied by a weight value which represents the probability of being chosen at each iteration. These weights are fixed before the algorithm runs, and can be tuned by off-the-shelf off-line automated algorithm configuration tools (e.g., SMAC). However, the large number of parameters might deteriorate the tuning tool's efficiency, especially in our case where each run of the algorithm is not computationally cheap, even when the number of parameters has been reduced by some intuition. In this work, we propose a systematic method to characterize each neighborhood's behaviours, representing them as a feature vector, and using cluster analysis to form similar groups of neighborhoods. The novelty of our characterization method is the ability of reflecting changes of behaviours according to hardness of different solution quality regions based on simple statistics collected during any algorithm runs. We show that using neighborhood clusters instead of individual neighborhoods helps to reduce the parameter configuration space without misleading the search of the tuning procedure. Moreover, this method is problem-independent and potentially can be applied in similar contexts.

Keywords: Algorithm configuration · Clustering · Multi-neighborhood local search

This paper is organized as follows. We describe the considered tuning problem in more detail in Sect. 1. The method for characterizing neighborhoods' behaviours and clustering them is explained in Sect. 2. Section 3 shows the advantage of using clustering in automated parameter tuning through out experimental results. Section 4 gives conclusion and discussion on future work.

1 Parameter Tuning for a Multi-neighborhood Local Search Algorithm

The local search algorithm considered in this work, which was developed by CODeS group's members of the University of Leuven (Belgium) [1], tackles the Swap-body Vehicle Routing problem. It is the winner of the Verolog Solver Challenge 2014 [2]. It is an iterated local search [3] algorithm that uses late

© Springer International Publishing AG 2016
P. Festa et al. (Eds.): LION 2016, LNCS 10079, pp. 234–239, 2016.
DOI: 10.1007/978-3-319-50349-3_17

acceptance hill climbing [4] as the local search component. At each iteration of the late acceptance hill climbing, a neighborhood N_k is randomly chosen with a probability proportional to its weight.

The algorithm consists of 18 neighborhood types. A number of them can be parameterized by their sizes, leading to a total number of 42 neighborhoods. Intuition can be used to reduce the number of weights to 28: some neighborhoods that belong to the same neighborhood type and have similar sizes can be grouped into one. This manual clustering was learned from the setting provided by the target algorithm developers, and is supposed to help the tuning better compared to the 42-weights version. The 42 neighborhoods and their corresponding groups of sizes (separated by slashes) are listed below:

– Non-parameterized   neighborhood   types: *Swap,   Intra-route-two-opt, Inter-route-two-opt, Change-swap-location, Merge-route, Split-to-sub-routes, Remove-route, Remove-sub-route, Remove-sub-route-with-cheapest-insertion, Convert-to-route, Convert-to-sub-route, Add-sub-route*
– Parameterized neighborhood types:
  - *Cheapest-insertion*: 1; 2; 3; 4; 5 / 10; 15 / 20; 25 / 35 / 50
  - *Ejection-chain*: 3; 4; 5 / 10 / 15 / 35
  - *Ruin-recreate*: 2; 3
  - *Remove-chain*: 1; 2; 3; 4 / 5; 6 / 7; 8
  - *Each-sequence-cheapest-insertion*: (2,5) / (5,2) / (4,4)

Parameter tuning is done on six (large) problem instances provided by the competition. An algorithm run on each instance takes 600 s. Note that the algorithm considered in this paper is actually not the same as the one that won the competition. The winning one is multi-threaded (4 independent parallel runs) while the one we use here is single-threaded. This is because the aim of our work is not to beat the winning algorithm, but to use this case study as a proof of concept for our characterization method.

2   Neighborhood Characterization and Clustering

Due to the limited space, in this section, the methodology is presented very briefly. An extended version of the paper with detail explanation and an R implementation of the method are available at https://sites.google.com/site/nguyenttdangxyz/.

Inspired by the idea from OSCAR [5], which is an automated approach for online selection of algorithm portfolio, we characterize each neighborhood N_k's behaviours on an instance I based on the following six observables:

– Probabilities that N_k improves, worsens or does nothing on a solution of I, denoted as $r_{improve}, r_{worsen}, r_{nothing}$ (they sum up to one)
– Magnitudes of improvement and worsening, denoted as $a_{improve}$ and a_{worsen}
– N_k's running time (used for tie-breaking)

Our method represents N_k using the estimated values of those observables on different *solution quality regions*, as they reflect changes of N_k's behaviours according to the *hardness* of the solution that it is dealing with. The methodology includes four steps [1]:

2.1   Step 1: Collect Raw Statistics During Algorithm Runs

When an algorithm configuration is run on an instance, a number of raw statistics representing each neighborhood's behaviours can be collected for the characterization. These statistics are accumulable among different algorithm runs. Given a problem instance, we assume that an upper bound and a lower bound of the optimal solution quality are available. We divide the range between the two bounds into a large number of small intervals (here we set it as 1000). Because higher quality solutions in general are harder to improve, we let the size of the intervals decrease exponentially. Each next interval has a size 0.99 the size of the previous interval. Now every time a neighborhood N_k is applied on a solution that has quality value belonging to an interval I_j, the following values are accumulatively collected for the pair (N_k, I_j):

- n_{iters}: the number of times N_k is applied,
- n_I, n_{SN}, n_W: the numbers of times N_k improves, does nothing, or worsens solutions, respectively,
- s_I, s_W: sums of the amount of improvement and worsening,
- s_{time}: sum of N_k's running time.

2.2   Step 2: Identify Solution Quality Regions Based on the Collected Statistics

Intervals are grouped into *frames* based on *sum_nIters*, which is the sum of all neighborhoods' n_{iters} values on each interval. This grouping tries to reflect the hardness of different regions in the solution quality space, based on the observation that the search of the algorithm normally stays the most in the regions around local optima or plateau.

2.3   Step 3: Characterize Neighborhood Behaviours as Feature Vectors

In this step, observable values for each frame are extracted from the collected raw statistics. For the first three observables, $r_{improve}$, r_{worsen} and $r_{nothing}$, we just simply sum the three values n_I, n_W and n_{SN} for all intervals belonging to the same frame. We then divide them by the sum of n_{iters} to get the ratios. For the other two observables $a_{improve}$ and a_{worsen}, aggregation is more complicated due to the incomparableness of their relevant raw statistics among

---

[1] Our provided R implementation does step 2, 3 and 4 automatically. The user only needs to do step 1 herself/himself.

different intervals. For each interval, neighborhoods are firstly ranked based on the averages of their corresponding s_I, s_W values (tie-breaking by the average running time). After that, these ranks are aggregated into frames using the R package *RobustRankAggreg* [6], a robust ranking aggregation method.

2.4 Step 4: Cluster Analysis on Neighborhoods

The first three observables, $r_{improve}$, r_{worsen} and $r_{nothing}$, sum up to one. As a result, their corresponding vector components belong to a special class named *compositional data*. Therefore, we apply the isometric log-ratio transformation proposed in [7] before applying a clustering method for high-dimensional data implemented in the R package HDclassif [8]. In the end, 42 neighborhoods are grouped into 9 clusters:

- *Ejection-chain 3, 4, 5; Remove-chain 1, 2, 3, 6, 7, 8; Remove-sub-route-with-cheapest-insertion;*
- *Swap; Inter-route-two-opt*
- *Cheapest-insertion 10, 15, 20, 25, 35, 50; Each-sequence-cheapest-insertion (2,5), (4,4), (5,2); Remove-chain 4*
- *Cheapest-insertion 1, 2, 3, 4, 5*
- *Change-swap-location; Merge-route*
- *Add-sub-route; Convert-to-sub-route*
- *Ejection-chain 10, 15, 35; Remove-chain 5; Intra-route-two-opt*
- *Ruin-recreate 2, 3*
- *Convert-to-route; Remove-sub-route; Remove-route; Split-to-sub-route*

3 Experimental Results

Our hypothesis is that the proposed characterization method does reflect neighborhood behaviours on the given set of instances, so that the generated feature vectors should correctly represent the neighborhoods and the clusters we obtained are meaningful. To test this hypothesis, we applied the automated tuning tool SMAC [9] to three configuration scenarios:

- *original*: the original 42 neighborhoods are treated independently, i.e., there is a weight value parameter for each neighborhood
- *basic*: the 28 groups of neighborhoods described in Sect. 1 are used.
- *clustered*: the 9 clusters of neighborhoods generated from our characterization method are used.

We carried out 18 runs of SMAC on each scenario. Each one has a budget of 2000 algorithm runs (13.9 CPU days). Due to the large CPU time each SMAC run requires, we use the shared-model-mode offered by SMAC with 10 cores (walltime is reduced to 1.39 days), and take the configuration which has the best training performance as the final one. Mean of optimality gaps (in percentage) on the instance set is used as tuning performance measure. Optimality gap on each instance is calculated by:

$$optimalityGap = 100 * (solutionCost - lowerBound)/lowerBound$$

where *lowerBound* is provided by the algorithm's authors, and is the best solution cost obtained after running the multi-threaded version of the algorithm on the corresponding instance in 6 h. The best algorithm configuration from each SMAC run is evaluated using test performance, which is the mean of optimality gaps obtained from 30 runs of the configuration on the instance set (5 runs/instance). Their corresponding test performance box-plots are shown in Fig. 1, in which the *clustered* scenario offers advantage over both of the others. *Paired t-tests* conducted on the two pairs (*clustered, original*) and (*clustered, basic*) give $p - values$ of 0.00216206 and 0.009258918 respectively, indicating statistical significance [2].

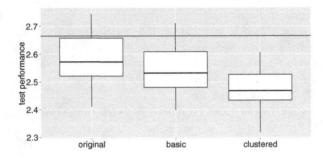

Fig. 1. The box-plots shows test performance of the three considered scenarios. The horizontal line is the default configuration.

4   Conclusion

In this paper, we propose a systematic method to characterize neighborhood behaviours in a multi-neighborhood local search framework, where the probability of choosing a neighborhood at each iteration is chosen in an off-line manner. The characterization is based on the probabilities that a neighborhood will improve, worsen or do nothing on a solution, on the magnitudes of its improvement and worsening, and on its running time. We have observed that these characteristics change according to hardness of different regions in solution quality space. As a result, we design our method such that it tries to detect these regions based on collected information and represent neighborhood behaviours on them as feature vectors. Cluster analysis is then applied to form groups of similar neighborhoods. A tuning experiment with the automated algorithm

---

[2] We also carried out another experiment to compare the tuned configurations with their corresponding Simple Random versions, as usually done in the Hyper-heuristic community. Interested readers are reffered to the extended version of this paper on http://arxiv.org/abs/1603.06459 for more details.

configuration tool SMAC [9] shows that using these clusters gives a statistically significant improvement on test performances of the obtained algorithm configurations over the non-clustered versions. It verifies the hypothesis that our characterization method is able to correctly reflect neighborhood behaviours on the given instance set. Since the information used in this method does not depend on a specific problem, the characterization and clustering procedure potentially can be applied in similar contexts. An R implementation of the methodology is publicly available.

For future work, since our current method are only limited to a small number of instances, we are seeking for the possibility of an extension to a large set of instances. We might want to exploit problem-specific expert knowledge, e.g., instance features, in such a case.

Acknowledgement. This work is funded by COMEX (Project P7/36), a BELSPO/IAP Programme. We thank Túlio Toffolo for his great help during the course of this research, Thomas Stützle and Jan Verwaeren for their valuable remarks. The computational resources and services used in this work were provided by the VSC (Flemish Supercomputer Center), funded by the Hercules Foundation and the Flemish Government department EWI.

References

1. Wauters, T., Toffolo, T., Christiaens, J., Van Malderen, S.: The winning approach for the verolog solver challenge 2014: the swap-body vehicle routing problem. In: Proceedings of ORBEL29 (2015)
2. Heid, W., Hasle, G., Vigo, D.: Verolog solver challenge 2014-vsc2014 problem description. In: VeRoLog (EURO Working Group on Vehicle Routing and Logistics Optimization) and PTV Group, pp. 1–6 (2014)
3. Lourenço, H.R., Martin, O.C., Stützle, T.: Iterated local search: framework and applications. In: Gendreau, M., Potvin, J.-Y. (eds.) Handbook of Metaheuristics. International Series in Operations Research and Management Science, vol. 146, pp. 363–397. Springer, Heidelberg (2010)
4. Burke, E.K., Bykov, Y.: A late acceptance strategy in hill-climbing for exam timetabling problems. In: PATAT 2008: Conference, Montreal, Canada (2008)
5. Mısır, M., Handoko, S.D., Lau, H.C.: OSCAR: online selection of algorithm portfolios with case study on memetic algorithms. In: Dhaenens, C., Jourdan, L., Marmion, M.-E. (eds.) LION 2015. LNCS, vol. 8994, pp. 59–73. Springer, Heidelberg (2015). doi:10.1007/978-3-319-19084-6_6
6. Kolde, R., Laur, S., Adler, P., Vilo, J.: Robust rank aggregation for gene list integration and meta-analysis. Bioinformatics **28**(4), 573–580 (2012)
7. Egozcue, J.J., Pawlowsky-Glahn, V., Mateu-Figueras, G., Barcelo-Vidal, C.: Isometric logratio transformations for compositional data analysis. Math. Geol. **35**(3), 279–300 (2003)
8. Bergé, L., Bouveyron, C., Girard, S.: HDclassif: an r package for model-based clustering and discriminant analysis of high-dimensional data. J. Stat. Softw. **46**(6), 1–29 (2012). http://www.jstatsoft.org/v46/i06/
9. Hutter, F., Hoos, H.H., Leyton-Brown, K.: Sequential model-based optimization for general algorithm configuration. In: Coello, C.A.C. (ed.) LION 2011. LNCS, vol. 6683, pp. 507–523. Springer, Heidelberg (2011). doi:10.1007/978-3-642-25566-3_40

Constraint Programming and Machine Learning for Interactive Soccer Analysis

Robinson Duque[1(✉)], Juan Francisco Díaz[1], and Alejandro Arbelaez[2]

[1] Universidad del Valle, Cali, Colombia
{robinson.duque,juanfco.diaz}@correounivalle.edu.co
[2] Insight Centre for Data Analytics, University College Cork, Cork, Ireland
alejandro.arbelaez@insight-centre.org

Abstract. A soccer competition consists of n teams playing against each other in a league or tournament system, according to a single or double round-robin schedule. These competitions offer an excellent opportunity to model interesting problems related to questions that soccer fans frequently ask about their favourite teams. For instance, at some stage of the competition, fans might be interested in determining whether a given team still has chances of winning the competition (i.e., finishing first in a league or being within the first k teams in a tournament to qualify to the playoff). This problem relates to the elimination problem, which is NP-complete for the actual FIFA pointing rule system (0, 1, 3), zero point to a loss, one point to a tie, and three points to a win. In this paper, we combine constraint programming with machine learning to model a general soccer scenario in a real-time application.

1 Introduction

Soccer fans usually have questions related to their favourite teams and most of the time they are subject to media speculations that are sometimes proved wrong. Many domestic leagues use a two-stage tournament structure with a single or double round-robin tournament for the regular season and a final knockout stage (aka playoffs). The first stage of the league typically features between 16 and 30 teams, each team faces each other team once or twice with *home* and *away* matches distributed evenly in the regular season. Depending on the results of the matches, every team is awarded some points under the FIFA three-point-rule (three points for a victory, one point for a draw, and zero points for a defeat), and the top k teams (typically eight) qualify for the playoffs.

The elimination problem is well-known in sports competitions, particularly from baseball [1,2] and consists in determining whether at some stage of the competition a given team still has the opportunity to be within the top teams to qualify for playoffs. The complexity of the problem depends on the score system for the results of the matches. In [3,4] the authors showed that the elimination problem is NP-complete for the current FiFA score system (0, 1, 3). However, interestingly [4] pointed out that with the old FIFA score system (0, 1, 2) from the 90's, the elimination problem could be solved in polynomial time using a

© Springer International Publishing AG 2016
P. Festa et al. (Eds.): LION 2016, LNCS 10079, pp. 240–246, 2016.
DOI: 10.1007/978-3-319-50349-3_18

network flow algorithms as first proposed by [5]. In this paper we attempt to present a general model to simulate scenarios and problems where fans can formulate queries about the positions of the teams at the end of a tournament, e.g., Will R. Madrid be in a better position than 3. To this end, we propose a combination of constraint programming (CP) with machine learning (ML) to answer soccer related queries.

2   CP Model for Soccer Queries

CP is a powerful technique to solve combinatorial problems which combines backtracking with constraint propagation. At each step a value is assigned to some variable. Each assignment is combined with a look-ahead process called constraint propagation which can reduce the domains of the remaining variables. In the following, we describe a CP formulation for soccer competitions, we start by offering a list of variables and notations for a basic soccer model.

- n: number of teams in the competition;
- T: set of team indexes in the competition;
- i, j: team indexes, such that $(i, j \in T)$;
- p_i: initial points of team i. If i has not played any games, then $p_i = 0$;
- F: number of fixtures left to be played in the competition. A fixture consists of one or more games between competitors;
- k: represents a fixture number, $(1 \leq k \leq F)$;
- G: set that represents the schedule of the remaining games to be played. Every game is represented as a triple $ng_e = (i, j, k) \wedge 0 \leq e \leq |G|$, where k is the fixture when both teams (i and j) meet in a game;
- pt_{ik}: represents the points that team i gets in fixture k, $(1 \leq k \leq F$ and $pt_{ik} \in \{0, 1, 3\})$. If team i is not scheduled to play fixture k, then $pt_{i,k} = 0$.
- tp_i: total points of team i at end of the competition;
- geq_{ij}: boolean variable indicating if team j has greater or equal total points as i: if $tp_j \geq tp_i$ then $geq_{ij} = 1$; otherwise $geq_{ij} = 0$, $(\forall i, j \in T)$;
- eq_{ij}: boolean variable indicating if two different teams i and j tie in points at the end of the competition: if $tp_j = tp_i$ and $i \neq j$ then $eq_{ij} = 1$; otherwise $eq_{ij} = 0$, $(\forall i, j \in T)$.
- pos_i: position of team i at the end of the competition;
- $worstPos_i$: upper bound for pos_i;
- $bestPos_i$: lower bound for pos_i;

Position in Ranking Queries: we use this set of variables to represent queries about positions of the teams at the end of the competition (e.g., R. Madrid will be in position 3).

- P: set of possible position in ranking queries, defined as a set of triples $np_b = (i, opr_i, ptn_i)$ and $0 \leq b \leq |P|$;
- opr_i: logical operator ($opr_i \in \{<, \leq, >, \geq, =\}$) to constrain team i;
- ptn_i: denoting the expected position for team i; $1 \leq ptn_i \leq n$;

2.1 CP Model Formulation

Basic Soccer Model: Constraints (1), (2), and (3) represent a valid game point assignment (0,3), (3,0) or (1,1) for each game $ng_e \in G$ between two teams i and j in a fixture k:

$$(0 \leq pt_{ik} \leq 3) \wedge (0 \leq pt_{jk} \leq 3) \quad \forall ng_e \in G \wedge ng_e = (i,j,k) \tag{1}$$

$$(pt_{ik} \neq 2) \wedge (pt_{jk} \neq 2) \quad \forall ng_e \in G \wedge ng_e = (i,j,k) \tag{2}$$

$$2 \leq pt_{ik} + pt_{jk} \leq 3 \quad \forall ng_e \in G \wedge ng_e = (i,j,k) \tag{3}$$

Constraint (4) corresponds to the final points tp_i of a team i. It is the addition of the initial points p_i and the points pt_{ik} obtained in every fixture k:

$$tp_i = p_i + \sum_{k=1}^{F} pt_{ik} \quad \forall i \in T \tag{4}$$

Constraints (5) to (8) are used to calculate final positions. All the final positions must be different and every position is bounded by $bestPos_i$ and $worstPos_i$:

$$geq_{ij} = \begin{cases} 1, & \text{if } tp_j \geq tp_i \\ 0, & otherwise \end{cases} \quad \forall i,j \in T$$

$$worstPos_i = \sum_{j=1}^{n} geq_{ij} \quad \forall i,j \in T \tag{5}$$

$$eq_{ij} = \begin{cases} 1, & \text{if } tp_j = tp_i \text{ and } i \neq j \\ 0, & otherwise \end{cases} \quad \forall i,j \in T$$

$$bestPos_i = worstPos_i - \sum_{j=1,j\neq i}^{n} eq_{ij} \quad \forall i,j \in T \wedge i \neq j \tag{6}$$

$$bestPos_i \leq pos_i \leq worstPos_i \quad \forall i \in T \tag{7}$$

$$alldifferent(pos_1, \ldots, pos_n) \tag{8}$$

Position in Ranking Queries: involves a set of constrained teams and indicates whether a given team can be above, below, or at a given position ptn_i, constraint (9) depicts the five possibilities:

$$\forall np_b \in P \wedge np_b = (i, opr_i, ptn_i) \begin{cases} pos_i = ptn_i, & \text{if } opr_i \text{ is } = \\ pos_i < ptn_i, & \text{if } opr_i \text{ is } < \\ pos_i \leq ptn_i, & \text{if } opr_i \text{ is } \leq \\ pos_i > ptn_i, & \text{if } opr_i \text{ is } > \\ pos_i \geq ptn_i, & \text{if } opr_i \text{ is } \geq \end{cases} \tag{9}$$

2.2 Variable/Value Selection

Generic heuristics (e.g., [6,7]) typically do not perform well for real-life problems as these heuristics do not exploit the structure of the problem. Therefore, in this paper we propose some query based heuristics for variable/value selection. First we introduce a set of required variables in order to describe a priority mechanism to select the team variables constrained in queries P:

- $spos_i$: starting position of team i before any branching strategy is applied;
- pri_i: denoting the priority of team i to be selected during branching, If team i does not appear in any query, then $pri_i = 0$;
- str_i: denoting the global branching strategy for the variables pt_{ik} of a particular team i in every fixture k. str_i starts with "tie" as a default value.

Heuristics for Position in Ranking Queries (P). Recall from (9) that we use the position pos_i to constrain a team to a wanted position ptn_i. Suppose we have the query ($pos_i < 8$). It's natural to try $str_i = win$ and assign a priority using the position ptn_i from the query. We depict in (10) some general rules for variable value selection:

$$\forall np_b \in P \wedge np_b = (i, opr_i, ptn_i) \begin{cases} \text{if } opr_i \text{ is } < \text{ or } \leq, & pri_i = n - ptn_i \wedge str_i = win \\ \text{if } opr_i \text{ is } > \text{ or } \geq, & pri_i = ptn \wedge str_i = lose \\ \text{if } opr_i \text{ is } =, & pri_i = ptn \wedge \begin{cases} str_i = lose, \text{ if } ptn_i > n/2 \\ str_i = win, \text{ otherwise} \end{cases} \end{cases} \quad (10)$$

Interestingly the defined heuristics in (10) for queries with the "=" operator seem to fail quite often (see Sect. 3). Suppose a scenario with a query ($pos_i = 7$) where $spos_i = 9$ with $F = 8$ fixtures to play. Given that the starting position is 9 and we have to reach position 7, the global branching strategy $str_i = win$ causes that pos_i overshoots position 7 and would require many backtracks of the search algorithm in order to reach such position, therefore, it might be useful to perform a bias search and in the following section we tackle this problem by using machine learning.

Machine Learning for Value Selection. For teams constrained with the "=" operator, we decided to assign a high priority ($pri_i = |spos_i - ptn_i| \cdot n$) for variable selection and to avoid position overshooting, we trained a classifier that selects among 9 branching strategies: S1 = [1,0,0], S2 = [0,1,0], S3 = [0,0,1], S4 = [0.5,0.5,0], S5 = [0.5,0,0.5], S6 = [0,0.5,0.5], S7 = [0.5,0.25,0.25], S8 = [0.25,0.5,0.25], S9 = [0.25,0.25,0.5]. Each strategy defines probabilities to select among [win, tie, lose] respectively, e.g., S7 means that for a team i, every variable pt_{ik} will be assigned *win* with a probability of 0.5, *tie* and *lose* with a probability of 0.25 each. We use the selected strategy with a restart-based search; therefore we restart the algorithm when some cutoff in the execution time is met. (3 s in this paper). Notice that we excluded a strategy [1/3, 1/3, 1/3] as preliminarily tests showed a poor performance for this alternative.

Training the Classifier: In order to train this classifier, we created a total of 500 P queries with the equality operator at different stages of a tournament (fixture 7, 9, 11, 14 and 16) with 18 teams, scheduled in a single round robin. We ran every query with each of the 9 branching strategies in order to get the strategy that solved the instance in the shortest time and created a data set with the following features: starting position ($spos_i$), wanted position (ptn_i), direction and distance ($spos_i - ptn_i$), fixtures to play (F), range rate ($|spos_i - ptn_i|/n$), best executing strategy $\in \{S1, S2, S3, S4, S5, S6, S7, S8, S9\}$.

In this paper we use J48 (the Weka v3.6.12 implementation of C4.5) to evaluate the performance of the algorithms. The objective is that for each query P with the equality operator "=", J48 assigns one of the nine branching strategies to the constrained team.

3   Empirical Evaluation

Tests Configuration. We evaluated our models using Mozart-Oz (V 1.4.0) as our CP reference solver. All the experiments were performed in a 4-core machine, featuring an Intel Core i5 processor at 2.3 Ghz and 4 GB of RAM. We focus our attention in the Colombian league (liga Postobón 2014-I) with 18 teams and 18 fixtures to play in a single round-robin schedule (17 fixtures + 1 extra fixture for the derbies). We provided five experimental scenarios (i.e., fixtures 7, 9, 11, 14, and 16). We also created a series of instances for each fixture (100 with 2 suppositions, 100 with 3 suppositions, and the same for 4, 5, 7, and 9 suppositions). Each instance (3000 in total) was executed with a time limit of 30 s. We recall that our models are implemented in SABIO, a Web based application where long answer times are not desirable. We experimented two scenarios: the basic CP implementation using the heuristics for position in ranking queries and the CP-ML implementation configured with 10 restarts (i.e., 3 s per restart), featuring the basic heuristics and the machine learning classifier for equality constraints.

Tests Results. Table 1 shows the number of unsolved instances and the average runtimes of the solved ones in our experiments. We observe 1069 unsolved instances with the CP model and we attribute this to 2 main reasons: first, the position bounds (i.e., $bestPos_i$ and $worstPos_i$) can only be computed after finding the total points (tp_i) for all the teams in the competition. As a result, position in ranking constraints standing are validated only when the search algorithm performs a complete game points assignment for all teams. Second, we observed that our variable/value selection heuristics struggle with queries related to the "=" operator and the lack of a biased search causes position overshooting. We also observed that our CP-ML implementation seems to perform better and the classifier improves the effectiveness of the algorithm by reducing the number of unsolved instances from 1069 to 627 while displaying a small trade-off in running time.

Table 1. Unsolved instances and average running times of CP and CP-ML

Fixture	Test	P Queries Sup.						Running times					
		2	3	4	5	7	9	2	3	4	5	7	9
Fixture 7	CP	24	28	38	46	50	52	.22	.48	.31	.41	.77	1.10
	CP-ML	6	9	14	21	27	35	1.19	1.34	1.83	2.40	3.28	3.09
Fixture 9	CP	23	28	35	46	49	44	.08	.41	.23	.65	.31	.72
	CP-ML	1	4	9	18	27	31	1.68	1.48	2.25	2.81	2.78	1.66
Fixture 11	CP	25	27	31	44	49	47	.56	.10	.75	.69	.86	.45
	CP-ML	2	7	11	18	31	34	1.53	1.09	2.01	2.67	2.56	2.46
Fixture 14	CP	23	33	38	41	51	45	.42	.13	.49	1.01	1.53	.74
	CP-ML	8	22	25	34	44	42	1.17	1.29	1.46	1.48	1.08	1.47
Fixture 16	CP	23	31	29	31	25	13	.12	.04	.89	.55	.45	.65
	CP-ML	19	25	25	33	28	17	.56	.50	.93	.05	.11	.03
CP results		Unsolved: 1069						Avg: 0.51 s					
CP-ML results		Unsolved: 627						Avg: 1.60 s					

4   Conclusions

In this paper we have combined the use of constraint programming and machine learning to solve general soccer fan queries at different stages of a competition and presented 2 alternative solutions CP and CP-ML. Our computational experiments showed that our CP-ML model improves CP effectiveness since it performs a query biased search. We also plan to extend our models to deal with more queries such as determining the maximum number of games that a team can afford to lose and still qualify to the playoffs and also explore the implementation of a MIP model, based on the work of Ribeiro and Urrutia [8].

We would like to thank Luis Felipe Vargas, María Andrea Cruz and Carlos Martínez for developing early versions of the CP model under the supervision of Juan Francisco Díaz. Robinson Duque is supported by Colciencias under the PhD scholarship program. Alejandro Arbelaez is supported by the DISCUS project (FP7 Grant Agreement 318137), and Science Foundation Ireland (SFI) Grant No. 10/CE/I1853.

References

1. Schwartz, B.L.: Possible winners in partially completed tournaments. SIAM Rev. 8(3), 302–308 (1966)
2. Hoffman, A., Rivlin, T.: When is a team mathematically eliminated? In: Princeton Symposium on Mathematical Programming, pp. 391–401. Princeton, NJ (1967)
3. Kern, W., Paulusma, D.: The new fifa rules are hard: complexity aspects of sports competitions. Discrete Appl. Math. 108(3), 317–323 (2001)

4. Bernholt, T., Gülich, A., Hofmeister, T., Schmitt, N.: Football elimination is hard to decide under the 3-point-rule. In: Kutyłowski, M., Pacholski, L., Wierzbicki, T. (eds.) MFCS 1999. LNCS, vol. 1672, pp. 410–418. Springer, Heidelberg (1999). doi:10.1007/3-540-48340-3_37
5. Wayne, K.D.: A new property and a faster algorithm for baseball elimination. SIAM J. Discrete Math. **14**(2), 223–229 (2001)
6. Arbelaez, A., Hamadi, Y.: Exploiting weak dependencies in tree-based search. In: SAC 2009, pp. 1385–1391 (2009)
7. Haralick, R.M., Elliott, G.L.: Increasing tree search efficiency for constraint satisfaction problems. In: IJCAI 1979, San Francisco, CA, USA, pp. 356–364 (1979)
8. Ribeiro, C.C., Urrutia, S.: An application of integer programming to playoff elimination in football championships. ITOR **12**(4), 375–386 (2005)

A Matheuristic Approach for the p-Cable Trench Problem

Eduardo Lalla-Ruiz$^{(\boxtimes)}$, Silvia Schwarze, and Stefan Voß

Institute of Information Systems, University of Hamburg, Von-Melle-Park 5,
20146 Hamburg, Germany
{eduardo.lalla-ruiz,silvia.schwarze,stefan.voss}@uni-hamburg.de

Abstract. The p-Cable Trench Problem is a telecommunications network design problem, which jointly considers cable and trench installation costs and addresses the optimal location of p facilities. In this work, a matheuristic approach based on the POPMUSIC (Partial Optimization Metaheuristic under Special Intensification Conditions) framework is developed. The inspected neighborhoods for building sub-problems include lexicographic as well as nearest neighbor measures. Using benchmark data available from literature it is shown that existing results can be outperformed.

1 Introduction

The Cable Trench Problem (CTP) reflects a scenario that appears in the installation of information technology infrastructure. In particular, it joins two cost types that appear in the construction of wire-based networks, namely cost for installation of cables and cost for preparing trenches. A trench may contain more than one cable such that a solution has to balance lengths of the cables on the one hand and the distance covered by the trenches on the other hand. As a result, the CTP combines the problems of finding a shortest path tree and of finding a minimum spanning tree. The CTP was proposed by [4] for the problem of connecting buildings to a central facility on a campus. Recent publications suggest further applications. In [5] a problem from bioinformatics, the representation of vascular network connectivity in medical image analysis is addressed by solving a Generalized CTP (GCTP). Moreover, [6] models the setup of a low-frequency radioastronomy station by applying a GCTP. An extension to the CTP is the p-CTP proposed by [1] and now introduced in more detail.

Let $G = (V, E)$ be a connected graph with nodes $i \in V$ and directed edges $e \in E$. For each edge (i, j) in E, the cost of installing one cable is given by $D_{ij} > 0$ and the cost of preparing a trench is denoted by $C_{ij} > D_{ij}$. In contrast to the CTP, where each node has to be connected to a given, single source node, the p-CTP requires to open exactly p facilities. The goal is to choose p of the $n = |V|$ nodes to act as facilities and to assign the remaining nodes to these p facilities, such that the total cost for cable and trenches is minimized.

A small example for a p-CTP is presented in Figs. 1 and 2. We consider a graph with $n = 11$ nodes and fix $p = 2$, i.e., two facilities shall be opened.

© Springer International Publishing AG 2016
P. Festa et al. (Eds.): LION 2016, LNCS 10079, pp. 247–252, 2016.
DOI: 10.1007/978-3-319-50349-3_19

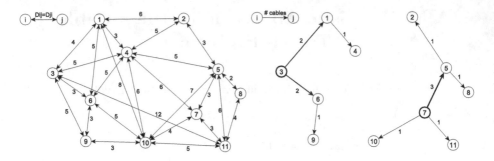

Fig. 1. Instance, $n = 11, p = 2$.   **Fig. 2.** Optimal solution: cost 97.

Moreover, Fig. 1 illustrates the graph G of the instance together with cable costs D_{ij} for each edge $(i, j) \in E$. The cost for preparing a trench on any edge (i, j) is fixed to $C_{ij} = 2D_{ij}$. An optimal solution is presented in Fig. 2. Nodes 3 and 7 are facility nodes and are indicated by bold circles. Moreover, each edge (i, j) is labeled with the number of installed cables (# cables). The cost of this solution is 97 and divides into trenching cost of 56 and cable cost of 41.

In [1], a mixed-integer programming formulation is proposed for the p-CTP and used to solve instances with up to 200 nodes. Nevertheless, when the dimensions of the instances increase, the solver runs out of memory. Furthermore, two heuristics based on Lagrangean relaxation are provided and tested for instances of up to 300 nodes. In order to improve the solution quality and avoid the memory-fault status, we propose in this work a matheuristic approach for the p-CTP. In particular, the Partial Optimization Metaheuristic under Special Intensification Conditions (POPMUSIC) [3] is applied. POPMUSIC addresses large instances by decomposing them into a set of parts. Subsets of parts are bundled and then used to form sub-problems for subsequently solving them.

The adaption of the POPMUSIC for the p-CTP is described in more detail in Sect. 2. Afterwards, numerical experiments are provided in Sect. 3 and the paper closes with concluding remarks in Sect. 4.

2   A POPMUSIC Approach for the p-CTP

The basic idea of POPMUSIC, proposed in [3], is to split an available solution S of the problem into t parts $part_1, part_2, \ldots, part_t$ and joining some of them to build a sub-problem R. To construct R, first a particular part, namely $part_{seed}$, is selected. Afterwards, r parts closest to $part_{seed}$ are merged with $part_{seed}$ to produce the sub-problem R. In order to determine the closeness of the parts, a distance measure is defined. Once a sub-problem R is constructed, it is solved by using an approximate or an exact solution approach. If parts and sub-problems are defined in an appropriate way, every improvement of a sub-problem corresponds to an improvement of the solution S. This process is repeated until the solution contains no sub-problem that can be improved.

Algorithm 1 depicts the POPMUSIC framework. An initial solution S is generated (line 1). Once it is build, the next step is to divide the solution into t parts (line 2). Then, a seed, $part_{seed}$, is selected (line 5). The sub-problem R is constructed by considering its r nearest parts according to a distance measure (line 6). In this regard, the unique parameter of this framework, r, is used for delimiting the size of the sub-problems. The sub-problem R is then solved by an approximate or exact procedure (line 7). In this framework, the set O gives the seed parts that correspond to sub-problems that have been unsuccessfully optimized. Once O contains all the parts of the complete solution (line 4), the process stops as all sub-problems have been examined without success.

Algorithm 1. POPMUSIC framework

1   Generate an initial solution S at random
2   Decompose S in t parts, $H = \{part_1, ..., part_t\}$
3   Set $O = \emptyset$
4   **while** $O \neq \{part_1, ..., part_t\}$ **do**
5   Select a seed $part_{seed} \notin O$
6   Build sub-problem R composed of r parts of S closest to $part_{seed}$
7   Optimize R by using an approximate or exact solution approach
8   **if** R *has been improved* **then**
9   Update solution S
10        $O \leftarrow \emptyset$
11   **end**
12   **else**
13        $O \cup \{part_{seed}\}$
14   **end**
15 **end**
16 return the improved solution S

In order to develop a POPMUSIC approach for the p-CTP, an initial solution is decomposed by considering those p nodes selected to be the facilities. All trenches and cables departing from each facility including the assigned nodes can be seen as a sub-network. Thus, in the context of POPMUSIC, the size of a solution is $t = p$ and each part is a sub-network induced by a facility. In the example provided in Fig. 2, the size of the solution structure is $t = 2$ and it is composed by nodes $part_1 = 3$ and $part_2 = 7$.

The sub-problems are built by means of the sub-networks represented by their starting nodes and the associated edges and nodes. Therefore, in the example provided in Fig. 2, when building a sub-problem of size $r = 2$ with the seed part $part_1 = 3$ all the nodes belonging to the corresponding sub-networks form the sub-problem. In the case of the aforementioned example, the new sub-problem may consider all the nodes from the network starting with $part_1 = 3$ and $part_2 = 7$. Moreover, for building the sub-problems, different measures or strategies can be used to indicate the closeness of the parts among themselves:

– Lexicographic: The sub-problems are grouped according to the indexes of the parts. That is, all the nodes belonging to $part_1$ are grouped to those belonging to $part_2$ if $r = 1$, also to those of $part_3$ if $r = 2$, and so on. For instance, for a solution divided into 4 parts and $r = 2$, we can have the following sub-problems, $R = \{part_1, part_2\}$, $R = \{part_2, part_3\}$, $R = \{part_3, part_4\}$, and $R = \{part_4, part_1\}$.
– Distance: This strategy takes into account the minimum distance between the facilities. For any $part_i \in H$, let i^* be the route node of $part_i$. Then the distance between $part_i$ and $part_j$ is given as $\bar{D}_{ij} = D_{i^*j^*}$, i.e., by the cable costs assigned to edge (i^*, j^*). If (i^*, j^*) does not exist, a high-enough value is assigned. The construction of the sub-problem is then performed in a greedy way. Once the seed $part_i \in H$ has been selected, that part with the minimum arc distance is assigned. That is, one $part_j = argmin(\bar{D}_{ij})_{part_j \in H, j \neq i}$ is chosen. For $r > 2$, the following parts are added taking into account the minimum average arc distance to the already assigned parts such that the next part to be added is calculated by means of $argmin(\sum_{part_i \in R} \bar{D}_{ij})_{part_j \in H \setminus R}$.

Once the sub-problem has been formed, it has to be solved by an approximate or exact method. In this work, we investigate the approach of applying a branch and cut method provided by a general-purpose solver such as CPLEX. The rationale behind this is (i) to provide flexibility in terms of not requiring to develop specific solvers for the problem itself, (ii) investigate the advantage of decomposing the problem for large-sized problem instances, and (iii) provide a competitive solution approach for addressing this problem in terms of solution quality. At this point, we may stress that dividing the problem into parts allows to address memory problems as the one indicated by Marianov et al. [1] for large-sized instances, where directly managing them may require high-amounts of computational memory.

3   Numerical Results

The computational experiments were conducted on a computer with an Intel i7 CPU 3.50 GHz and 6 GB of RAM, restricted to use one CPU. The model was implemented in CPLEX 12.6. The instances used in this work are those large-sized ones from [1] for the p-CTP.

Table 1 shows the results provided by the best approach reported in the literature based on a Lagrangean relaxation [1] and the results of our POPMUSIC approach with $t = p$ and $r = 0.5p$, for both measures distance (dist) and lexicographic (lex). Moreover, with the aim of reducing the computational time, a modified stopping criterion is realized in rPOPMUSIC. In this version, the algorithm stops if the set O, see Algorithm 1, line 4, contains r elements. That is, in the experiments, rPOPMUSIC stops if $0.5p$ parts have been unsuccessfully examined. The relative error of each approach is calculated by means of the lower bound provided by the Lagrangean approach. Moreover, it should be mentioned that the Lagrangean approach is executed until the step size is lower than 0.0001. Therefore, due to the fact that both approaches reach their respective stopping

Table 1. Numerical results for the large-sized instances provided in [1]

Instance	p	Lagrangean relax. heur. [1]				POPMUSIC - dist			POPMUSIC - lex			rPOPMUSIC - dist			rPOPMUSIC - lex		
		UB	LB	Gap (%)	Time (s)	UB	Gap (%)	Time (s)	UB	Gap (%)	Time (s)	UB	Gap (%)	Time (s)	UB	Gap (%)	Time (s)
Pmed11	30	4566	4375.9	4.3	503	**4532**	3.57	3914.65	4533	3.59	2346.04	4533	3.59	336.31	4544	3.84	378.11
Pmed11	60	3295	2907.7	13.3	493.2	**3277**	12.70	4885.59	**3277**	12.70	2688.89	3283	12.91	314.15	3290	13.15	314.85
Pmed11	90	2403	2169.8	10.7	499.1	2372	9.32	6125.81	**2366**	9.04	3915.45	2369	9.18	327.67	2374	9.41	357.65
Pmed12	30	4650	4451	4.5	612.6	4641	4.27	1780.31	**4636**	4.16	1730.32	4648	4.43	361.73	4646	4.38	303.98
Pmed12	60	3432	2924.5	17.4	762.1	**3385**	15.75	4891.89	**3385**	15.75	2211.76	3391	15.95	322.93	3402	16.33	347.29
Pmed12	90	2498	2228.2	12.1	738.7	2481	11.35	8807.69	**2478**	11.21	3935.17	2486	11.57	347.43	2487	11.61	315.09
Pmed13	30	4584	4288.1	6.9	883.2	4575	6.69	2054.39	**4574**	6.67	1604.44	4594	7.13	349.53	4636	8.11	318.38
Pmed13	60	3364	2873.3	17.1	524.3	3346	16.45	5820.94	**3344**	16.38	2426.28	3357	16.83	333.71	3357	16.83	334.94
Pmed13	90	2475	2223.8	11.3	703.8	**2466**	10.89	9492.88	**2466**	10.89	3002.48	2468	10.98	355.67	2469	11.03	302.50
Pmed14	30	4712	4438.7	6.2	347.4	**4676**	5.35	4812.77	**4676**	5.35	2825.34	4720	6.34	317.73	4716	6.25	332.67
Pmed14	60	3416	2752.8	24.1	918.9	**3392**	23.22	3820.63	**3392**	23.22	3076.78	**3392**	23.22	360.93	3410	23.87	340.96
Pmed14	90	2483	2210.5	12.3	880.1	2455	11.06	5514.69	**2453**	10.97	3325.85	2459	11.24	333.69	2454	11.02	342.00
Pmed15	30	4469	4293.9	4.1	577.5	4448	3.59	5771.30	**4437**	3.33	1622.57	4451	3.66	351.36	4450	3.64	356.95
Pmed15	60	3281	2807.9	16.8	748.4	**3266**	16.31	3626.79	**3266**	16.31	2366.72	3302	17.60	342.03	3282	16.88	304.09
Pmed15	90	2456	2236.6	9.8	542.9	**2427**	8.51	10996.87	**2427**	8.51	3472.56	2430	8.65	346.38	**2427**	8.51	346.62
		3472.27	3145.51	11.39	649.01	3449.27	10.60	5487.81	3447.33	10.54	2703.38	3458.87	10.89	340.08	3462.93	10.99	333.07

criteria without running out of memory, the quality of the solutions provided by them is analyzed.

Independent from the measurement used to determine the closeness among the parts, the results are similar in terms of average gap, see Table 1. In terms of computational time, however, differences are observed. In particular, rPOP-MUSIC allows to provide high-quality solutions in less computational time than required for the Lagrangean heuristic. Moreover, new best values for the inspected instances are reported in bold font. It can be highlighted that for all instances, POPMUSIC and its variants are able to provide new best values.

4 Conclusion

In this work, a matheuristic approach for the p-CTP based on the POPMUSIC template is introduced. Two different ways for building the sub-problems and stopping criteria are proposed and assessed. Moreover, solving the sub-problems is done by means of an available mathematical programming formulation and using the standard solver CPLEX. Under this approach, the complete problem is decomposed and can be treated by the solver, while for the full problem, depending on computer performance, it can reach an out-of-memory status. Thus, the POPMUSIC-based approach provides new best values for all the large-sized problem instances considered for this problem.

For future research, we are going to perform an extensive analysis of different configurations for the POPMUSIC parameters (including various options for choosing seed parts) as well as study other stopping criteria and neighborhood measures.

References

1. Marianov, V., Gutiérrez-Jarpa, G., Obreque, C., Cornejo, O.: Lagrangean relaxation heuristics for the p-cable-trench problem. Comput. Oper. Res. **39**, 620–628 (2012)
2. Nielsen, R.H., Riaz, M.T., Pedersen, J.M., Madsen, O.B.: On the potential of using the cable trench problem in planning of ICT access networks. In: 50th International Symposium ELMAR, pp. 585–588 (2008)
3. Taillard, É.D., Voß, S.: Popmusic - partial optimization metaheuristic under special intensification conditions. In: Ribeiro, C.C., Hansen, P. (eds.) Essays and Surveys in Metaheuristics. Operations Research/Computer Science Interfaces Series, vol. 15, pp. 613–629. Springer, New York (2002)
4. Vasko, F.J., Barbieri, R.S., Rieksts, B.Q., Reitmeyer, K.L., Stott Jr., K.L.: The cable trench problem: combining the shortest path and minimum spanning tree problems. Comput. Oper. Res. **29**, 441–458 (2002)
5. Vasko, F.J., Landquist, E., Kresge, G., Tal, A., Jiang, Y., Papademetris, X.: A simple and efficient strategy for solving very large-scale generalized cable-trench problems. Networks **67**(3), 199–208 (2015)
6. Zyma, K., Girard, J.N., Landquist, E., Schaper, G., Vasko, F.J.: Formulating and solving a radio astronomy antenna connection problem as a generalized cable-trench problem: an empirical study. Int. Trans. Oper. Res. (2016). doi:10.1111/itor.12312

An Empirical Study
of Per-instance Algorithm Scheduling

Marius Lindauer(✉), Rolf-David Bergdoll, and Frank Hutter

University of Freiburg, Freiburg im Breisgau, Germany
`lindauer@cs.uni-freiburg.de`

Abstract. Algorithm selection is a prominent approach to improve a system's performance by selecting a well-performing algorithm from a portfolio for an instance at hand. One extension of the traditional algorithm selection problem is to not only select one single algorithm but a schedule of algorithms to increase robustness. Some approaches exist for solving this problem of selecting schedules on a per-instance basis (e.g., the *Sunny* and *3S* systems), but to date, a fair and thorough comparison of these is missing. In this work, we implement *Sunny*'s approach and dynamic schedules inspired by *3S* in the flexible algorithm selection framework *flexfolio* to use the same code base for a fair comparison. Based on the algorithm selection library (*ASlib*), we perform the first thorough empirical study on the strengths and weaknesses of per-instance algorithm schedules. We observe that on some domains it is crucial to use a training phase to limit the maximal size of schedules and to select the optimal neighborhood size of k-nearest-neighbor. By modifying our implemented variants of the *Sunny* and *3S* approaches in this way, we achieve strong performance on many *ASlib* benchmarks and establish new state-of-the-art performance on 3 scenarios.

Keywords: Algorithm selection · Algorithm schedules · Constraint solving

1 Introduction

A common observation in many areas of AI (e.g., SAT or CSP solving) and machine learning is that no single algorithm dominates the performance of all others. To exploit this complementarity of algorithms, algorithm selection systems [6,8,11] are used to select a well-performing algorithm for a new given instance. Algorithm selectors, such as *SATzilla* [12] and *3S* [7], demonstrated in several SAT competitions that they can outperform pure SAT solvers by a large margin (see, e.g., the results of the SAT Challenge 2012[1]).

An open problem in algorithm selection is that the machine learning model sometimes fails to select a well-performing algorithm, e.g., because of uninformative instance features. An extension of algorithm selection is to select a schedule of multiple algorithms at least one of which performs well.

---

[1] http://baldur.iti.kit.edu/SAT-Challenge-2012/.

© Springer International Publishing AG 2016
P. Festa et al. (Eds.): LION 2016, LNCS 10079, pp. 253–259, 2016.
DOI: 10.1007/978-3-319-50349-3_20

To date, a fair comparison of such algorithm schedule selectors is missing, since every publication used another benchmark set and some implementations (e.g., *3S*) are not publicly available (because of license reasons). To study the strengths and weaknesses of such schedulers in a fair manner, we implemented well known algorithm schedule approaches (i.e., *Sunny* [1] and dynamic schedules inspired by *3S* [7]) in the flexible framework of *flexfolio* (the successor of *claspfolio 2* [5]) and studied them on the algorithm selection library (*ASlib* [3]).

2   Per-instance Algorithm Scheduling

Similar to the per-instance algorithm selection problem [11], the per-instance algorithm scheduling problem is defined as follows:

Definition 1 (Per-instance Algorithm Scheduling Problem). *Given a set of algorithms $\mathcal{P}$, a set of instances $\mathcal{I}$, a runtime cutoff κ, and a performance metric $m : \Sigma \times \mathcal{I} \rightarrow \mathbb{R}$, the* per-instance algorithm scheduling problem *is to find a mapping $s : \mathcal{I} \rightarrow \Sigma$ from an instance $\pi \in \mathcal{I}$ to a (potentially unordered) algorithm schedule $\sigma_\pi \in \Sigma$ where each algorithm $\mathcal{A} \in \mathcal{P}$ gets a runtime budget $\sigma_\pi(\mathcal{A})$ between 0 and κ such that $\sum_{\mathcal{A} \in \mathcal{P}} \sigma_\pi(\mathcal{A}) \leq \kappa$ and $\sum_{\pi \in \mathcal{I}} m(s(\pi), \pi)$ will be minimized.*

The algorithm scheduler *aspeed* [4] addresses this problem by using a static algorithm schedule; i.e., *aspeed* applies the same schedule to all instances. The schedule is optimized with an answer set programming [2] solver to obtain a timeout-minimal schedule on the training instances. The scheduler *aspeed* either uses a second optimization step to determine a well-performing ordering of the algorithms or sorts the algorithms by their assigned times, in ascending order (such that a wrongly selected solver does not waste too much time).

Systems such as *3S* [7], *SATzilla* [12] and *claspfolio 2* [5] combine static algorithm schedules (also called pre-solving schedules) and classical algorithm selection. All these systems run the schedule for a small fraction of the runtime budget κ (e.g., *3S* uses 10% of κ), and if this pre-solving schedule fails to solve the given instance, they apply per-instance algorithm selection to run an algorithm predicted to perform well. *3S* and *claspfolio 2* use mixed integer programming and answer set programming solvers, respectively, to obtain a timeout-minimal pre-solving schedule. *SATzilla* uses a grid search to obtain a pre-solving schedule that optimizes the performance of the entire system.

The algorithm scheduler *Sunny* [1] determines the schedule for a new instance π by first determining the set of k training instances $\mathcal{I}_k$ closest to π in instance feature space, and then assigns each algorithm a runtime proportional to the number of instances in $\mathcal{I}_k$ it solved. The algorithms are sorted by their average PAR10 scores on $\mathcal{I}_k$, in ascending order (which corresponds to running the algorithm with the best expected performance first).

3   Instance-Specific Aspeed (ISA)

Kadioglu et al. [7] proposed a variant of $3S$ that uses per-instance algorithm schedules instead of a fixed split between static pre-solving schedule and algorithm selection. In order to evaluate the potential of per-instance timeout-optimized scheduling, we developed the scheduler ISA, short for *instance-specific aspeed*. Inspired by Kadioglu et al. [7], our implementation uses k-nearest neighbor (k-NN) to identify the set $\mathcal{I}_k$ of training instances closest to a given instance π and then applies *aspeed* to obtain a timeout-minimal schedule for them.

During offline training, we have to determine a promising value for the neighborhood size k. In our experiments, we evaluated different k values between 1 and 40 by running cross-validation on the training data and stored the best performing value to use online. We chose this small upper bound for k to ensure a feasible runtime of the scheduler[2] (in our experiments less than 1 second). Furthermore, to optimize the runtime of the scheduler, we reduced the set of training instances, omitting all instances that were either solved by every algorithm or solved by none within the cutoff time.

For each new instance, ISA first computes the k nearest neighbor instances from the reduced training set. This instance set is passed to *aspeed* [4], which returns a timeout-minimal unordered schedule for the neighbor set. The schedule is finally aligned by sorting the time slots in ascending order.

4   Trained Sunny ($TSunny$)

To offer a form of scheduling with less overhead in the online stage than ISA, we implemented a modified version of *Sunny* [1] by adding a training phase. For a new problem instance *Sunny* first selects a subset of k training instances $\mathcal{I}_k$ using k-NN. Then time slots are assigned to each candidate algorithm: Each solver gets one slot for each instance of $\mathcal{I}_k$ it can solve within the given time. Additionally, a designated backup solver gets one slot for each instance of $\mathcal{I}_k$ that cannot be solved by any of the algorithms. Having this slot assignment, the actual size of a single time slot is computed by dividing the available time by the total number of slots. Finally, the schedule is aligned by sorting the algorithms by their average PAR10 score on $\mathcal{I}_k$, thereby running the most promising solver first.

Preliminary experiments for our implementation of this algorithm produced relatively poor results. Examining the schedules, we found that *Sunny* tends to employ many algorithms per schedule, which we suspected to be a weakness. Thus, we enhanced the algorithm by limiting the number of algorithms used in a single schedule to a specified number λ.

Originally, *Sunny* is defined as lazy, i.e. not applying any training procedures after the benchmark data is gathered. However, to obtain better values for our new parameter λ, and also to improve the choice of the neighborhood size k, we

---

[2] Optimizing a schedule is NP-hard; thus the size of the input set, defined by k, must be kept small to make the process applicable during runtime.

implemented a training process for *Sunny*. Similar to *ISA*, different configurations for λ (range 1 to the total number of solvers) and k (range 1 to 100) are evaluated by cross-validation on the training data. To distinguish this enhanced algorithm from the original *Sunny*, we dubbed this trained version *TSunny*.

5 Empirical Study

To compare the different algorithm scheduling approaches of *ISA* and *Sunny*, we implemented them in the flexible algorithm selection framework *flexfolio*[3] and compared them to various other systems: The static algorithm scheduling system *aspeed* [4], the default configuration of *flexfolio* (which is similar to *SATzilla* [12] and *claspfolio 2* [5] and includes a static-presolving schedule), as well as the per-instance algorithm selector *AutoFolio* [9] (an automatically-configured version of *flexfolio* without consideration of per-instance algorithm schedules). If not mentioned otherwise, we used the default parameter values of *flexfolio*. The comparison is based on the algorithm selection library (*ASlib* [3]), which is specifically

Table 1. Gap metric on PAR10: 1.0 corresponds to a perfect oracle score and 0.0 corresponds to the single best score. The best score for each scenario is highlighted with bold face and all system performances have a star that are not significantly worse than the best system (permutation test with 100 000 random permutations and $\alpha = 0.05$; "Equal to Best"). All systems are implemented in *flexfolio*, except *Sunny* which is the original version.

	flexfolio	*AutoFolio*	*aspeed*	*Sunny*	*TSunny*	*ISA*
ASP-POTASSCO	0.78*	0.80*	0.34	0.69	**0.81***	0.72
CSP-2010	**0.80***	0.75*	0.05	0.68	0.77*	0.74*
MAXSAT12-PMS	0.67	0.90*	0.65	0.87	0.93*	**0.94***
PREMAR-2013	0.70	0.74*	0.74*	0.71	0.62	**0.78***
PROTEUS-2014	0.82	0.87	0.87	0.88	**0.94***	0.91
QBF-2011	0.90	0.91	0.80	0.90	**0.94***	0.92
SAT11-HAND	0.73*	0.71*	**0.74***	0.54	0.52	0.69*
SAT11-INDU	0.29*	0.36*	0.06	0.19	0.37*	**0.43***
SAT11-RAND	0.93*	**0.95***	0.80	0.59	0.87	0.95*
SAT12-ALL	0.69*	0.69*	0.10	0.58	0.69*	**0.71***
SAT12-HAND	0.68	0.71	0.46	0.57	0.72	**0.78***
SAT12-INDU	0.39	0.46*	−0.22	0.01	0.53*	**0.54***
SAT12-RAND	0.17	0.24*	−0.28	−0.14	**0.32***	0.12
Average	0.66	0.70	0.39	0.54	0.69	0.71
Equal to Best	6	10	2	0	9	9

---

[3] The source code and all benchmark data are available at http://www.ml4aad.org/algorithm-selection/flexfolio/.

designed to fairly measure the performance of algorithm selection systems. Version 1.0 of *ASlib* consists of 13 scenarios from a wide range of different domains (SAT, MAXSAT, CSP, QBF, ASP and operations research).

Table 1 shows the performance of the systems as the fraction of the gap closed between the static single best algorithm and the oracle (i.e., the performance of an optimal algorithm selector), using performance metric PAR10[4]. As expected, the per-instance schedules (i.e., *Sunny* and *ISA*) performed better on average than *aspeed*'s static schedules. However, *aspeed* still establishes the best performance on SAT11-HAND. By comparing *Sunny* and *TSunny*, we see that parameter tuning substantially improved performance. Comparing *TSunny* and *ISA*, we note that their overall performance is similar but that either has advantages on different scenarios; thus, there is still room for improvement by selecting the better of the two on a per-scenario basis. Surprisingly, the per-instance schedules had a similar performance (*ISA* with 0.71) to the state-of-the-art procedure *AutoFolio* (0.70); however, *AutoFolio* performed slightly more robustly, being amongst the best systems on 10/13 scenarios. Nevertheless, *ISA* establishes new state-of-the-art performance on PREMAR-2013

Table 2. Statistics of schedules: neighborhood size k, average size $\varnothing|\sigma|$ of schedules, average position $\varnothing suc$ of successful solver in schedule for our systems *aspeed*, *ISA*, *Sunny'* (a reimplementation of the lazy version of *Sunny*), and *TSunny* (the non-lazy trained version of *Sunny'*)

	aspeed		ISA			Sunny'			TSunny		
	$\varnothing|\sigma|$	$\varnothing suc$	k	$\varnothing|\sigma|$	$\varnothing suc$	k	$\varnothing|\sigma|$	$\varnothing suc$	k	$\varnothing|\sigma|$	$\varnothing suc$
ASP-POTASSCO	5.9	1.96	14.4	1.6	1.07	34.0	10.7	1.15	19.6	1.0	1.01
CSP-2010	2.0	1.2	5.8	1.1	1.0	43.0	1.9	1.01	12.8	1.9	1.0
MAXSAT12-PMS	3.0	1.98	7.3	1.2	1.02	28.0	5.4	1.04	6.4	3.0	1.01
PREMAR-2013	4.0	1.75	32.6	2.3	1.3	22.0	4.0	1.22	9.0	3.6	1.21
PROTEUS-2014	18.3	7.27	30.6	3.2	1.41	60.0	13.9	1.77	26.6	12.5	1.5
QBF-2011	4.9	2.2	27.8	1.9	1.26	35.0	4.5	1.1	14.1	3.3	1.06
SAT11-HAND	5.9	2.96	27.8	3.1	1.92	16.0	13.5	1.6	10.2	1.7	1.02
SAT11-INDU	4.6	2.82	3.8	1.3	1.03	16.0	16.5	1.55	4.2	1.4	1.02
SAT11-RAND	3.8	1.94	14.4	1.8	1.13	23.0	7.8	1.04	18.3	1.5	1.02
SAT12-ALL	12.6	5.24	8.8	1.6	1.12	38.0	24.4	1.72	4.2	1.0	1.0
SAT12-HAND	10.9	5.45	4.8	1.5	1.09	26.0	26.2	1.68	4.6	1.0	1.01
SAT12-INDU	6.2	3.64	6.1	1.2	1.04	32.0	22.5	1.75	4.3	1.0	1.0
SAT12-RAND	5.2	2.27	18.3	1.8	1.07	35.0	15.7	1.12	67.2	1.0	1.0

---

[4] PAR10 is the penalized average running time where timeouts are counted as 10 times the running time cutoff.

(short for PREMARSHALLING-ASTAR-2013) and *TSunny* on PROTEUS-2014 and QBF-2011 according to the on-going evaluation on *ASlib*[5].

Table 2 gives more insights into our systems' behavior. It also includes our implemented version of *Sunny* without training, dubbed *Sunny'*. *Sunny* (and also *Sunny'*) sets the neighborhood size k as the square root of the number of instances, whereas *TSunny* optimizes k on the training instances. The reason for *TSunny*'s better performance in comparison to *Sunny* is probably its much smaller values for k on all scenarios except on SAT12-RAND. Also *TSunny*'s average schedule size was smaller on nearly all scenarios (except CSP-2010).

Comparing the static *aspeed* and the instance-specific *aspeed* (*ISA*), the average schedule size of *aspeed* is rather large since *aspeed* has to compute a single static schedule that is robust across all training instances and not only on a small subset. Surprisingly, the values of k for *ISA* and *TSunny* differ a lot, indicating that the best value of k depends on the scheduling strategy.

6   Conclusion and Discussion

We showed that per-instance algorithm scheduling systems can perform as well as algorithm selectors and even establish new state-of-the-art performance on 3 scenarios of the algorithm selection library [3]. Additionally, we found that the performance of the algorithm schedules strongly depends on the adjustment of their parameters for each scenario, here the neighborhood size of the k-nearest neighbor and the maximal size of the schedules.

In our experiments we did not tune all possible parameters of *Sunny* and *ISA* in the flexible *flexfolio* framework; e.g., we fixed the pre-processing strategy of the instance features. Therefore, a future extension of this line of work would be to extend the search space of the automatically-configured algorithm selector *AutoFolio* [9] to also cover per-instance algorithm schedules. Another extension could be to allow communication between the algorithms in the schedule [10].

References

1. Amadini, R., Gabbrielli, M., Mauro, J.: SUNNY: a lazy portfolio approach for constraint solving. TPLP **14**(4–5), 509–524 (2014)
2. Baral, C.: Knowledge Representation, Reasoning and Declarative Problem Solving. Cambridge University Press, Cambridge (2003)
3. Bischl, B., Kerschke, P., Kotthoff, L., Lindauer, M., Malitsky, Y., Frechétte, A., Hoos, H., Hutter, F., Leyton-Brown, K., Tierney, K., Vanschoren, J.: ASlib: a benchmark library for algorithm selection. AIJ **237**, 41–58 (2016)
4. Hoos, H., Kaminski, R., Lindauer, M., Schaub, T.: aspeed: Solver scheduling via answer set programming. TPLP **15**, 117–142 (2015)
5. Hoos, H., Lindauer, M., Schaub, T.: claspfolio 2: Advances in algorithm selection for answer set programming. TPLP **14**, 569–585 (2014)

---

[5] www.aslib.net.

6. Huberman, B., Lukose, R., Hogg, T.: An economic approach to hard computational problems. Science **275**, 51–54 (1997)
7. Kadioglu, S., Malitsky, Y., Sabharwal, A., Samulowitz, H., Sellmann, M.: Algorithm selection and scheduling. In: Lee, J. (ed.) CP 2011. LNCS, vol. 6876, pp. 454–469. Springer, Heidelberg (2011). doi:10.1007/978-3-642-23786-7_35
8. Kotthoff, L.: Algorithm selection for combinatorial search problems: a survey. AI Mag. **35**, 48–60 (2014)
9. Lindauer, M., Hoos, H., Hutter, F., Schaub, T.: Autofolio: an automatically configured algorithm selector. JAIR **53**, 745–778 (2015)
10. Malitsky, Y., Sabharwal, A., Samulowitz, H., Sellmann, M.: Boosting sequential solver portfolios: knowledge sharing and accuracy prediction. In: Nicosia, G., Pardalos, P. (eds.) LION 2013. LNCS, vol. 7997, pp. 153–167. Springer, Heidelberg (2013). doi:10.1007/978-3-642-44973-4_17
11. Rice, J.: The algorithm selection problem. Adv. Comput. **15**, 65–118 (1976)
12. Xu, L., Hutter, F., Hoos, H., Leyton-Brown, K.: SATzilla: portfolio-based algorithm selection for SAT. JAIR **32**, 565–606 (2008)

Dynamic Strategy to Diversify Search Using a History Map in Parallel Solving

Seongsoo Moon and Mary Inaba[✉]

Tokyo University, Bunkyo-ku Yayoi 1-1-1, Tokyo, Japan
logic85@hotmail.com, mary@is.s.u-tokyo.ac.jp

Abstract. Diversification plays an important role in portfolio-based parallel SAT solvers. To maintain diversity, state-of-the-art solvers allocate different search policies and share learned clauses. However, the possibility of similarities between search space areas remains as learning progresses. In this paper we attempt to avoid frequently visited areas. We divide a search space into areas and convert each area into an index. We propose a heuristic to dynamically change the search space area using a history map of these indexes. The proposed heuristic was evaluated experimentally using the benchmarks from the application tracks of SAT-Race 2015.

1 Introduction

The satisfiability (SAT) problem is a well-known NP-complete problem. Problem instances from domains such as puzzles, circuit verification, and planning can be easily encoded into SAT problems. Throughout the evolution of multicore hardware, many types of parallel SAT solvers have been proposed. The initial approach was divide-and-conquer [1,2]. However, finding a successful load-balancing solution is difficult. In recent years, a portfolio-based approach [3,4] has become mainstream for parallel SAT solvers. In portfolio approaches, maintaining the diversification and intensification tradeoff is very important [5]. To balance diversification and intensification, many solvers allocate different combinations of policies and share learned clauses among the workers. For example, in ManySAT [3], each worker has different strategies for restart, decision heuristics, polarity, and clause sharing. However, finding a good combination of policies is difficult and clause sharing does not ensure diversity because of limited resources. In this paper, we propose a method for diversifying search among different workers by sharing phase saving statuses. To share these statuses efficiently, we devised a metric that represents the current phase saving status, i.e., Polarity Search Space Index (PSSI). By accumulating PSSI data as a history map, and sharing the history map among workers, each worker can dynamically walk toward the sparsely visited areas. We implemented this method in our solver ParaGlueminisat, and evaluated our method experimentally using the benchmarks from the application tracks of the SAT-Race 2015.

The remainder of this paper is organized as follows. In Sect. 2, we define PSSI and describe preliminary PSSI experiments. In Sect. 3, we demonstrate

© Springer International Publishing AG 2016
P. Festa et al. (Eds.): LION 2016, LNCS 10079, pp. 260–266, 2016.
DOI: 10.1007/978-3-319-50349-3_21

our sparsely visited area walking on search space (SaSS) heuristic using a PSSI history map and its implementation. We show effects of the PSSI experimentally in Sect. 4 and present conclusions and suggestions for future work in Sect. 5.

2   Conversion of Current Area into an Index

In this section, we propose converting the current area of the search space into an index. We use this metric primarily for the dynamic changes of the current area in our heuristic. Details are given in Sect. 3.

2.1   Polarity Search Space Index (PSSI)

Many state-of-the-art solvers use phase saving to reuse a previous phase for intensive search after restarts. This phase has a strong relationship with learned clauses found by the current worker. However, clauses imported from other workers may not fit the current phase. By changing only a small part of the phase, we expect to maintain an intensive search and may be able to use imported clauses.

For this, we convert the current phase to a PSSI as follows. First, we divide the variable set into k-blocks $(B_1, B_2, ..., B_k)$. Second, we calculate the ratio $(r_1, r_2, ..., r_k)$ of variables currently allocated to TRUE, and divide the ratio into uniform m sections where each section has a value between 0 on the far left to $m - 1$ on the far right. For each block B_i, the ratio is converted to integer b_i. For B_i, $b_i = p$ if $p/m \leq r_i < (p+1)/m$, where $p \in \{0, 1, ..., m - 1\}$. After calculating each b_i, we calculate the PSSI as follows.

$$PSSI = \sum_{i=1}^{k} b_i \times m^{i-1}$$

PSSI is now an integer; thus, we can easily, although roughly, compare the areas in the search space among the workers. Consider a problem with n variables x_1, x_2, x_3 and x_n. We can solve this problem using the parallel SAT solver with two workers w_1 and w_2. Let p_i be the current phase of w_i. If we simply calculate the hamming distance between the workers, it takes only $O(n)$ time. However, to compute the distance between the workers, they must be synchronized. This method would be unwieldy when the number of workers increases.

When $p_1 = 0, 1, 1, ..., 0$ has been visited and we fail to find a model, then w_1 and w_2 should avoid the same status in future. However, memoization for this needs a lot of memory.

In PSSI, for example, if we suppose $k = 4$ and $m = 2$, ratio(w_1) = (0.3, 0.7, 0.0, 1.0) and ratio(w_2) = (0.6, 0.7, 1.0, 0.2), then we get $PSSI_1 = (0 * 2^0 + 1 * 2 + 0 * 2^2 + 1 * 2^3) = 10$ and $PSSI_2 = (1 * 2^0 + 1 * 2 + 1 * 2^2 + 0 * 2^3) = 7$. We can compare their areas using PSSI based on bitwise XOR. In this case (0101 $\veebar$ 1110) = 1011, and we count the number of 1 s to obtain distance = 3. However, we do not compare these areas directly because of the synchronization problem. We prefer considering past PSSI results (Sect. 3).

PSSIs have k-blocks, and this structure helps in maintaining search intensification. We can easily modify fractional changes by selecting a block and inverting each polarity in a phase. Using this structure, we dynamically change a phase (Sect. 3).

Currently, we are dividing k-blocks by simply using the indexes of variables, which means if there are kx variables in the original formula, we allocate $1 \sim x$ to block 1, $x + 1 \sim 2x$ to block 2, and so on. We have adopted this policy to attempt to minimize sudden changes in the current area.

2.2 Block Division Policy

Optimal division of variable sets into blocks is difficult. Using a community detection algorithm such as the Louvain method [6] would be a good idea, because when we pick a block representing a community and change the polarities in a phase, these changes will have fractional effects on the entire search because a community have sparse connections to other communities. However, at this time, we simply divide the blocks according to their indexes. We have adopted this policy based on our experiments. We performed 300 tests with a benchmark to produce different models, and observed their polarity trends. The benchmark chosen was 002-80-8.cnf from the application problems in SAT competition 2014 because a model for this benchmark can usually be found rapidly. For each test, we obtained a model and checked its final polarities. The polarity distribution counts for each variable from 300 repeated tests are shown in Fig. 1. Each point indicates the number of TRUEs assigned to the model. From this data, we concluded that there is a high probability of strong relations among proximate variables by index. For example, in Fig. 1, variables with an index between 0 and 2,500 have similar TRUE assigned numbers. Also, a group of variables with less than 30 TRUE assigned numbers at indexes between 9,500 and 13,300 can be seen in Fig. 1 relatively frequently. From the results of this experiment, we suspect that some hidden structures based on the indexes of variables might exist;, this is why we divided blocks using the index order.

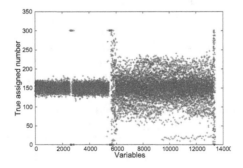

Fig. 1. Polarity distribution of each variable from 300 tests (Benchmark: 002-80-8.cnf)

2.3   Representability of PSSI

To check how well PSSI differentiates between models, we chose four satisfiable benchmarks from the application problems in SAT competition 2014 that can relatively easily find a model using our solver. The benchmarks are 008-80-8.cnf, 002-80-8.cnf, 004-80-8.cnf and 004-22-160.cnf. We performed 20 tests per benchmark (i.e., a total of 80 tests) with a time limit of 3,600 s. Parameter settings of the solver were the same for all tests. In 10 tests, we reached a time limit. Models were obtained for the remaining 70 tests, and we checked their PSSIs. PSSI is calculated under the conditions k = 10 and m = 4;, thus, there are 4^{10} different PSSIs. Sixty-seven different PSSIs were found in 70 models. These results indicate that models for a CNF-formula could be found in many different areas in the search space, and these areas might be differentiated through the PSSI.

3   Walk Towards the Sparsely Visited Areas Using a History Map

We proposed the use of PSSI to represent the current area of the search space in Sect. 2. Using this metric, we diversify the search space areas. Each worker periodically calculates PSSI, and we accumulate these data as a history map of PSSIs. Our main idea is to avoid the frequently visited areas, and to walk towards the sparsely visited areas. The history map is a one-dimensional array comprising the PSSI counts. Each element counts how many times this area is visited. We can walk from the current area to a sparsely visited area by sharing the history map among workers and by dynamically changing a phase. However, we can not anticipate whether we will reach the sparsely visited area. It depends on the block division policy and the structure of the problem. Therefore we refer to this sparsely visited area as the target area.

Algorithms 1 and 2 show the pseudo-code of the SaSS heuristic. In Algorithm 1, after every c conflicts (line 1), each worker calculates the current area as PSSI (line 2), updates the history map of the PSSIs (shared for all workers) and obtains a target area as a PSSI (line 3). If the target area differs from the current area (line 4), it changes polarities to walk towards the target area (line 5). A block is selected by calculating the bitwise XOR of p and $p\prime$, and the polarity of each variable is updated. If the selected block is B_i, then each variable's polarity is allocated to TRUE b_i in m. In Algorithm 2, we obtain a target area using a history map. The history map is updated (line2), but the

---

Algorithm 1. SaSS heuristic: *changeCurrentArea()*

1: **if** *conflicts* % interval == 0 **then**
2:     $p := getCurrArea();$
3:     $p\prime := updateHistoryMap(p);$
4:     **if** $p \mathrel{!=} p\prime$ **then**
5:         $changeBlockPolarities(p, p\prime);$

Algorithm 2. SaSS heuristic: *updateHistoryMap(p)*

Input: PSSI p
Output: PSSI $p\prime$
1: $p\prime := p$;
2: historyMap[$p\prime$]++;
3: **if** $p\prime <$ *c-threshold* $\times$ *thread number* **then**
4:    **return** $p\prime$;
5: $p\prime := checkNearestAreas(p, d)$;
6: **return** $p\prime$;

---

area is not changed in the early stages (lines 3 and 4). When the early stages end, the target area is searched on the basis of the current area (line 5). Areas within hamming distance d from the current area are identified, and the area with the minimal count in the history map is selected as the target area.

4   Experimental Results

In this section, we present an experimental evaluation of the SaSS heuristic. SaSS is implemented in our parallel solver ParaGlueminisat and evaluated by the number of solved instances and the runtime for both SAT/UNSAT. Currently our parallel solver can be executed in two different environments. We used both environments to maximize our test numbers because results from parallel solvers are unstable, particularly for SAT problems.

- Work Station (WS): Xeon X5680 3.3 GHz (12 physical + 12 hyper-threading cores) with 140 GB RAM
- VMware (VM): Intel(R) Xeon(R) CPU E5-2650 v2 2.60 GHz (16 physical + 16 hyper-threading cores) with 128 GB RAM

Parallel track benchmarks from SAT-Race 2015 were used for the evaluation. As a default setting, each worker updates its history map of PSSIs in shared memory every 50,000 conflicts. PSSI is calculated for $k = 10$ and $m = 4$, i.e., for 4^{10} different PSSIs. SaSS only checks proximate areas with hamming distance $= 1$ from the current area. To assess scalability, we performed tests with 12 and 64 workers. We compared the results of SaSS and non-SaSS and, also of the test with 12 and 64 workers. For comparison, we set the CPU time limit to $3600x$ (where x is the number of workers) seconds. We only have 24 or 32 threads including hyper-threading; therefore when we performed tests with 64 workers their actual time limit is 9,600 s in WS and 6,400 s in VM.

We obtained the best results using SaSS with 64 workers. With 64 workers, SaSS, solved 73 instances in 100 hard benchmarks from SAT-Race 2015. Overall, SaSS solved more problems than non-SaSS for SAT problems with 12 and 64 workers. For UNSAT, SaSS did not deteriorate the results; however, we cannot, claim that it provided better results. The implementation of SaSS provided a speed-up of 7% over non-SaSS in the total running time over the SAT benchmarks and solved 6.5 more instances on an average within the time limit (Fig. 2).

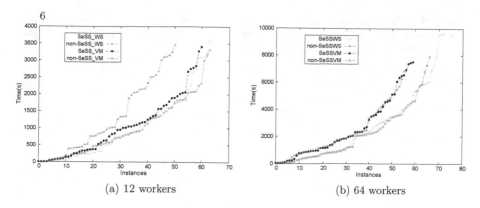

(a) 12 workers   (b) 64 workers

Fig. 2. Time required to solve benchmarks within $3600 \times worker\ number$ seconds. Time limit is set by CPU time, not real time. Measured time (Y-axis) is real time.

5   Conclusions

In this study, we focused on the current phase and dynamically changed search space areas using a history map to diversify search by walking towards the sparsely visited areas. We experimentally evaluated this using benchmarks. The proposed heuristic can be applied regardless of the size of the problem. However, several questions remain unanswered. We used static parameters to divide search space into blocks; however this approach may not suitable for some problems. To address this issue, we could dynamically change the blocks. Our SaSS heuristic deterministically chooses a sparsely visited area as the target area. We could consider a biased random walk to increase the robustness of SaSS. We could also consider the time factor because our map only counts the number of visits to each area, and does not consider the time at which the area was last visited.

References

1. Zhang, H., Bonacina, M.P., Hsiang, J.: PSATO: a distributed propositional prover and its application to quasigroup problems. J. Symb. Comput. **21**, 543–560 (1996)
2. Chrabakh, W., Wolski, R.: GrADSAT: a parallel SAT solver for the grid. Technical report, UCSB CS TR N (2003)
3. Hamadi, Y., Jabbour, S., Sais, L.: ManySAT: a parallel SAT solver. JSAT **6**(4), 245–262 (2009)
4. Audemard, G., Hoessen, B., Jabbour, S., Lagniez, J.-M., Piette, C.: Revisiting clause exchange in parallel SAT solving. In: Cimatti, A., Sebastiani, R. (eds.) SAT 2012. LNCS, vol. 7317, pp. 200–213. Springer, Heidelberg (2012). doi:10.1007/978-3-642-31612-8_16
5. Guo, L., Hamadi, Y., Jabbour, S., Sais, L.: Diversification and intensification in parallel SAT solving. In: Cohen, D. (ed.) CP 2010. LNCS, vol. 6308, pp. 252–265. Springer, Heidelberg (2010). doi:10.1007/978-3-642-15396-9_22

6. Blondel, V.D., Guillaume, J.-L., Lambiotte, R., Lefebvre, E.: Fast unfolding of communities in large networks. J. Stat. Mech.: Theory Exp. (2008). doi:http://dx.doi.org/10.1088/1742-5468/2008/10/P10008

Faster Model-Based Optimization Through Resource-Aware Scheduling Strategies

Jakob Richter[1](✉), Helena Kotthaus[2](✉), Bernd Bischl[3], Peter Marwedel[2],
Jörg Rahnenführer[1], and Michel Lang[1]

[1] Department of Statistics, TU Dortmund University, Dortmund, Germany
{richter,rahnenfuehrer,lang}@statistik.tu-dortmund.de
[2] Department of Computer Science 12, TU Dortmund University,
Dortmund, Germany
{helena.kotthaus,peter.marwedel}@tu-dortmund.de
[3] Department of Statistics, LMU München, Munich, Germany
bernd.bischl@stat.uni-muenchen.de

Abstract. We present a Resource-Aware Model-Based Optimization
framework RAMBO that leads to efficient utilization of parallel computer
architectures through resource-aware scheduling strategies. Conventional
MBO fits a regression model on the set of already evaluated configurations and their observed performances to guide the search. Due to its
inherent sequential nature, an efficient parallel variant can not directly
be derived, as only the most promising configuration w.r.t. an infill criterion is evaluated in each iteration. This issue has been addressed by
generalized infill criteria in order to propose multiple points simultaneously for parallel execution in each sequential step. However, these
extensions in general neglect systematic runtime differences in the configuration space which often leads to underutilized systems. We estimate
runtimes using an additional surrogate model to improve the scheduling
and demonstrate that our framework approach already yields improved
resource utilization on two exemplary classification tasks.

Keywords: Black-box optimization · Hyperparameter tuning · Model
selection · Model-based optimization · Resource-aware scheduling ·
Performance management · Parallelization

1 Introduction

In the field of hyperparameter optimization for machine learning methods, efficient black-box optimization is often necessary to obtain a well-performing
hyperparameter configuration for a given data set. A state-of-the-art optimization strategy for expensive black-box functions is the model-based optimization
(MBO) [6]. MBO is an iterative optimization algorithm that starts on an initial
set of already evaluated configurations. In each step a regression model is fitted

---

J. Richter and H. Kotthaus—These authors are contributed equally.

© Springer International Publishing AG 2016
P. Festa et al. (Eds.): LION 2016, LNCS 10079, pp. 267–273, 2016.
DOI: 10.1007/978-3-319-50349-3_22

on the so far available evaluations. It serves as a surrogate model to predict the
outcome of the black-box on yet unseen configurations. The infill criterion of the
model guides the search to a new configuration which is usually a compromise
between good predicted performance and uncertainty of the search space region –
expected improvement is a popular choice. The new configuration is evaluated,
appended to the current data and the next iteration step starts until the budget
of evaluations is depleted. Many extensions to the basic MBO algorithm have
been suggested for parallel point proposal [3].

One popular application for MBO is hyperparameter tuning [10,12] where
the objective function is defined as a resampled performance measure of a
machine learning algorithm. Here, resource requirements like CPU utilization
or memory usage heavily vary depending on the type and configuration of the
applied machine learning algorithm. Heterogeneous runtimes have already been
addressed in [11] where the authors suggest to model these with an additional
surrogate leading to an "expected improvement per second" which favors less
expensive configurations. We also use surrogate models to estimate resource
requirements but instead of adapting the infill criterion, we use them for efficient scheduling of parallel point evaluations. Resource-aware scheduling is an
active field of research which is often tailored specifically for different hardware
platforms, from small embedded systems [13] up to heterogeneous clusters [4]. In
contrast to these classical scheduling problems, we are in control of the job generation as we can query the resource model for jobs with suitable resource requirements and postpone or skip suggested jobs if deemed not promising enough.

2   Resource-Aware Model-Based Optimization

Our framework (RAMBO) is shown in Fig. 1. In the first of three steps, the **MBO
Method** proposes a set of promising configurations w.r.t. the infill criterion.
Each configuration forms a job with different resource demands. Based on all
previous evaluations, we build surrogate regression models to predict the computational resources for arbitrary configurations. Such a model is called *Job
Utility Estimator* and is used to create *Job Profiles*. Configurations to evaluate
are selected in the **Job Selection** step. Jobs are prioritized depending on their
estimated usefulness for optimization and their predicted resource requirements.
The **Scheduling** step uses the estimated *Job Profiles* and a *System Description*
(e.g., number of CPUs and free memory) to efficiently map the jobs to the available resources. The jobs are started and can be monitored by a **Job Tracker**.
Since job profiles are only estimated, a job whose resource utilization deviates
from its predicted requirements might need to be rescheduled or stopped to guarantee efficient resource utilization. We propose two possibilities to update the
model with results. One way is the *synchronous* feedback, where the results of
all jobs within one iteration are gathered before each model update. The other
way is to update the model each time a job has finished its computation in an
asynchronous fashion. Either way, the updated model is then used to propose
new candidate points.

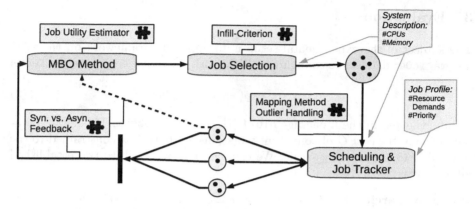

Fig. 1. Ressource-Aware Model-Based Optimization Framework.

To demonstrate our general framework, we show a simple exemplary setup in this work. We pick kriging as surrogates to model the misclassification error and the logarithmic runtime. We opt for a multipoint *Lower Confidence Bound* (LCB), which is an optimistic estimate of the objective function, similar to [5] as infill criterion, which we call qLCB. qLCB simultaneously generates q configurations by drawing q random values λ_i ($i = 1, \ldots, q$) from the exponential distribution with a mean of 2. Each λ_i results in a different trade-off between exploitation ($\lambda_i \downarrow$) and exploration ($\lambda_i \uparrow$) and thus leads to a different optimal configuration $\mathbf{x}_i^*$ after solving

$$\mathbf{x}_i^* := \arg\min_{\mathbf{x}} \left[\text{LCB}(\mathbf{x}, \lambda_i) \right] = \arg\min_{\mathbf{x}} \left[\hat{y}(\mathbf{x}) - \lambda_i \hat{s}(\mathbf{x}) \right]. \tag{1}$$

Here, $\hat{y}(\mathbf{x})$ denotes the posterior mean and $\hat{s}(\mathbf{x})$ the root of the posterior standard deviation of the regression model at point $\mathbf{x}$, respectively. Unfortunately, there is no direct ordering of the set of obtained candidates $\mathbf{x}_i^*$. Therefore, we assign candidates with a balanced exploration-exploitation trade-off a higher priority: $p_i = -|\log(\lambda_i) - \log(2)|$ is inversely proportional to the absolute distance of λ_i to its expected value 2 on a log-scale.

For scheduling, we use the synchronous approach. In each iteration we generate a list of $q = 3m$ proposed jobs with the help of qLCB. We then determine the job j_{i^*}, $i^* := \arg\max_i p_i$, with highest priority and run it CPU$_1$ exclusively. Accordingly, on a system with m homogenous CPUs the remaining jobs are scheduled on CPU$_2, \ldots,$ CPU$_m$, limited by the upper time bound $\hat{t}_{i^*}$, which is directly derived from the estimated runtime of job j_{i^*}. Jobs which have an estimated runtime $\hat{t}_i \leq \hat{t}_{i^*}$ are mapped in decreasing order of their priorities to the remaining CPUs in a greedy first fit manner. A job j_i is mapped on CPU$_k$ if its runtime $\hat{t}_i \leq \hat{t}_{i^*} - \sum_{i \in J_k} \hat{t}_i$ where J_k is the set of jobs already mapped to CPU$_k$. Jobs of the inital list that do not fit on any CPU are discarded. If any CPU is left without a job we query the surrogate model for a new job for each CPU with a runtime smaller or equal to $\hat{t}_{i^*}$ to fill the gaps. When all scheduled jobs are evaluated the surrogate model is updated and the iteration starts over.

3   Evaluation

The subject of the experimental setup is to apply our framework on the w6a[1] and magic04[2] data set to configure an SVM with the radial basis function kernel

$$k(\mathbf{x}, \mathbf{x}') = \exp(-\gamma \|\mathbf{x} - \mathbf{x}'\|^2) \tag{2}$$

as implemented in the R package e1071 [7], based on libsvm. The kernel parameter γ and the cost C of constraint violations are both box-constrained to the interval $[-15, 15]$ on a $\log_2$-scale. We compare our approach to two established alternatives:

Random Search (RS): A simple parallelized random search. This relatively naive yet often effective [1] approach does not need a synchronization step like MBO, therefore the next random point will be scheduled immediately after each function evaluation which guarantees maximum load of all CPUs.

qLCB: A simple MBO approach with a multipoint LCB infill criterion [3], using a kriging model and naive scheduling. At each sequential step, $q = $ ncores points are selected minimizing the LCB (1) w.r.t. random $\lambda_i \sim \text{Exp}(\frac{1}{2})$ $(i = 1, \ldots, q)$.

Since the concept of a fixed budget of evaluations does not translate well into a scenario with heterogeneous runtimes, we define the budget via the elapsed time. We use a 3-fold cross validation to define the objective function for the tuner and an outer 10-fold cross validation to evaluate the optimization results. All variants start with an initial latin hypercube design with 10 points. To increase comparability, initial designs are fixed per outer cross-validation fold.

The software is implemented in R using mlr[3] to interface the machine learning algorithms and mlrMBO[4] as optimization toolbox. BatchExperiments [2] is used to parallelize the experiments on high performance computing cluster. The traceR framework [8,9] guarantees reliable measures of computational resources.

Figure 2 shows the mean misclassification errors (MMCE) of the best configuration after 1, 10, 120 and 180 min. The left hand side displays the tuning error, i.e. the over-optimistic error on the internal tuning set. The right hand side shows the MMCE on the outer cross-validation. Unfortunately, on these data sets only marginal improvements are achieved after evaluation of the initial design. Yet our RAMBO approach seems to perform well, yielding comparable performance and sometimes slightly less variance. The reasons for this can be found in Fig. 3 which visualizes the mapping of parallel jobs. We can observe unused CPU time for

---

[1] Platt: http://www.csie.ntu.edu.tw/~cjlin/libsvmtools/datasets/binary/w6a.

[2] Bock: https://archive.ics.uci.edu/ml/datasets/MAGIC+Gamma+Telescope.

[3] Bischl, B., Lang, M., Kotthoff, L., Schiffner, J., Richter, J., Studerus, E., Casalicchio, G., Jones, Z.M.: mlr: machine learning in R. J. Mach. Learn. Res. **17**(170), 1–5 (2016). http://jmlr.org/papers/v17/15-066.html

[4] Bischl et al., mlrMBO: Model-Based Optimization for mlr. https://github.com/berndbischl/mlrMBO.

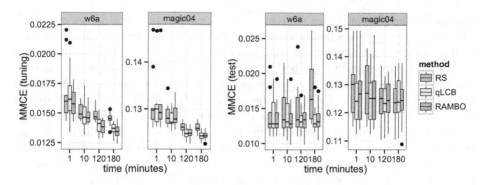

Fig. 2. Averaged misclassification errors (MMCE): tuning (left) and test data (right) for the best observed configuration after a given time budget.

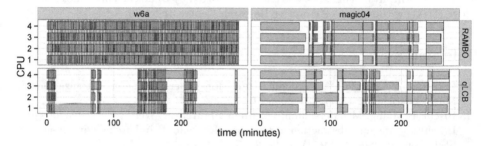

Fig. 3. Scheduling visualization for one run: The boxes show the mapping of jobs on CPUs. Less empty spaces indicate higher CPU utilization. Vertical lines indicate the end of one MBO iteration.

qLCB whereas RAMBO balances long execution times more evenly. The estimation of runtimes reliably estimates the runtimes so that only 2.3% of the evaluations exceed $\hat{t} + 2 \cdot s(\hat{t})$. qLCB often schedules four jobs with vastly different runtimes and hence wastes available CPU time idling. Thus our results demonstrate that RAMBO achieves higher **CPU utilization**, meaning more evaluations which yields better knowledge of the hyperparameter space and thus higher confidence in the optimization result. It also shows on magic04 that it not only prefers short jobs but is also able to schedule long jobs more efficiently. On the w6a dataset RAMBO is capable of evaluating twice as many configurations as the unscheduled baseline method qLCB. In contrast it only yields 25% more evaluations on the magic04 dataset which indicates that promising configurations have longer runtimes then average and vice versa for w6a.

4 Conclusion

With our RAMBO framework we present a novel approach to perform a faster model-based optimization through resource-aware scheduling. We demonstrate that our yet heuristic mapping approach already leads to improved resource utilization and thus to more evaluations within the same time budget. This potentially yields a better knowledge of the hyperparameter space and thus higher confidence in the optimization result. In order to efficiently use hardware resources, we are planning further improvements. Firstly, further work will concentrate on integrating memory profiles since memory usage heavily influences runtime if the amount of RAM in the system is too small to hold all required data. Secondly, we aim to improve the resource estimation. Thirdly, we are planning to implement dynamic scheduling of jobs for cases of remaining deviations. Fourthly, we plan to implement a multi-objective approach with respect to hardware costs, memory, runtime and priority for performance optimization for an more optimized resource-aware scheduling strategy. This is especially important for an efficient utilization of heterogeneous architectures.

Acknowledgments. This work was partly supported by Deutsche Forschungsgemeinschaft (DFG) within the Collaborative Research Center SFB 876, A3.

References

1. Bergstra, J., Bengio, Y.: Random search for hyper-parameter optimization. J. Mach. Learn. Res. **13**(1), 281–305 (2012)
2. Bischl, B., Lang, M., Mersmann, O., Rahnenführer, J., Weihs, C.: BatchJobs and BatchExperiments: abstraction mechanisms for using R in batch environments. J. Stat. Comput. Simul. **64**(11), 1–25 (2015)
3. Bischl, B., Wessing, S., Bauer, N., Friedrichs, K., Weihs, C.: MOI-MBO: multi-objective infill for parallel model-based optimization. In: Pardalos, P.M., Resende, M.G.C., Vogiatzis, C., Walteros, J.L. (eds.) LION 2014. LNCS, vol. 8426, pp. 173–186. Springer, Heidelberg (2014). doi:10.1007/978-3-319-09584-4_17
4. Delimitrou, C., Kozyrakis, C.: Quasar: resource-efficient and QoS-aware Cluster Management. In: ASPLOS 2014, pp. 127–144. ACM (2014)
5. Hutter, F., Hoos, H.H., Leyton-Brown, K.: Parallel algorithm configuration. In: Hamadi, Y., Schoenauer, M. (eds.) LION 6. LNCS, vol. 7219, pp. 55–70. Springer, Heidelberg (2012). doi:10.1007/978-3-642-34413-8_5
6. Jones, D.R., Schonlau, M., Welch, W.J.: Efficient global optimization of expensive black-box functions. J. Global Optim. **13**(4), 455–492 (1998)
7. Karatzoglou, A., Meyer, D., Hornik, K.: Support vector machines in R. J. Stat. Softw. **15**(1), 1–28 (2006)
8. Kotthaus, H., Korb, I., Lang, M., Bischl, B., Rahnenführer, J., Marwedel, P.: Runtime and memory consumption analyses for machine learning R programs. J. Stat. Comput. Simul. **85**(1), 14–29 (2015)
9. Kotthaus, H., Korb, I., Marwedel, P.: Performance analysis for parallel R programs: towards efficient resource utilization. Technical report 01/2015, Department of Computer Science 12, TU Dortmund University (2015). SFB876 Project A3
10. Lang, M., Kotthaus, H., Marwedel, P., Weihs, C., Rahnenführer, J., Bischl, B.: Automatic model selection for high-dimensional survival analysis. J. Stat. Comput. Simul. **85**(1), 62–76 (2015)
11. Snoek, J., Larochelle, H., Adams, R.P.: Practical bayesian optimization of machine learning algorithms. In: NIPS Workshop on Bayesian Optimization, Sequential Experimental Design, and Bandits, pp. 2960–2968 (2012)

12. Thornton, C., Hutter, F., Hoos, H.H., Leyton-Brown, K.: Auto-WEKA: combined selection and hyperparameter optimization of classification algorithms. In: Proceedings of ACM SIGKDD, pp. 847–855 (2013)
13. Tillenius, M., Larsson, E., Badia, R.M., Martorell, X.: Resource-aware task scheduling. ACM Trans. Embed. Comput. Syst. 14(1), 5:1–5:25 (2015)

Risk-Averse Anticipation for Dynamic Vehicle Routing

Marlin W. Ulmer[1]([⊠]) and Stefan Voß[2]

[1] Technische Universität Braunschweig, Mühlenpfordtstr. 23,
38106 Braunschweig, Germany
m.ulmer@tu-braunschweig.de
[2] Universität Hamburg, Von-Melle-Park 5, 20146 Hamburg, Germany
stefan.voss@uni-hamburg.de

Abstract. In the field of dynamic vehicle routing, the importance to integrate stochastic information about possible future events in current decision making increases. Integration is achieved by anticipatory solution approaches, often based on approximate dynamic programming (ADP). ADP methods estimate the expected mean values of future outcomes. In many cases, decision makers are risk-averse, meaning that they avoid "risky" decisions with highly volatile outcomes. Current ADP methods in the field of dynamic vehicle routing are not able to integrate risk-aversion. In this paper, we adapt a recently proposed ADP method explicitly considering risk-aversion to a dynamic vehicle routing problem with stochastic requests. We analyze how risk-aversion impacts solutions' quality and variance. We show that a mild risk-aversion may even improve the risk-neutral objective.

Keywords: Dynamic vehicle routing · Anticipation · Risk-aversion · Approximate dynamic programming · Stochastic customer requests

1 Introduction

Many service providers dispatch a fleet of vehicles during the day to transport goods or passengers and to conduct services at customers. Factors like e-commerce, digitization, and urbanization lead to an increase in uncertainty dispatchers have to consider in their plans, e.g., in travel times, service times, or customer demands [1]. Especially, customer requests often occur spontaneously during the day. In many cases, new requests require significant adaptions of the current plan [2]. These are enabled by real-time computational resources. Practical routing applications are generally modeled as dynamic vehicle routing problems (DVRPs, compare [1]). For many DVRPs, static approaches applied on a rolling horizon are not suitable [3]. Anticipation of possible future events and decisions is mandatory to allow reliable, flexible, and effective plans.

Anticipation can be achieved by approximate dynamic programming [4]. ADP for DVRPs is widely established, especially for stochastic requests [2]. ADP

© Springer International Publishing AG 2016
P. Festa et al. (Eds.): LION 2016, LNCS 10079, pp. 274–279, 2016.
DOI: 10.1007/978-3-319-50349-3_23

methods evaluate decisions regarding the expected future rewards (or costs). The expected future rewards are usually approximated via simulation. Generally, a tradeoff between current and future rewards can be experienced. High immediate rewards may lower the expected future rewards. Dispatchers aim for an "optimal" balance between immediate and future rewards.

All ADP approaches applied to DVRPs maximize the sum of immediate and expected future rewards. In practice, decisions also depend on the variance of the expected future rewards, i.e., the service provider's *risk-aversion* [5]. A risk-averse provider may discount the expected future rewards if a high variance, i.e., a high uncertainty of a decision's success is given. In some cases, practitioners are able to quantify their risk-aversion. In other cases, the degree of risk-aversion can be derived by analyzing historical decisions [6]. The derived properties then have to be integrated in a suitable anticipatory DVRP-approach.

Work on risk-aversion for vehicle routing problems is limited. In (static) vehicle routing with stochastic travel times explicit inclusion of risk-aversion is, e.g., achieved by [7]. [8] evaluate plans by risk for a dynamic orienteering problem. Until now, the ADP methods applied to DVRPs are not able to integrate practitioners' risk-aversion. Anticipation is based on mean values. Especially low probability - high impact incidences are not sufficiently considered [9]. Recently, Jiang and Powell [10] proposed a general ADP method integrating quantiles of the expected value-distribution and therefore the variance in the anticipation. In this paper, we adapt the proposed method to an ADP approach of anticipatory time budgeting (ATB, [2]) for a DVRP with stochastic customer requests. We analyze the impact on rewards and variances for different instance settings and degrees of risk-aversion.

This paper is the first work integrating risk-aversion in an ADP approach for dynamic vehicle routing. We show that an explicit inclusion of risk-aversion in DVRPs is possible and that a mild risk-aversion even strengthens the approximation process resulting in higher rewards and lower variances compared to the risk-neutral equivalent.

2   Dynamic VRP with Stochastic Requests

In this section, we define the DVRP with stochastic requests via a Markov decision process (MDP, [11]). The problem is an extension of [12]. An uncapacitated vehicle serves customers in a service area considering a time limit. The tour starts and ends in a depot. A set of known early request customers (ERC) has to be served. During the day, new requests occur. If the vehicle is located at a customer, the dispatcher has to decide about the subset of occurred requests to be confirmed and the next customer to visit. Waiting is permitted. The dispatcher aims on maximizing the confirmed late request customers (LRC). Modeling the problem as MDP, a decision point k occurs if the vehicle is located at a customer. A state S_k consists of the point of time, the vehicle's position, the set of not yet served ERC and confirmed LRC, and the set of new LRC. Decisions x are made about the subset to be confirmed and the next customer to visit, respectively, waiting. The immediate reward $R(S_k, x)$ is the number of newly confirmed LRC. A post-decision state S_k^x consists of the point of time, the vehicle's position, the not yet served ERC and confirmed LRC, and the next customer to visit. The transition results from the vehicle's travel and provides a new set of requesting LRC. The process terminates in state S_K when no customers remain to be served and the vehicle has returned to the depot. A solution for the DVRP is a policy π, a sequence of decision rules $(X_0^\pi, \ldots, X_\pi^K)$ assigning a decision $x = X_k^\pi(S_k)$ to every state S_k. The objective is to derive an optimal policy π maximizing the expected sum of rewards over all decision points. Notably, the objective is defined for a risk-neutral dispatcher.

3 Risk-Averse Time Budgeting

In this section, we extend ATB by [2] to ATB$^\lambda$ allowing the integration of risk-aversion. ATB draws on the ADP method of approximate value iteration (AVI, [4]) to evaluate post-decision states (PDSs) S^x regarding the expected number of future confirmations, i.e., their *value* $V(S^x)$. To be more specific, AVI represents ways of using past experience about the algorithm behavior to improve future performance. Tuning refers to the update of values. Due to the curses of dimensionality, PDSs are aggregated to vectors containing the point of time and the remaining free time budget. The resulting vector space is then partitioned to a lookup table (LT). Every entry of the LT contains a set of vectors. AVI starts with initial tuning and entry values $\hat{V}_0$ inducing a policy π_0. Then, AVI iteratively simulates a problem's realization i and tune the values $\hat{V}_{i-1}$ regarding the algorithms performance. Within each approximation run i, policy π_{i-1} is applied based on Bellman's Equation [11] depicted in Eq. (1). The values for the new policy π_i are tuned by the realized values of approximation run i.

$$X_k^{\pi_i}(S_k) = \operatorname*{argmax}_{x \in \mathcal{X}(S_k)} \left\{ R(S_k, x) + \hat{V}_i(S^x) \right\} \tag{1}$$

$V(S^x)$ is a random variable. A risk-averse policy aims on avoiding highly volatile $V(S^x)$. Notably, $V(S^x)$ is the sum of a sequence of interdependent random variables $R(S_{k+i}, x), 0 < i < K - k$, i.e., the volatility and the impact of the volatility may change over the subsequent decision points. A straightforward evaluation of the variance of $V(S^x)$ is not sufficient to consider risk-aversion in dynamic decision problems. [13] describe *dynamic risk measures* $\rho(S^x)$ considering the risk over the subsequent decision points. [10] present an algorithm to approximate $\rho(S^x)$ for every post-decision state by ρ^α via ADP methods. They use the quantiles of ρ^α as an approximation of the real value distribution of ρ. For ATB$^\lambda$, we draw on the concept of conditional value at risk (CVaR, [14]). The considered dynamic risk measure ρ^α is induced by the one-step conditional risk measure ρ^λ as depicted in Eq. (2).

$$\rho^\lambda(S^x) = (1 - \lambda)V(S^x) + \lambda \rho^\alpha(S^x) \tag{2}$$

The value of ρ^λ is then used instead of $V(S^x)$ in Eq. (1). To achieve ρ^α, ρ^λ is recursively applied over the subsequent decision points. For an efficient approximation, we simplistically assume that $V(S^x)$ follows a uniform distribution. This avoids an extensive estimation of the distribution for every value. As a result, parameter $\lambda \in [0,1]$ directly determines the dispatcher's risk-aversion. $\lambda = 0$ results in risk-neutrality and ATB. $\lambda = 1$ results in a myopic policy. For the tuning of ATB^λ, we approximate both $V(S^x)$ and $\rho^\alpha(S^x)$ via AVI.

4   Computational Studies

In this section, we define the settings of ATB^λ, briefly describe the instances, and analyze the results. For ATB^λ, we follow the parameter settings of [2]. We use a (static) LT with interval length of one. The values are updated via moving average. We consider the tuning after 1 million approximation runs. The instances base on [2]. The time limit is set to 360 min. The vehicle travels with monotone speed $v = 15\,\mathrm{km/h}$ in a service area of $20 \times 20\,\mathrm{km}^2$. The depot is located in the center of the service area. The expected number of customers is 100. The percentage of LRC is 75%. Customer requests follow a Poisson distribution over time. We consider three spatial customer distributions. Customers are distributed uniformly ($\mathcal{F}_U$), equally grouped in two clusters ($\mathcal{F}_{2C}$), or distributed in three clusters ($\mathcal{F}_{3C}$). For $\mathcal{F}_{2C}$, the cluster centers are $(5,5)$ and $(15,15)$. For $\mathcal{F}_{3C}$, the cluster centers are $(5,5)$, $(5,15)$, and $(15,10)$. Within the clusters, the request probability is normally distributed with standard deviation of 1 km.

Table 1. Results; best values are depicted in bold.

λ	Confirmations			Variance		
	$\mathcal{F}_U$	$\mathcal{F}_{2C}$	$\mathcal{F}_{3C}$	$\mathcal{F}_U$	$\mathcal{F}_{2C}$	$\mathcal{F}_{3C}$
0.0	34.3	49.6	44.9	23.2	34.4	31.9
0.1	**34.5**	50.2	45.3	22.7	33.0	29.7
0.2	34.4	50.6	45.7	22.1	32.2	29.3
0.3	34.2	**50.7**	**46.0**	21.3	32.3	27.7
0.4	33.8	**50.7**	**46.0**	20.8	30.7	25.9
0.5	33.3	50.3	45.9	**20.4**	30.2	25.6
0.6	32.1	49.8	45.3	21.1	**28.8**	**24.5**
0.7	30.4	49.0	44.4	21.7	29.2	25.4
0.8	28.6	48.0	43.3	22.0	29.6	26.1
0.9	27.4	47.2	42.5	23.0	31.0	27.1
1.0	26.8	46.8	42.0	24.2	31.1	28.4

For each instance setting, we run 1,000 test runs for $\lambda = 0.0, 0.1, \dots, 1.0$. The average number of confirmations and the variance are depicted in Table 1. Notably, a mild risk-aversion leads to a higher risk-neutral objective value.

This can be explained by the impact of risk-aversion on the tuning process. For a high λ, only the (relatively certain) outcomes of the next few decision points define the decision policy leading to a fast and more reliable tuning process. A low λ results in an equal consideration of all subsequent decision points and outcomes. The according tuning process requires a high number of approximation runs to be accurate. This is especially the case for the clustered customer distributions [2]. Further, ATB is based only on temporal attributes and may provide a less reliable tuning for clustered distributions compared to $\mathcal{F}_U$ [15]. As a result, the highest amount of confirmations is achieved for $\lambda = 0.3$ and $\lambda = 0.4$ for the clustered distributions. As expected, we experience a constant decrease of the variances between $\lambda = 0.0$ and $\lambda = 0.5$. Afterwards, the variance increases, since a high λ is similar to a myopic policy.

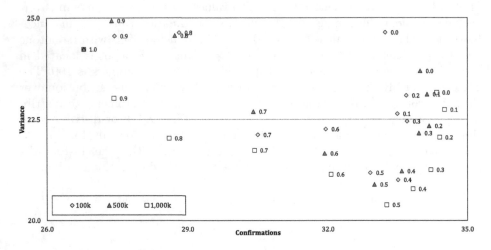

Fig. 1. Solution quality and standard deviation for varying λ and $\mathcal{F}_U$

We now analyze the tuning process and the tradeoff between number of confirmations and variance in more detail. Figure 1 shows the number of confirmations and variance for varying λ and $\mathcal{F}_U$ for 1,000 test runs and policies achieved by 100 k, 500 k and 1,000 k approximation runs. For 1,000 k, $\lambda = 0.1$ to $\lambda = 0.5$ span a Pareto-front for both dimensions similar to the theoretical results of [10]. For 100 k, the tuning for ATB$^\lambda$ with $\lambda = 0.1$ is (still) not sufficient. During the tuning process, we experience an increase in the number of confirmations for low λ, and a decrease in the variance for high λ. Hence, a directed tuning of ATB$^\lambda$ (and AVI) to the two different objectives can be achieved. The integration of risk-aversion further results in a faster and more reliable AVI-tuning.

5 Conclusion

In this paper, we applied an ADP method to a DVRP with stochastic customer requests enabling anticipation and the inclusion of service provider's risk-aversion. Even though we simplistically assume the expected values V to follow a uniform distribution, results show that the integration is not only possible, but also strengthens the tuning process and even improves the overall (risk-neutral) objective. In this paper, we considered a "vanilla" DVRP. Future work may focus on more real-world related problems and problems containing unlikely events with significant impacts (e.g., vehicle breakdowns). For a more efficient tuning process, risk directed sampling may be included in the approach as proposed in [10]. Further, historical data about previous decision making may be analyzed to quantify service providers' risk-aversion. For a more accurate approximation, the distribution of V could be explicitly considered by a set of quantiles. Finally, a mild risk-aversion improves the (risk-neutral) objective. Hence, it may be beneficial for many ADP methods to include a dynamic risk measure for a strengthened and more reliable tuning process. The risk-aversion may decrease during the tuning process once a more reliable approximation is achieved.

References

1. Psaraftis, H.N., Wen, M., Kontovas, C.A.: Dynamic vehicle routing problems: three decades and counting. Networks **67**(1), 3–31 (2016)
2. Ulmer, M.W., Mattfeld, D.C., Köster, F.: Budgeting time for dynamic vehicle routing with stochastic customer requests. Transp. Sci. (2016, to appear)
3. Powell, W.B., Towns, M.T., Marar, A.: On the value of optimal myopic solutions for dynamic routing and scheduling problems in the presence of user noncompliance. Transp. Sci. **34**, 67–85 (2000)
4. Powell, W.B.: Approximate Dynamic Programming: Solving the Curses of Dimensionality. Wiley, New York (2011)
5. Dyer, J.S., Sarin, R.K.: Relative risk aversion. Manag. Sci. **28**, 875–886 (1982)
6. Jackwerth, J.C.: Recovering risk aversion from option prices and realized returns. Rev. Financ. Stud. **13**, 433–451 (2000)
7. Adulyasak, Y., Jaillet, P.: Models and algorithms for stochastic and robust vehicle routing with deadlines. Transp. Sci. **50**(2), 608–626 (2016)
8. Lau, H.C., Yeoh, W., Varakantham, P., Nguyen, D.T., Chen, H.: Dynamic stochastic orienteering problems for risk-aware applications. In: Proceedings of the Conference on Uncertainty in Artificial Intelligence, pp. 448–458 (2012)
9. Taniguchi, E., Thompson, R.G., Yamada, T.: Incorporating risks in city logistics. Procedia-Soc. Behav. Sci. **2**, 5899–5910 (2010)
10. Jiang, D.R., Powell, W.B.: Approximate dynamic programming for dynamic quantile-based risk measures. Technical report, Princeton University (2015)
11. Puterman, M.L.: Markov Decision Processes: Discrete Stochastic Dynamic Programming. Wiley, New York (2014)
12. Thomas, B.W.: Waiting strategies for anticipating service requests from known customer locations. Transp. Sci. **41**, 319–331 (2007)
13. Ruszczyński, A.: Risk-averse dynamic programming for Markov decision processes. Math. Program. **125**, 235–261 (2010)
14. Rockafellar, R.T., Uryasev, S.: Optimization of conditional value-at-risk. J. Risk **2**, 21–42 (2000)
15. Ulmer, M.W., Mattfeld, D.C., Hennig, M., Goodson, J.C.: A rollout algorithm for vehicle routing with stochastic customer requests. In: Mattfeld, D., Spengler, T., Brinkmann, J., Grunewald, M. (eds.) Logistics Management. Lecture Notes in Logistics, pp. 217–227. Springer, Cham (2015). doi:10.1007/978-3-319-20863-3_16

GENOPT Papers

Solving GENOPT Problems
with the Use of ExaMin Solver

Konstantin Barkalov$^{(\boxtimes)}$, Alexander Sysoyev,
Ilya Lebedev, and Vladislav Sovrasov

Lobachevsky State University of Nizhny Novgorod, Nizhny Novgorod, Russia
{konstantin.barkalov,alexander.sysoyev,ilya.lebedev,
vladislav.sovrasov}@itmm.unn.ru

Abstract. This paper describes an algorithm for solving multidimensional multiextremal optimization problems. This algorithm uses Peanotype space-filling curves for dimension reduction. It has been used for solving problems at GENeralization-based contest in global OPTimization (GENOPT). Computational experiments are carried out on 1800 multidimensional problems.

Keywords: Global optimization · Multiextremal functions · Space-filling curves · Mixed global-local algorithm · GENOPT

1 Introduction

A well-known approach to the investigation and comparing of the multiextremal optimization algorithms is based on testing these methods by solving a set of test problems, chosen randomly from some specially designed class. Each test problem can be considered as a particular instance of a random function generated by a special generator. Application of multiextremal optimization algorithms to large sets of such functions allows estimating the characteristics of the methods and evaluating the efficiency of each particular algorithm.

Among such generators for the one-dimensional problems, there are samples from Fourier series proposed by Hill [1]. A generator proposed by Shekel [2] generates another well-known class of the one-dimensional test problems. For the investigation of various one-dimensional algorithms using the samples of the functions generated by Hill and Shekel generators Globalizer software has been developed. A comprehensive description of the capabilities of this system and the examples of its application are given in [3]. Note also that Hill and Shekel functions have been successfully used in the construction of one-dimensional constrained problems (with controlled measure of the feasible domain) [4].

A generator for a random sampling of two-dimensional test functions has been proposed in [5]. A generator for the functions of arbitrary dimensionality with known positions of the local and global minima (GKLS generator) has been proposed in [6]. Its application for the studying of some multidimensional algorithms has been described in [7,8].

© Springer International Publishing AG 2016
P. Festa et al. (Eds.): LION 2016, LNCS 10079, pp. 283–295, 2016.
DOI: 10.1007/978-3-319-50349-3_24

Approach to the comparison of the algorithms by solving a set of test problems has been used by the organizers of GENeralization-based contest in global OPTimization (GENOPT). The functions to be optimized are broadly divided into three function families: GKLS, conditioned transforms of classical benchmarks (Rosenbrock (unimodal, narrowing bending valley), Rastrigin (strongly multimodal), Zakharov (unimodal)), and a composition of classical benchmarks. In their turn, each family is subdivided into six function types, which differ by their analytical definition and by the number of dimensions. The functions belonging to the same type share the majority of properties and can be assumed to have the same difficulty. Finally, every function type of every family is realized into an unlimited number of function instances which differ by some randomly generated parameters. In particular, every instance will have a randomly generated offset $c \in [-1, 1]$ added to the function's output in order to make the global minimum value unpredictable. A detailed description of the competition problems can be found at the website http://genopt.org.

An efficient approach for solving global optimization problems has been developed under the supervision by prof. R.G. Strongin at Lobachevsky State University of Nizhni Novgorod [10–23]. Within the framework of this approach, solving multidimensional global optimization problems is reduced to solving a set of the corresponding one-dimensional ones. The corresponding dimension reduction is based on the use of Peano space-filling curves (also called *evolvents*), unambiguously mapping the unit interval of the real axis onto a hypercube, as well as the generalization of these ones, which can be applied to solving the problems using the multiprocessor systems. The proposed algorithms have been implemented in ExaMin solver applied to solving the problems within GENOPT competition. In the present work, a brief description of the global optimization algorithm applied and of its modifications are given, and the results of the computational experiments with the problems of the competition are presented.

2   Problem Statement

Let us consider the problem of search for a global minimum of an N-dimensional function $\varphi(y)$ within a hyperinterval D

$$\varphi(y^*) = \min \{\varphi(y) : y \in D\}, \tag{1}$$
$$D = \left\{y \in R^N : a_i \le y_i \le b_i, 1 \le i \le N\right\}.$$

Let us assume that the function φ satisfies the Lipschitz condition with an a priori unknown constant L

$$|\varphi(y_1) - \varphi(y_2)| \le L \|y_1 - y_2\|, \ y_1, y_2 \in D, \ 0 < L < \infty.$$

In this study, we will use the approach based on the idea of dimension reduction by means of a Peano curve $y(x)$, which continuously and unambiguously maps the unit interval [0,1] onto the n-dimensional cube

$$\left\{y \in R^N : -2^{-1} \le y_i \le 2^{-1}, 1 \le i \le N\right\} = \{y(x) : 0 \le x \le 1\}.$$

Problems of numerical construction of Peano-type space filling curves and the corresponding theory are considered in detail in [3,8]. Here we will note that a numerically constructed curve is 2^{-m} accurate approximation of the theoretical Peano curve in L_∞ metric, where m is an evolvent construction parameter. Examples of the evolvent with different m in two dimensions are given in Fig. 1.

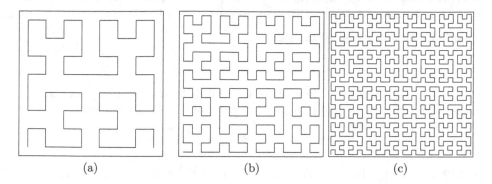

(a)   (b)   (c)

Fig. 1. Evolvents in two dimensions with (a) $m = 3$, (b) $m = 4$ and (c) $m = 5$

By using this kind of mapping it is possible to reduce the multidimensional problem (1) to a univariate problem

$$\varphi(y^*) = \varphi(y(x^*)) = \min\left\{\varphi(y(x)) : x \in [0,1]\right\}.$$

An important property of such mapping is preservation of boundedness of function relative differences (see [3,9]): if the function $\varphi(y)$ in the domain D satisfies the Lipschitz condition, then the function $\varphi(y(x))$ on the interval $[0,1]$ will satisfy a uniform Hölder condition

$$|\varphi(y(x_1)) - \varphi(y(x_2))| \le H\,|x_1 - x_2|^{1/N},$$

where the Hölder constant H is linked to the Lipschitz constant L by the relation

$$H = 2L\sqrt{N+3}.$$

Therefore, it is possible, without loss of generality, to consider minimization of univariate function

$$f(x) = \varphi(y(x)), \quad x \in [0,1],$$

satisfying the Hölder condition.

3 Global Search Algorithm

The considered algorithm for solving this problem (here, according to [3]) involves constructing a sequence of points x^k, where the values of minimized function $z^k = f(x^k) = \varphi(y(x^k))$ are calculated. Let us call the process of calculating the function value (including the construction of an image $y^k = y(x^k)$) the "trial", and the pair (x^k, z^k), the "trial result". The set of pairs $\{(x^k, z^k)\}, 1 \leq k \leq n$, makes up the search data collected using the method after carrying out n steps. The rules that determine the work of the global search algorithm are as follows.

At the first iteration of the method the trial is carried out at an arbitrary internal point x^1 of the interval $[0, 1]$. The point of trial at the next iteration $(k + 1)$ is determined according to the rules presented below.

Rule 1. Renumber points of the set

$$X_k = \{x^1, \dots, x^k\} \cup \{0\} \cup \{1\},$$

which includes boundary points of the interval $[0, 1]$ and the points of the previous trials, with subscripts in increasing order of coordinate values, i.e.,

$$0 = x_0 < x_1 < \cdots < x_k < x_{k+1} = 1.$$

Rule 2. Supposing that $z_i = f(x_i) = \varphi(y(x_i))$, $1 \leq i \leq k$, calculate values

$$\mu = \max_{2 \leq i \leq k} \frac{|z_i - z_{i-1}|}{\Delta_i}, \tag{2}$$

$$M = \begin{cases} r\mu, & \mu > 0, \\ 1, & \mu = 0, \end{cases}$$

where $r > 1$ is a preset parameter of the method, and $\Delta_i = (x_i - x_{i-1})^{1/N}$.

Rule 3. Calculate a *characteristic* for every interval (x_{i-1}, x_i), $1 \leq i \leq k + 1$, according to the following formulae

$$R(1) = 2\Delta_1 - 4\frac{z_1}{M},$$

$$R(i) = \Delta_i + \frac{(z_i - z_{i-1})^2}{M^2 \Delta_i} - 2\frac{z_i + z_{i-1}}{M}, 1 < i < k + 1, \tag{3}$$

$$R(k + 1) = 2\Delta_{k+1} - 4\frac{z_k}{M}.$$

Rule 4. Find interval (x_{t-1}, x_t) with the maximum characteristic

$$R(t) = \max\{R(i) : 1 \leq i \leq k + 1\}. \tag{4}$$

Rule 5. Carry out a trial at the point $x^{k+1} \in (x_{t-1}, x_t)$, calculated using the following formulae

$$x^{k+1} = \frac{x_t + x_{t-1}}{2}, \text{ if } t = 1 \text{ or } t = k + 1,$$

$$x^{k+1} = \frac{x_t + x_{t-1}}{2} - \text{sign}(z_t - z_{t-1})\frac{1}{2r}\left[\frac{|z_t - z_{t-1}|}{\mu}\right]^N, \text{ if } 1 < t < k + 1. \tag{5}$$

The algorithm terminates if the condition $\Delta_t < \epsilon$ is satisfied; here $\epsilon > 0$ is the preset accuracy. As the *current estimate* of an optimum at the step k we accept the value

$$\varphi_k^* = \min_{1 \leq i \leq k} z^i, \tag{6}$$

and the vector

$$y_k^* = \arg \min_{y \in \{y^1, \ldots, y^k\}} \varphi(y).$$

This *global search algorithm* (GSA) was developed in the framework of information - statistical approach (see [3]). From this point of view the normalized characteristics $R(i)$ from (3) can be considered as probabilities of locating the global minimum within the interval $(x_{i-1}, x_i), 1 \leq i \leq k + 1$. Thus, at each iteration a new trial point is selected inside the interval, which has the greatest probability of finding the global minimum. At the same time, use of running estimates (2) of Lipschitz constant L allows us to apply GSA to the problems with a priori unknown values of L.

Let us illustrate the work of GSA for minimization of a multiextremal function of two variables generated by GKLS generator [6]. The experiment used the method parameters $r = 3$, $\epsilon = 10^{-3}$, and the evolvent construction parameter $m = 10$. Fig. 2 shows the level lines of the function and the points of 421 trials carried out by the method before the required accuracy was obtained.

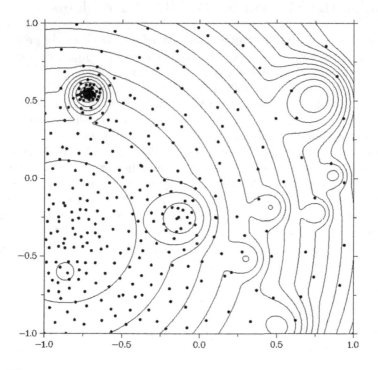

Fig. 2. Minimization of a GKLS function of two variables by GSA

The following theorem from [3] determines sufficient convergence conditions of the global search algorithm.

Theorem. Let the point $\bar{y}$ be the limit point of the sequence y^k generated by the rules of GSA while minimizing the Lipschitzian with the constant L function $\varphi(y), y \in D$. Then:

1. If side by side with $\bar{y}$ there exists another limit point y' of the sequence y^k, then $\varphi(\bar{y}) = \varphi(y')$.
2. For any $k \geq 1$ $x^k = \varphi(y^k) \geq \varphi(\bar{y})$.
3. If at some step of the search process the value μ from (2) satisfies the condition

$$r\mu > 2^{3-1/N} L\sqrt{N+3},$$

than $\bar{y}$ is a global minimizer of the function $\varphi(y)$ over D and any global minimizer y^* from () is also a limit point of the sequence y^k.

Remark. As follows from the theorem, the limit points of the trial sequence y^k generated by GSA are global minimizers. This property makes GSA substantially different from search techniques one way or another based on the random search concept which generates trial sequences everywhere dense in the search domain.

4   Tuning the Method for GENOPT Problems

The organizers of GENOPT competition have offered 18 classes of problems (see Table 1, where last two rows correspond to six classes of composite functions with $n = 0, 1, 2$). In each class, 100 functions have been given.

The limit of 1 million computations of the function (trials) imposed by the organizers is essential for finding the global minimum of the functions with given dimensionalities. Thus, even for the 10-dimensional problems, having built a uniform grid with four points in each dimension, one goes already out of the 1 million points limit. The global optimization algorithms that build non-uniform grid in the search domain put the points more efficiently. However, for the 30-dimensional problems such an efficiency appears to be not enough as well. In this section, we describe four modifications of the core method used in solving various problems within the competition. The first two modifications have been applied to all problems, the third one – to the problems of GKLS class, and the last one has been used for solving the problems with Rastrigin, Rosenbrock, and Zakharov functions.

4.1   Mixed Global-Local Algorithm

One scheme aiming to accelerate the search process is to be outlined in this subsection (other schemes are described, for example, in [24,25]). The idea of acceleration is to magnify the characteristics of the intervals containing the best current estimates by introducing some factor depending on the function values

Table 1. The classes of problems to be solved within the competition

Index	Family	Type	Dimension
0	GKLS	non-differentiable	10
1			30
2		differentiable	10
3			30
4		twice differentiable	10
5			30
6	High condition	Rosenbrock	10
7			30
8		Rastrigin	10
9			30
10		Zakharov	10
11			30
$12 + 2n$	Composite		10
$13 + 2n$			30

estimated at the end-points of the corresponding interval. Following this idea, we introduce the modified characteristics

$$R_\alpha(i) = \frac{R(i)}{\sqrt{(z_i - \varphi_k^*)(z_{i-1} - \varphi_k^*)} + \mu(1.5)^{-\alpha}}, \tag{7}$$

where μ, $R(i)$ and φ_k^* are respectively from (2), (3), (6), and α is some integer parameter for setting the desired level of localization.

Next, it is possible to build a scheme *mixing* the 'global' and the 'local' decision rules, i.e. switching between formulae (3) and (7) in some systematic way. In our experiments we use *global-to-local* ratio q, specifying the number of global trials preceding each local trial.

The mixed strategy has the following features. Firstly, both decision rules are based on the same information, so that each decision action (no matter local or global) uses the outcomes of all the trials performed. Secondly, non-stop global search assures the global convergence; the aim of the local refinement is to accelerate the attainment of low function values.

Let us use a mixed global-local algorithm for solving a problem from Sect. 3 with the global-to-local ratio $q = 4$ and level of localization $\alpha = 15$. With this parameters the number of iterations was 203. Fig. 3 shows level lines of the same function from Fig. 2 with points of the trials performed by the mixed method.

4.2   Local Refinement of the Best Current Estimate

The second implemented modification of the core method consisted in a direct utilizing of a local optimization method, namely Hooke–Jeeves method [26] (see also [27,28]). Schematically, the combined method works as follows:

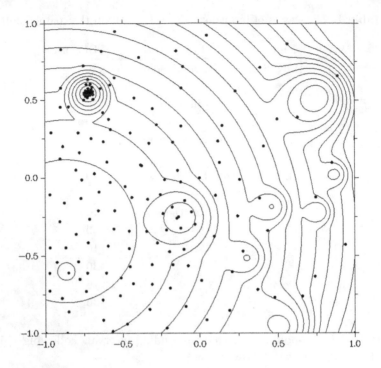

Fig. 3. Minimization of a GKLS function of two variables by mixed GSA

- The global phase
    - Perform the GSA iterations until the current optimum (the minimal value of the objective function in the trial points computed already) is renovated.
- The local phase
    - Start Hooke-Jeeves method from the current optimum point with given exit condition according to precision.
    - Add all the trial points of the local method into the GSA trial points database.
    - Upon achieving given accuracy by the local method, go to the global phase.

The accuracy of the local method (the termination condition) was taken as 10^{-5} for all problem classes of GKLS as well as for the Rastrigin and composite ones, 10^{-7} for Rosenbrock functions, and 10^{-6} for Zakharov ones.

4.3   Utilizing the Random Search for GKLS-30

The modification of the core method considered above have allowed solving almost all problems from the 10-dimensional GKLS classes (see more details in Sect. 5). However, these ones appeared to be insufficient for the 30-dimensional problems of this class. The multistart scheme was the next variant of the modification. Using Sobol quasi-random number generator, we have selected 500 start points. A local optimization method with the limitation not only in precision (stated above) but also in the number of iterations (1000) has been started in each point. Thus, 500 thousand trials could be spent in the local phase in the worst case. All the points of the trials executed at the local phase are stored in the database of the global method. Then, the global phase was started, when the rest of 1 million trials were executed using GSA.

4.4   Use of Separable Search as an Initial Stage

Since Rastrigin function is a separable one, one can search for its global optimum by performing the optimization in each coordinate separately. Thus, the following scheme has been employed in order to solve Rastrigin problem:

- The separable stage
  - Select a start point.
  - For each coordinate to perform:
    * Fix all coordinated except the current one.
    * Perform the optimization by one-dimensional GSA.
  - Store all the trial points of the local method into the trial point database of GSA.
- The local stage
  - Start Hooke-Jeeves method with given exit condition according to precision from the point of current optimum.
  - Store all the trial points of the local method into the trial point database of GSA.
  - Upon achievement given precision by the local method, go to the global stage.
- The global stage
  - Perform the iterations of GSA, until the current optimum (the minimum value of the objective function in the trial points computed already) is renovated.

The scheme considered above has allowed us to solve all the problems with 10-dimensional Rastrigin functions as well as with the 30-dimensional ones.

The same approach has been applied to solving the problems based on the unimodal functions (Rosenbrock and Zakharov ones) as well. In these cases, the separable stage provides a good initial approximation for the local method. Without the use of the separability, the local method starts from a point located far away from the global minimum and performs too many iterations until the termination conditions are satisfied.

5   Numerical Experiments

The methods considered in Sects. 3 and 4 and the modifications of these ones have been implemented in ExaMin solver intended for the parallel solving of the multidimensional multiextremal global optimization problems developed in Lobachevsky State University of Nizhni Novgorod. Global search algorithm and block nested optimization scheme [20] make the algorithmic basis for ExaMin solver. According to the competition conditions, the sequential mode of the solver execution only has been used when solving the problems. However, ExaMin supports the systems with distributed memory (using MPI) as well as with shared memory (using OpenMP). Moreover, NVIDIA graphic processors and Intel Xeon Phi coprocessors are supported.

In the final stage of the competition, ExaMin solver has taken the 3rd prize in the overall ranking and the 1st one in the total number of the solved problems (Fig. 4). The distribution of the classes of the solved tasks is presented in Table 2. The parameter of the evolvents building was $m = 10$. The parameter of the method was $r = 2.5$ for all problems besides the ones from GKLS class. For GKLS, r varied from 2.5 to 20.

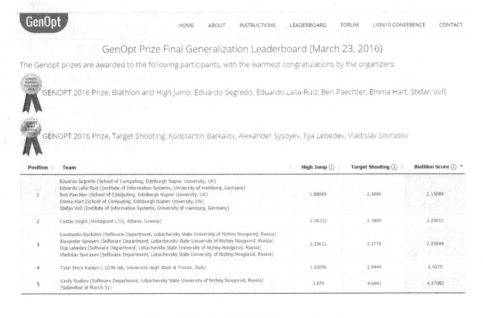

Fig. 4. Final GENOPT leaderboard

The modification with the separable stage with the precision 0.02 for the stop condition of one-dimensional GSA was used for all classes except GKLS. When solving the GKLS problems, the modifications from Subsects. 4.1 and 4.2 have been used first, then modification from Subsect. 4.3 was applied. The number of the solved problems with and without random search is presented in Table 3.

Table 2. Total number of solved tasks from different classes

Class	Tasks solved
GKLS–nd–10	99
GKLS–nd–30	15
GKLS–cd–10	96
GKLS–cd–30	1
GKLS–td–10	94
GKLS–td–30	0
Rosenbrock–10	100
Rosenbrock–30	100
Rastrigin–10	100
Rastrigin–30	100
Zakharov–10	100
Zakharov–30	100
Composite–10	100
Composite–30	100

Table 3. Total number of solved tasks from different classes

Class	Solved without random mode	Solved with random mode
GKLS–nd–10	78	99
GKLS–nd–30	0	15
GKLS–cd–10	67	96
GKLS–cd–30	0	1
GKLS–td–10	65	94
GKLS–td–30	0	0

6   Conclusion

In this work, the results of solving of the 10- and 30-dimensional problems of the unconditional global optimization from GenOpt 2016 competition are presented. The optimization methods used and the modifications of these ones directed onto the obtaining of a solution at given limitation of 1 million trials (computations of the objective function) are described. All the modifications considered are implemented in ExaMin solver employed in the conducting of the experiments. The numerical experiments have been carried out using Lobachevsky supercomputer [29].

Acknowledgements. This study was supported by the Russian Science Foundation, project No 15-11-30022 "Global optimization, supercomputing computations, and applications".

References

1. Hill, J.D.: A search technique for multimodal surfaces. IEEE Trans. Syst. Sci. Cybern. **5**(1), 2–8 (1969)
2. Shekel, J.: Test functions for multimodal search technique. In: Proceedings of the 5th Princeton Conference on Information Science Systems, pp. 354–359. Princeton University Press, Princeton (1971)
3. Strongin, R.G., Sergeyev, Y.D.: Global Optimization with Non-convex Constraints. Sequential and Parallel Algorithms. Kluwer Academic Publishers, Dordrecht (2000)
4. Barkalov, K.A., Strongin, R.G.: A global optimization technique with an adaptive order of checking for constraints. Comput. Math. Math. Phys. **42**(9), 1289–1300 (2002)
5. Grishagin, V.A.: Operation characteristics of some global optimization algorithms. Probl. Stoch. Search **7**, 198–206 (1978). (in Russian)
6. Gaviano, M., Kvasov, D.E., Lera, D., Sergeyev, Y.D.: Software for generation of classes of test functions with known local and global minima for global optimization. ACM Trans. Math. Softw. **29**(4), 469–480 (2003)
7. Kvasov, D., Sergeyev, Y.D.: Multidimensional global optimization algorithm based on adaptive diagonal curves. Comput. Math. Math. Phys. **43**(1), 40–56 (2003)
8. Sergeyev, Y.D., Strongin, R.G., Lera, D.: Introduction to Global Optimization Exploiting Space-filling Curves. Springer, New York (2013)
9. Lera, D., Sergeyev, Y.D.: Lipschitz and holder global optimization using space-filling curves. Appl. Numer. Math. **60**(1–2), 115–129 (2010)
10. Sergeyev, Y.D., Grishagin, V.A.: A parallel method for finding the global minimum of univariate functions. J. Optim. Theory Appl. **80**(3), 513–536 (1994)
11. Sergeyev, Y.D., Grishagin, V.A.: Sequential and parallel global optimization algorithms. Optim. Methods Softw. **3**, 111–124 (1994)
12. Gergel, V.P.: A method of using derivatives in the minimization of multiextremum functions. Comput. Math. Math. Phys. **36**(6), 729–742 (1996)
13. Gergel, V.P.: A global optimization algorithm for multivariate functions with lipschitzian first derivatives. J. Glob. Optim. **10**(3), 257–281 (1997)
14. Grishagin, V.A., Sergeyev, Y.D., Strongin, R.G.: Parallel characteristical algorithms for solving problems of global optimization. J. Glob. Optim. **10**(2), 185–206 (1997)
15. Gergel, V.P., Sergeyev, Y.D.: Sequential and parallel algorithms for global minimizing functions with Lipschitzian derivatives. Comput. Math. Appl. **37**(4–5), 163–179 (1999)
16. Sergeyev, Y.D., Grishagin, V.A.: Parallel asynchronous global search and the nested optimization scheme. J. Comput. Anal. Appl. **3**(2), 123–145 (2001)
17. Strongin, R.G., Sergeyev, Y.D.: Global optimization: fractal approach and non-redundant parallelism. J. Glob. Optim. **27**(1), 25–50 (2003)
18. Gergel, V.P., Strongin, R.G.: Parallel computing for globally optimal decision making on cluster systems. Future Gener. Comput. Syst. **21**(5), 673–678 (2005)

19. Barkalov, K., Polovinkin, A., Meyerov, I., Sidorov, S., Zolotykh, N.: SVM regression parameters optimization using parallel global search algorithm. In: Malyshkin, V. (ed.) PaCT 2013. LNCS, vol. 7979, pp. 154–166. Springer, Heidelberg (2013). doi:10.1007/978-3-642-39958-9_14
20. Barkalov, K.A., Gergel, V.P.: Multilevel scheme of dimensionality reduction for parallel global search algorithms. In: Proceedings of the 1st International Conference on Engineering and Applied Sciences Optimization - OPT-i 2014, pp. 2111–2124 (2014)
21. Gergel, V., Grishagin, V., Israfilov, R.: Local tuning in nested scheme of global optimization. Procedia Comput. Sci. **51**(1), 865–874 (2015)
22. Gergel, V., Grishagin, V., Gergel, A.: Adaptive nested optimization scheme for multidimensional global search. J. Global Optim. **66**, 35–51 (2016)
23. Barkalov, K., Gergel, V.: Parallel global optimization on GPU. J. Global Optim. **66**, 3–20 (2016)
24. Sergeyev, Y.D., Kvasov, D.E.: Global search based on efficient diagonal partitions and a set of Lipschitz constants. SIAM J. Optim. **16**(3), 910–937 (2006)
25. Paulavicius, R., Sergeyev, Y., Kvasov, D., Zilinskas, J.: Globally-biased DISIMPL algorithm for expensive global optimization. J. Global Optim. **59**(2–3), 545–567 (2015)
26. Hooke, R., Jeeves, T.A.: "Direct search" solution of numerical and statistical problems. J. ACM. **8**(2), 212–229 (1961)
27. Wilde, D.J.: Optimum Seeking Methods. Prentice-Hall, Engelwood Cliffs (1964)
28. Himmelblau, D.M.: Applied Nonlinear Programming. McGraw-Hill, New York (1972)
29. http://www.top500.org/system/178472

Hybridisation of Evolutionary Algorithms Through Hyper-heuristics for Global Continuous Optimisation

Eduardo Segredo$^{1(\boxtimes)}$, Eduardo Lalla-Ruiz^2, Emma Hart^1, Ben Paechter^1, and Stefan Voß^2

^{1} School of Computing, Edinburgh Napier University, Edinburgh, Scotland, UK
{e.segredo,e.hart,b.paechter}@napier.ac.uk
^{2} Institute of Information Systems, University of Hamburg, Hamburg, Germany
{eduardo.lalla-ruiz,stefan.voss}@uni-hamburg.de

Abstract. Choosing the correct algorithm to solve a problem still remains an issue 40 years after the *Algorithm Selection Problem* was first posed. Here we propose a hyper-heuristic which can apply one of two meta-heuristics at the current stage of the search. A scoring function is used to select the most appropriate algorithm based on an estimate of the improvement that might be made by applying each algorithm. We use a differential evolution algorithm and a genetic algorithm as the two meta-heuristics and assess performance on a suite of 18 functions provided by the *Generalization-based Contest in Global Optimization* (GENOPT). The experimental evaluation shows that the hybridisation is able to provide an improvement with respect to the results obtained by both the differential evolution scheme and the genetic algorithm when they are executed independently. In addition, the high performance of our hybrid approach allowed two out of the three prizes available at GENOPT to be obtained.

Keywords: Global search · Differential evolution · Genetic algorithm · Global continuous optimisation · Hyper-heuristic

1 Introduction

A significant amount of real-world applications requires finding global optima over continuous decision spaces. Examples from diverse domains such as economics and finance [28], circuit design [21], control theory [1], chemistry [23], and electricity [26], among others, highlight the importance of properly addressing them in order to provide satisfactory solutions. Due to this, the development of efficient algorithms has been of increasing interest for researchers, also accompanied by the urgency from the side of practitioners requiring high-quality feasible and fast solutions for their difficult problems at hand.

In this context, various are the approaches that have been recently proposed for non-differentiable global optimisation. For instance, in [13], a predictive approach to the reproduction phase of new individuals for a well-known

© Springer International Publishing AG 2016
P. Festa et al. (Eds.): LION 2016, LNCS 10079, pp. 296–305, 2016.
DOI: 10.1007/978-3-319-50349-3_25

meta-heuristic was proposed. Another example is given by [14], where a derivative-free global heuristic, which deals with constraints by static and dynamic penalty function techniques, was presented. Finally, a modification over an existing exact penalty algorithm for making it derivative-free, which in addition makes use of a local search procedure, was introduced in [7].

Evolutionary Computation (EC) is a relevant field with many applications within global optimisation [5,25]. Its main goal is to study, develop, and analyse algorithms following the biological notion of evolution within the Darwinian principles. The above has motivated the development of a wide variety of algorithms. In this regard, some of the most frequently used methods, which belong to the family of *Evolutionary Algorithms* (EAs), are *Genetic Algorithms* (GAs) [10], due to their easy and flexible implementation, as well as their exhibited performance. Furthermore, during the last two decades, another EA called *Differential Evolution* (DE), proposed in [22], has been successfully applied not only to benchmark problems but also to several real-world applications [4].

Another field of research that has gained a significant popularity during last years is that of *Hyper-heuristics* (HH). A HH can be defined as a search method or a learning mechanism for *selecting* or *generating* meta-heuristics or tailored heuristics to solve computational search problems [2]. Therefore, they function at a higher level of abstraction when compared to meta-heuristics and heuristics, and usually have no knowledge about the domain of the problem at hand. In this context, HH based on selection try to address the *Algorithm Selection Problem* [15] by iteratively identifying and selecting the most promising low-level meta-heuristics or heuristics, from a set of candidates, for solving a particular instance of an optimisation problem [3]. This can be done by means of a scoring function that is used for assessing the performance of each low-level approach.

In this work, we propose a hybridised EA that uses a selection-based HH to address the set of global continuous optimisation problems provided for the *Generalization-based Contest in Global Optimization* (GENOPT)[1] organised in the field of the *Learning and Intelligent Optimization Conference* (LION 10). The HH selects the most suitable meta-heuristic to be applied at the current stage of the search procedure, choosing between a DE scheme and a GA. If both algorithms are applied in isolation, then for some problems, DE is the best performing approach, with the GA failing to converge to high-quality solutions, while for other instances, the opposite situation is observed. By combining the EAs by means of a HH, we are able to produce a more powerful approach that overcomes the weaknesses of the individual algorithms on the majority of considered problems.

The remainder of this paper is organised as follows. Section 2 describes our hybridisation of EAs through the use of a HH. Section 3 describes the experimental evaluation and provides a discussion of the results obtained. Finally, Sect. 4 draws the main conclusions extracted from the work and provides several lines for further research.

---

[1] The manifesto of the contest, including its instructions and rules, can be found in the following URL: http://genopt.org/genopt.pdf.

2   Hybridisation of Evolutionary Algorithms

This section is devoted to the description of the hybridisation, through the use of a HH (Sect. 2.4), of two meta-heuristics. Section 2.1 introduces an adaptive version of DE, while Sect. 2.2 presents the implementation of the GA applied. Additionally, with the aim of increasing the convergence speed of the whole optimisation scheme, a *Global Search* (GS) procedure is described in Sect. 2.3, which is incorporated into both meta-heuristics.

2.1   Adaptive Differential Evolution

DE is a stochastic direct search method particularly suited for continuous global optimisation [22]. In DE, the decision variables of a given problem are defined by a vector $X = [x_1, x_2, \ldots, x_i, \ldots, x_D]$, being D the number of decision variables or the dimensionality of the problem, and every x_i a real number. The quality of each vector X is given by the objective function $f(X)(f : \Omega \subseteq \mathbb{R}^D \to \mathbb{R})$. The goal of global optimisation, considering a minimisation problem, is thus to find a vector $X^* \in \Omega$ where $f(X^*) \leq f(X)$ holds for all $X \in \Omega$. In the particular case of box-constrained optimisation problems, the feasible region Ω is defined by particular values for the lower (a_i) and upper (b_i) bounds of each variable, i.e. $\Omega = \prod_{i=1}^{D}[a_i, b_i]$.

In this work, we apply an adaptive version of the approach DE/current-to-pbest/1/bin, which uses JADE [27] as the control scheme. We selected this variant since it showed to be one of the best exploitative schemes in [19]. JADE is responsible for adapting the values of the *mutation scale factor F* and the *crossover rate CR* of DE, which will be introduced in the following lines.

The operation of this DE scheme is as follows. First of all, a population P with NP individuals $(P = [X_1, X_2, \ldots, X_j, \ldots, X_{NP}])$, also called vectors in the scope of DE, is initialised by using a particular strategy. Each individual comprises D decision variables. The value of the decision variable i belonging to the individual X_j is denoted by $x_{j,i}$. Then, successive iterations are evolved by executing the following steps, until a stopping criterion is satisfied. For each vector in the current population, referred to as *target vector* (X_j), a new *mutant vector* (V_j) is created using a *mutant vector generation strategy*. In our case, we apply the current-to-pbest/1 scheme. Any vector in the population different from the target vector is randomly selected as the *base vector*. The mutant vector V_j for target vector X_j is thus created as shown in Eq. 1, where r_1 and r_2 are mutually exclusive integers different from the index j chosen at random from the range $[1, NP]$. Moreover, the individual X_{r_3} is randomly selected from the fittest $p \times 100\%$ individuals. Another parameter K is also introduced, but in order to facilitate the parameterisation of the whole scheme, $K = F$ is usually considered, with F the mutation scale factor allowing the exploration and exploitation abilities of DE to be balanced.

$$V_j = X_j + K \times (X_{r_3} - X_j) + F \times (X_{r_1} - X_{r_2}) \tag{1}$$

After applying the mutant vector generation strategy, the mutant vector is combined with the target vector to generate a *trial vector* (U_j) through a crossover operator. The combination of the mutant vector generation strategy and the crossover operator is usually referred to as the *trial vector generation strategy*. The most commonly applied operator for combining the target and mutant vectors, and the one considered herein, is the *binomial crossover* (*bin*). The crossover operation is controlled by means of the crossover rate CR. The binomial crossover generates a trial vector as shown in Eq. 2. A uniformly distributed random number in the range $[0, 1]$ is given by $rand_{j,i}$, and $i_{rand} \in [1, 2, ..., D]$ is an index selected in a random way that ensures that at least one variable is propagated from the mutant vector to the trial one. For the remaining cases, the probability of the variable being inherited from the mutant vector is CR. Otherwise, the variable of the target vector is considered.

$$u_{j,i} = \begin{cases} v_{j,i} \; if \; (rand_{j,i} \; \leq \; CR \; or \; i \; = \; i_{rand}) \\ x_{j,i} \; otherwise \end{cases} \tag{2}$$

The trial vector generation strategy, as described above, might generate vectors outside the feasible region Ω. One of the most widely used schemes is based on randomly reinitialising the infeasible values in their corresponding feasible ranges, and it is the one applied herein. After generating NP trial vectors, each one is compared against its corresponding target vector. For each pair, the one that minimises the objective function is selected to survive. In case of a tie, in our version the trial vector survives. Finally, the GS depicted in Sect. 2.3 is applied to the surviving population.

2.2 Genetic Algorithm

The other approach we selected for our hybridisation is a generational GA with elitism preservation. This was selected as it has previously been demonstrated to be the best performing mono-objective approach when solving continuous optimisation problems with different dimensions [20]. The operation of this algorithm follows the typical scheme of a GA. First of all, an initial population with NP individuals is randomly generated. Then, for each generation, $NP - 1$ offspring are created. Parents are selected by using the well-known *Binary Tournament* [8], while offspring are obtained by applying the *Uniform Mutation* operator [8] and the *Simulated Binary Crossover* operator [6], with mutation and crossover rates p_m and p_c, respectively. Afterwards, during the replacement stage, all parents, except the fittest one, are discarded, and they are replaced by the generated offspring. Finally, the last step of the algorithm is the application of the GS described in Sect. 2.3 to the surviving population. The above process is repeated until a given stopping criterion is achieved. In order to complete the definition of this GA, we should note that individuals are represented by a vector of D real numbers, being D the number of decision variables of the problem considered.

2.3 Global Search Procedure

In order to address potential slow convergence in both DE and GA that arises when addressing difficult problems, and to improve the quality of the solutions provided, a GS procedure based on the one proposed in [11], is applied to both algorithms. It is defined as follows. Given an individual X_k randomly selected from the current population, a new individual V is generated by means of Eq. 3,

$$V = a_1 \times X_k + a_2 \times X_{Best} + a_3 \times (X_{r_1} - X_{r_2}), \tag{3}$$

with a_1, a_2, and a_3 being three numbers randomly selected from the range $[0, 1]$, and for which the condition $a_1 + a_2 + a_3 = 1$ is satisfied. X_{Best} is the best individual in the current population, i.e. the one with the lowest objective value, and X_{r_1} and X_{r_2} represent two different individuals randomly selected from the current population. We should note that indexes k, r_1, and r_2 are mutually exclusive. Once the new individual V is generated, it is evaluated and compared to the individual X_k. In case $f(V) < f(X_k)$, V replaces X_k in the current population, i.e. $X_k = V$, and the GS starts another iteration for trying to improve X_k. Otherwise, individual V is discarded and the GS is stopped. The main novelty in our work is that the GS is iteratively applied to individual X_k until it cannot be improved anymore. This contrasts to earlier work in [11] in which the GS is only applied once to individual X_k.

2.4 Hyper-heuristic

A variant of the selection HH firstly proposed in [24] is used to select between the two aforementioned EAs. The said variant was proposed and has been successfully applied by the authors in previous work [16,17,20]. It is based on using a scoring and a selection strategy for choosing the most suitable low-level configuration. Once a low-level configuration is selected, only that strategy is executed until a local stopping criterion is achieved. When this happens, another low-level configuration is selected and executed. The final population of the last low-level configuration becomes the initial population of the new low-level configuration. This process continues until a global stopping criterion is satisfied. In the particular case of the current work, a fixed number of evaluations, established by the GENOPT rules, is considered as the global stopping criterion.

The low-level configuration that must be executed is selected as follows. First, the *scoring strategy* assigns a score to each of the two EAs. This score estimates the improvement that each configuration might achieve starting from the current population. Larger values are assigned to more promising approaches, based on their historical performance. To calculate this estimate, the previous improvements in the objective value achieved by each configuration are used. The improvement γ is defined as the difference, in terms of the objective value, between the best individual found so far, and the best initial individual. Given a configuration $conf$ that has been executed j times, the score $s(conf)$ is calculated as a weighted average of its last k improvements. This is shown in Eq. 4, where $\gamma[conf][j - i]$ represents the improvement achieved by the configuration *conf* in execution number $j - i$. The adaptation level of HH, *i.e.* the amount of historical knowledge considered to perform its decisions, can be varied depending on the value of k. Finally, the weighted average assigns a greater importance to the most recent executions, with the aim of better adapting decisions to the current stage of the search procedure, thus discarding old information.

$$s(conf) = \frac{\sum_{i=1}^{min(k,j)} (min(k,j) + 1 - i) \cdot \gamma[conf][j-i]}{\sum_{i=1}^{min(k,j)} i} \tag{4}$$

The HH is elitist, namely, it selects the low-level configuration that maximises the score $s(conf)$. However, some selections are randomly performed by following a uniform distribution: this is tuned by means of a parameter β, which represents the minimum selection probability that should be assigned to each low-level configuration. If n_h is the number of low-level configurations involved, then a random selection is performed in $\beta \cdot n_h$ percentage of the cases.

3   Experimental Evaluation

This section is focused on describing the experiments conducted with the optimisation scheme introduced in Sect. 2.

Experimental Method. The EAs, as well as the HH framework, were implemented using the *Meta-heuristic-based Extensible Tool for Cooperative Optimisation* (METCO) [12]. Tests were run on a Debian GNU/Linux computer with four AMD® Opteron™ processors (model number 6164 HE) at 1.7 GHz and 64 GB RAM. Since all experiments used stochastic algorithms, each execution was repeated 100 times with different initial seeds. With respect to the former, comparisons between algorithms were carried out by applying the following statistical analysis [18]. First, a *Shapiro-Wilk test* was performed to check whether the values of the results followed a normal (Gaussian) distribution. If so, the *Levene test* checked for the homogeneity of the variances. If the samples had equal variance, an ANOVA *test* was done. Otherwise, a *Welch test* was performed. For non-Gaussian distributions, the non-parametric *Kruskal-Wallis* test was used. For all tests, a significance level $\alpha = 0.05$ was considered.

Problem Set. Experiments were carried out using the set of continuous optimisation problems proposed for the GENOPT. The set is composed of three families of functions with different features, and a particular function is defined by its identifier. For the contest, 6 functions created by the GKLS generator [9] (f_1-f_6), 6 conditioned transforms of classical benchmarks (f_7-f_{12}), and 6 composite functions ($f_{13}-f_{18}$), were proposed. Functions f_1-f_{12} were defined by identifiers 0–11, while functions $f_{13}-f_{18}$ were defined by identifiers 1586038869–1586038874. Initial seeds were fixed by the GENOPT organisation

Table 1. Parameterisation of the genetic algorithm

Parameter	Value	Parameter	Value
Stopping criterion	$1 \cdot 10^6$ evals.	Mutation rate (p_m)	1/D
Population size (NP)	5	Crossover rate (p_c)	1

Table 2. Parameterisation of the differential evolution scheme

Parameter	Value	Parameter	Value
Stopping criterion	$1 \cdot 10^6$ evals.	Mutation scale factor (F)	JADE
Population size (NP)	32	Crossover rate (CR)	JADE
% of best individuals (p)	0.1		

Table 3. Parameterisation of the hyper-heuristic

Parameter	Value	Parameter	Value
Local stopping criterion	$1.2 \cdot 10^4$ evals.	Minimum selection rate (β)	0.1
Low-level configs. (n_h)	2	Historical knowledge (k)	5

to values 1586038869–1586038968. Finally, following the instructions given for the contest, in the current work, for those functions with an even identifier, the number of decision variables D was fixed to 10. For the remaining functions, 30 decision variables were considered.

Parameters. Tables 1 and 2 show the parameterisation for the GA and DE, respectively. Parameter values for both EAs were selected based on previous knowledge of the authors [19,20]. However, in order to fix parameter values for HH, different parameterisations were considered, which did not present statistically significant differences among them. The above means that HH is robust from the point of view of its parameters, since altering them is not going to significantly affect the performance of the whole optimisation scheme. Table 3 shows the particular configuration of HH that we applied for the set of functions considered. Regarding the number of low-level configurations n_h, we should note that different values were also tested, by taking into account different parameterisations for DE and GA as the candidate set of HH. Nevertheless, the usage of more than two low-level configurations, i.e. $n_h > 2$, started to degrade somewhat the performance of the whole optimisation scheme. The reader should recall that HH makes some random decisions. If some candidate configurations do not perform properly, some function evaluations might be lost due to the random selection of one of those configurations, with the consequent decrease in performance of the whole search procedure. This is the main reason why we selected only two low-level configurations, one based on DE and the other one based on GA.

Results. Table 4 shows, for each considered problem, a statistical comparison among HH and each of both EAs executed independently. Particularly, it shows if

Table 4. Statistical comparison between HH and EAs considering problems f_1–f_{18}

f	Alg.	p-value	Dif.	f	Alg.	p-value	Dif.	f	Alg.	p-value	Dif.
f_1	DE	1.53e−26	↑	f_2	DE	2.51e−28	↑	f_3	DE	1.81e−29	↑
	GA	6.69e−12	↑		GA	1.38e−10	↑		GA	1.28e−12	↑
f_4	DE	2.69e−21	↑	f_5	DE	2.19e−28	↑	f_6	DE	2.23e−18	↑
	GA	4.75e−05	↑		GA	3.07e−13	↑		GA	1.25e−03	↑
f_7	DE	2.40e−16	↑	f_8	DE	1.48e−01	↔	f_9	DE	8.09e−02	↔
	GA	2.52e−34	↑		GA	2.52e−34	↑		GA	3.07e−34	↑
f_{10}	DE	5.53e−39	↑	f_{11}	DE	6.94e−14	↑	f_{12}	DE	7.70e−12	↑
	GA	5.54e−39	↑		GA	2.52e−34	↑		GA	2.52e−34	↑
f_{13}	DE	9.98e−09	↑	f_{14}	DE	3.27e−02	↑	f_{15}	DE	1.98e−05	↓
	GA	2.52e−34	↑		GA	2.66e−33	↑		GA	2.52e−34	↑
f_{16}	DE	5.18e−34	↑	f_{17}	DE	2.87e−30	↑	f_{18}	DE	5.17e−05	↑
	GA	2.52e−34	↑		GA	2.52e−34	↑		GA	2.52e−34	↑

HH statistically outperformed DE or GA (↑), if HH was statistically outperformed by DE or GA (↓), and cases for which statistically significant differences did not appear between HH and DE or GA (↔). We should note that HH statistically outperforms a particular EA if there exist statistically significant differences between them, *i.e.* if the p-value is lower than $\alpha = 0.05$, and if at the same time, the *Vargha Delaney effect size* between HH and the given EA is lower than 0.5, since we are dealing with minimisation problems.

It can be observed that HH was statistically better in 33 out of 36 statistical comparisons. In 15 out of 18 problems, HH was able to provide statistically better solutions than DE and GA. For problems f_8 and f_9, DE did not present statistically significant differences with HH, but the latter was able to statistically outperform GA. Finally, considering the problem f_{15}, HH was statistically outperformed by DE. Bearing the above in mind, the superiority of HH when compared to DE and GA executed independently is clear. Using HH removes the issue of algorithm selection from the user, with the HH autonomously selecting the most appropriate algorithm at the current stage of the search for a given instance, and enabling a hybridisation of both algorithms.

Additionally, HH is able to provide even better solutions than those obtained by DE or GA executed independently for most of the considered problems, thus showing that it is able to properly combine the benefits of both EAs for solving global continuous optimisation problems.

4 Conclusions and Future Work

In this work, a hyper-heuristic solution approach, HH, enabling hybridisations of EAs for solving global continuous optimisation problems was proposed. The approach hybridised a differential evolution DE, and genetic algorithm GA.

Furthermore, it included the use of a stochastic global search following the selection of the surviving population. The HH selects the most appropriate method to use at each point based on a scoring function that estimates potential improvement. The method is evaluated using a set of continuous optimisation problems proposed for the *Generalization-based Contest in Global Optimization* (GENOPT).

The computational results show that the use of our proposed hyper-heuristic framework leads to an overall enhancement when compared to the evolutionary algorithms executed in isolation. This highlights the capability of HH for switching the best evolutionary algorithm along the search. Moreover, in the majority of the cases, the improvement exhibited by HH goes further the best performing EA for each given problem, suggesting its use instead of using the embedded methods independently. Finally, it is worth mentioning that the high performance of our hybridisation through HH was recognised with two out of the three prizes available at the GENOPT, including the best overall approach prize.

On the basis of the findings presented in this paper, the next stage of our research will be focused on extending the numerical experimentation including the assessment of the different parameters of HH and its integrated EAs, as well as studying the performance of HH with additional algorithms and/or problems. Another promising line of research would be to analyse the impact that different scoring functions have over the performance of the whole optimisation scheme.

References

1. Bingül, Z., Karahan, O.: A Fuzzy Logic Controller tuned with PSO for 2 DOF robot trajectory control. Expert Syst. Appl. **38**(1), 1017–1031 (2011)
2. Burke, E.K., Gendreau, M., Hyde, M., Kendall, G., Ochoa, G., Ozcan, E., Qu, R.: Hyper-heuristics: a survey of the state of the art. J. Oper. Res. Soc. **64**(12), 1695–1724 (2013)
3. Burke, E., Kendall, G., Newall, J., Hart, E., Ross, P., Schulenburg, S.: Hyper-heuristics: an emerging direction in modern search technology. In: Glover, F., Kochenberger, G.A. (eds.) Handbook of Metaheuristics. International Series in Operations Research & Management Science, vol. 57, pp. 457–474. Springer US, New York (2003)
4. Das, S., Mullick, S.S., Suganthan, P.: Recent advances in differential evolution - an updated survey. Swarm Evol. Comput. **27**, 1–30 (2016)
5. Dasgupta, D., Michalewicz, Z.: Evolutionary Algorithms in Engineering Applications. Springer Science & Business Media, Berlin (2013)
6. Deb, K., Agrawal, R.B.: Simulated binary crossover for continuous search space. Complex Syst. **9**, 115–148 (1995)
7. Di Pillo, G., Lucidi, S., Rinaldi, F.: A derivative-free algorithm for constrained global optimization based on exact penalty functions. J. Optim. Theory Appl. **164**(3), 862–882 (2015)
8. Eiben, A.E., Smith, J.E.: Introduction to Evolutionary Computing. Springer, Heidelberg (2015)
9. Gaviano, M., Kvasov, D.E., Lera, D., Sergeyev, Y.D.: Algorithm 829: software for generation of classes of test functions with known local and global minima for global optimization. ACM Trans. Math. Softw. **29**(4), 469–480 (2003)

10. Goldberg, D.E.: Genetic Algorithms in Search, Optimization and Machine Learning. Addison-Wesley Publishing Company, Reading (1989)
11. Guo, Z., Liu, G., Li, D., Wang, S.: Self-adaptive differential evolution with global neighborhood search. Soft Comput., 1–10 (2016, in press)
12. León, C., Miranda, G., Segura, C.: METCO: a parallel plugin-based framework for multi-objective optimization. Int. J. Artif. Intell. Tools **18**(4), 569–588 (2009)
13. Li, Y.L., Zhan, Z.H., Gong, Y.J., Chen, W.N., Zhang, J., Li, Y.: Differential evolution with an evolution path: a DEEP evolutionary algorithm. IEEE Trans. Cybern. **45**(9), 1798–1810 (2015)
14. Liu, J., Teo, K.L., Wang, X., Wu, C.: An exact penalty function-based differential search algorithm for constrained global optimization. Soft Comput. **20**(4), 1305–1313 (2016)
15. Rice, J.R.: The algorithm selection problem. Adv. Comput. **15**, 65–118 (1976). Elsevier
16. Segredo, E., Segura, C., León, C.: Memetic algorithms and hyperheuristics applied to a multiobjectivised two-dimensional packing problem. J. Glob. Optim. **58**(4), 769–794 (2013)
17. Segredo, E., Segura, C., León, C.: Fuzzy logic-controlled diversity-based multi-objective memetic algorithm applied to a frequency assignment problem. Eng. Appl. Artif. Intell. **30**, 199–212 (2014)
18. Segura, C., Coello, C.A.C., Segredo, E., Aguirre, A.H.: A novel diversity-based replacement strategy for evolutionary algorithms. IEEE Trans. Cybern. 1–14 (2015, in press)
19. Segura, C., Coello Coello, C.A., Segredo, E., León, C.: On the adaptation of the mutation scale factor in differential evolution. Optim. Lett. **9**(1), 189–198 (2015)
20. Segura, C., Segredo, E., León, C.: Analysing the robustness of multiobjectivisation approaches applied to large scale optimisation problems. In: Tantar, E., Tantar, A.-A., Bouvry, P., Del Moral, P., Legrand, P., Coello Coello, C.A., Schütze, O. (eds.) EVOLVE- A Bridge between Probability, Set Oriented Numerics and Evolutionary Computation. SCI, vol. 447, pp. 365–391. Springer, Heidelberg (2013)
21. Storn, R.: On the usage of differential evolution for function optimization. In: 1996 Biennial Conference of the North American Fuzzy Information Processing Society (NAFIPS), pp. 519–523. IEEE (1996)
22. Storn, R., Price, K.: Differential evolution - a simple and efficient heuristic for global optimization over continuous spaces. J. Glob. Optim. **11**(4), 341–359 (1997)
23. Thomsen, R.: Flexible ligand docking using differential evolution. In: 2003 IEEE Congress on Evolutionary Computation (CEC), vol. 4, pp. 2354–2361. IEEE (2003)
24. Vinkó, T., Izzo, D.: Learning the best combination of solvers in a distributed global optimization environment. In: Proceedings of Advances in Global Optimization: Methods and Applications (AGO), Mykonos, Greece, pp. 13–17, June 2007
25. Yao, X.: Evolutionary Computation: Theory and Applications. World Scientific, Singapore (1999)
26. Yuan, X., Zhang, Y., Wang, L., Yuan, Y.: An enhanced differential evolution algorithm for daily optimal hydro generation scheduling. Comput. Math. Appl. **55**(11), 2458–2468 (2008)
27. Zhang, J., Sanderson, A.: JADE: adaptive differential evolution with optional external archive. IEEE Trans. Evol. Comput. **13**(5), 945–958 (2009)
28. Zhu, H., Wang, Y., Wang, K., Chen, Y.: Particle swarm optimization (PSO) for the constrained portfolio optimization problem. Expert Syst. Appl. **38**(8), 10161–10169 (2011)

Subject Index

Author Index

Printed in the United States
By Bookmasters